# Overview of Low-Carbon Development

Xiangwan Du · Dadi Zhou · Qingchen Chao ·
Zongguo Wen · Taoli Huhe · Qiang Liu

# Overview of Low-Carbon Development

CEPG

Springer

Xiangwan Du
National Energy Expert Advisory Committee
Beijing, China

Dadi Zhou
Energy Research Institute of the National Development and Reform Commission
Beijing, China

Qingchen Chao
National Climate Center
Beijing, China

Zongguo Wen
Tsinghua University
Beijing, China

Taoli Huhe
Changzhou University
Changzhou, China

Qiang Liu
National Center for Climate Change Strategy and International Cooperation
Beijing, China

ISBN 978-981-13-9249-8　　ISBN 978-981-13-9250-4 (eBook)
https://doi.org/10.1007/978-981-13-9250-4

Jointly published with China Environment Publishing Group Co., Ltd.
The print edition is not for sale in China. Customers from China please order the print book from: China Environment Publishing Group Co., Ltd.
ISBN of the China edition: 978-7-5111-1454-9



This Springer imprint is published by the registered company Springer Nature Singapore Pte Ltd.
The registered company address is: 152 Beach Road, #21-01/04 Gateway East, Singapore 189721, Singapore

# Preface I

The report of the 18th National Congress of the Communist Party of China stated that "we should strive for green, circular and low-carbon development, preserve our geographical space and improve our industrial structure, way of production and way of life in the interest of conserving resources and protecting the environment, address the root cause of deterioration of the ecological environment so as to reverse this trend, create a sound working and living environment for the people, and contribute our share to global ecological security." The *Opinions of the CPC Central Committee and the State Council on Accelerating the Building of Ecological Civilization,* released on April 25, 2015, further specifies the development path of "green, circular and low-carbon development." In fact, low-carbon development, the green development, and circular development are similar in nature and consistent in the work. Low-carbon development is not only a strategy for tackling climate change but also the only way for global sustainable development. It has strategic significance for current practices and long-term development of China.

The Chinese society has a better understanding of "green development" and "circular development" than "low-carbon development." Today, as "low-carbon development" has become a global trend and valued by the CPC and the state government, it is necessary to popularize relevant knowledge, reach consensuses, and take stronger actions so that China can take the lead in this global endeavor and make greater contributions to the progress of human civilization.

In this context, China Environmental Science Press has organized the publication of the book series *Low Carbon Development in China*, which has received great supports from relevant government departments and research scholars.

This series is designed as an advanced science read for the general public. The readers we have in mind include civil servants, business leaders, science and technology educators, university students, graduate students, and those who are interested in learning more about low-carbon development, as they will be the advocates and participants of China's low-carbon development. And hopefully, these books will be of some help to them.

The editorial committee of this series of books is comprised of well-known experts and scholars in related fields, who organize and preside over the design and writing of each volume in the series. This series is structured around various aspects of low-carbon development, involving climate change, manufacturing industry, transportation, construction, low-carbon cities, agriculture and forestry, energy, and relevant international experience. This series aims to provide scientific, systematic, novel, and readable information.

The publication of this series is an objective requirement and response to the call for green and low-carbon development. It is the fruit of the hard work of many experts, scholars, and editors of China Environmental Science Press. This series inevitably contains flaws and errors due to the limits of time and our abilities. Your comments and suggestions are greatly appreciated.

Beijing, China
December 2015

Xiangwan Du

# Preface II

"Development" has always been an eternal theme as the human society continues to evolve and advance. After a long period of primitive civilization and early agricultural civilization, industrialization began about 200 years ago, a civilization built upon the discovery of fossil energy such as coal and oil. By using the accumulated scientific and technological knowledge, human began to develop and utilize the energy that was buried under the ground of the Earth. We can say that elevating from the primitive and poor living conditions and pursuing a wealthy and culturally advanced life drive the progress of human society.

However, the total population of the world has increased exponentially since ancient times, and the speed of development is no longer the same. Yet, the earth human call home remains as big as before. The extensive exploitation of nature and the excessive use of fossil energy not only threaten the mankind with the supply shortage but also create serious environmental crises. The basic conditions for human survival—the quality of the air, water, and soil—are significantly degraded, directly harming people's health and even life. The bad results that go against the original intention for development give a warning about development: development can be a "double-edged sword." Therefore, China has put forward a series of new development concepts that are both theoretical and practical: "taking a new path to industrialization" and "building a resource-conserving and environment-friendly society," specifying the "unbalanced, uncoordinated, and unsustainable development," advocating "outlook on scientific development" and "the urgent needs for change of growth model," identifying the path of "green development, circular development, low-carbon development," gradually deepening the concept and content of "ecological civilization" and prioritizing such development as a the national strategy.

In this context, this publication aims to present an overview of thoughts on low-carbon development, from the evolution of human civilization, the emergence of low-carbon concept, to China's choice of low-carbon path. This series details the concept, its significance, the necessity of change of growth model for China, as well as the mode, focuses, and goals of low-carbon development program. These books also address "what is low carbon?" "Why should we implement low-carbon development?" "How does China proceed on the path of low-carbon development" and a series of issues closely related to the present social development.

The book consists of 11 chapters following the introduction: Chap. 1 on the evolution of human civilization; Chap. 2 on the study on climate change and the concept of low carbon; Chap. 3 detailing the concept of low-carbon development and the relationship between low carbon, circular, and green development; Chap. 4 disputing high-carbon development as the only way to modernization by citing the development of different countries; Chap. 5 validating the urgent needs to change the growth model of China by analyzing a series of problems brought about by the past development modes; Chap. 6 detailing the strategic objectives of China's low-carbon development; Chap. 7 highlighting the importance of low-carbon energy and its development direction and focus in China; Chap. 8 on the importance of low carbon production methods and the development trend; Chap. 9 on the key areas in China that need guidance for low-carbon development and low carbon consumption; Chap. 10 detailing the importance of science and technology support for low-carbon development, and the current situation of such development; Chap. 11 detailing China's policies on low-carbon development, their existing problems and suggestions for improvement.

This monograph is the collaborative result of the following authors:

Introduction: Xiangwan Du

Chapter 1: Zongguo Wen, Xiaojun Chen, Ning Wang, Xin Cao, Qian Tang

Chapter 2: Qingchen Chao, Hongbin Liu, Lei Huang

Chapter 3: Xiangwan Du, Ying Cao, Tao Ma, Leilei Cui

Chapter 4: Xiangwan Du, Xiaolong Liu, Bo Yang, Leilei Cui

Chapter 5: Dan Hu, Shiji Gao, Taoli Huhe, Tao Hong, Haiqin Wang, Haoran Yuan, Xiaolin Wang, Yuanzheng Li

Chapter 6: Qiang Liu, Yi Chen

Chapter 7: Zhiyu Tian, Xiang Gao

Chapter 8: Zhiyu Tian, Guanyun Fu

Chapter 9: Dadi Zhou

Chapter 10: Hongwei Yang, Guanyun Fu

Chapter 11: Ding Ding, Xiu Yang

Zongguo Wen and Taoli Huhe finalized the copy editing; Xiangwan Du revised and proofread the manuscript.

The writing of this monograph has received the full support from relevant experts and scholars, and I would like to express my heartfelt gratitude to them. At the same time, I hope that this publication can serve as a bridge among the experts, scholars, and the general public on related issues, and contribute to the disseminating and publicizing knowledge about low-carbon development in China.

In view of the authors' limitation, any feedback and correction of the inevitable mistakes contained in the book are greatly appreciated.

Beijing, China — Xiangwan Du
Beijing, China — Dadi Zhou
Beijing, China — Qingchen Chao
Beijing, China — Zongguo Wen
Changzhou, China — Taoli Huhe
Beijing, China — Qiang Liu
December 2015

# Introduction

## Promote Low-Carbon Development for Ecological Progress

### *Theories on Development and the Development of Theories*

Human history is but a short moment relative to the history of the universe and the history of the earth. However, humanity progresses from a primitive state to the modernized civilization, a prolonged process that takes tens of thousands of years. At the beginning, such an evolution was basically natural and spontaneous, with human's limited understanding of themselves and the external environment in which they survived. For a long time, people lived with no concept of "human being"; the earth existed, but people had no idea about this planet. It was only a few hundred years ago that people gained initial knowledge about the Earth that they had a preliminary understanding of how humans evolved.

One characteristic that distinguishes humans from other species is the former possess higher intelligence. Humanity continues to advance through productive practices, interaction with nature, constant improvement of living conditions, and social development. At the same time, humanity also develops through deliberative thinking. From the symbolic records of ancient civilizations to the textual documentation of history, human beings present a group of early thinkers, philosophers, educators, and scientists, among others. In different regions, due to the differences in geographical and historical conditions, various forms of civilization that are marked by both unique and universal characteristics of humanity have gradually formed. However, the intellectual achievements of humans in different historical periods obviously bear the limitations of the times. The thoughts that concern the global development and all mankind have just come into being in the past few centuries.

After a long period of primitive lifestyle and farming culture, human civilization entered industrialization about 200 years ago. This is a historic progress, to which an important material foundation is the discovery of fossil energy, such as coal and petroleum. Combined with the accumulated scientific and technological knowledge, the discovery enables human beings to develop and utilize these energy sources that are once dormant inside the Earth. Steam engines and internal combustion engines were created one after another, and then electricity was put to use. Labor productivity has been significantly increased, which spurs development of a number of countries. People who enjoy better quality of life brought about by the development are full of confidence in "modifying nature", and even desire to "conquer nature". However, facing the ancient and profound universe, the human beings are still young, to whom there are more unknown than known, and whose understandings of the inner working of the objective world are quite superficial and contain many blind spots in the "battlefield" without sufficient rationality. The extensive exploitation of nature and the excessive use of fossil energy not only make people risk the potential supply shortage but also cause serious environmental crisis. The basic conditions for human survival—the quality of the air, water, and soil are significantly degraded, directly harming people's health and even life. This kind of vicious result that violates the original intention of development raises a warning about the way of development: development is also a "double-edged sword".

In this context, after the mid-twentieth century, thoughts on development show more insights. In 1962, the book *Silent Spring*, by American biologist Rachel Carson, reveals the conflict between man and nature behind industrial prosperity and sounds the alarm of environmental crisis in the industrial society. According to *The Limits to Growth*, published by the Club of Rome in 1972, the "exponential growth of population and pollution, sharply reduced resources and limited self-purification" require that "humans must change production and lifestyle, otherwise the Earth's existing resources won't sustain the development of mankind". In the same year, the Human Environment Conference was held in Stockholm, Sweden, issued a programmatic document "Declaration of the United Nations Conference on the Human Environment", pointing out the challenges that environmental destruction poses to the people across the world. It emphasizes that since we have only one planet, protecting the environment requires the joint efforts of all countries in the world. In 1987, the World Commission on Environment and Development published the report "Our Common Future", which officially coined the concept of "sustainable development", further clarifying the connotation of sustainable development: "To meet the needs of modern people while not to harm the needs of future generations". It is necessary to combine environmental protection with human development. The concept of sustainable development marks an important leap in human thoughts on environment and development.

In the nineteenth and twentieth centuries, the impact of greenhouse gases emission created by human activities on global climate change has gradually formed a subject area on the basis of observation and theoretical analysis. Based on previous studies on the radiation physical properties of the molecules and the laws of radiation transport in the atmosphere, the Swedish chemist Arrhenius proposed in 1896 that the massive burning of fossil fuels would increase the concentration of carbon dioxide in the atmosphere, a cause of global warming. This idea is later supported by multitude of scientific data. The United Nations Conference on Environment and Development, held in Rio de Janeiro in June 1992, presented the

*United Nations Framework Convention on Climate Change* (hereinafter referred to as the Convention) to member countries for signature. The Convention clearly states that developed and developing countries have "common but differentiated responsibilities" and the ultimate goal is to "stabilize the greenhouse gases concentration in the atmosphere at a level that would prevent dangerous anthropogenic interference with the climate system". Based on the high attention to climate change issues, the concept and consensus of "low-emission development" characterized by reducing greenhouse gases such as carbon dioxide are established, and active actions begin. Subsequently, other treaties such as the *Kyoto Protocol*, which set emission targets for developed countries, were signed. In 2003, the British government issued the White Paper "The Future of Our Energy—Creating a Low-Carbon Economy", which coins the concept of "low-carbon economy" and proposed the goal "to fundamentally turn the UK into a low-carbon economy". These concepts and goals have made a global impact quickly.

After adopting the policy of reform and opening up, China took economic development as the central task and has attracted worldwide attention for its accelerated economic growth and remarkable achievements. However, it shows limited understanding of how to develop according to the objective laws of economic and social development. At the turn of the century, the negative effects of extensive development have gradually emerged, with increasingly severe constraints on resources and the environment. Therefore, the "transformation of growth model" was explicitly put on the agenda, and the state proposed a series of new development measures that are both theoretical and practical: "to take a new path of industrialization" "to build a resource-conserving and environment-friendly society", "to identify unbalanced, uncoordinated, unsustainable development", "to propose the Scientific Outlook on Development", "to immediately transform the growth model", "to specify the path to green development, circular development, low-carbon development", and "to gradually deepen the concept and connotation of ecological civilization and raise it to the height of the national strategy". These development concepts have contributed to the treasure house of philosophical thinking about development. The problem is that many actual practices are far removed from these new development concepts.

"Development" has always been a central theme of human society. Getting rid of the primitive state and poverty, and further pursuing higher living standards in both material and cultural life are the driving force for social progress. However, in the initial stage of development, humans had no systematic and in-depth thinking about the "development path". Since ancient times, the global population has grown by tens of thousands, and the level of development is no longer the same. However, the earth human call home remains the same. As a result, the concept of "environmental capacity" began to be embedded in the concept of development. Before an inhabitable "new home" is discovered, human beings can only discuss problems in the context of "we have only one earth" and have to face the black confusion and increasingly sharp environmental conflicts brought about by the highly developed industrial civilization. Development can solve many problems but also generates new problems. The pros and cons are related to the way of development. What is more important than the economic growth rate is the growth model. People began to realize that "just like science and technology, development is also a 'double-edged sword'", which is a dialectical materialist understanding of the laws governing the development of the objective world, and awareness of culture of "development". The theoretical milestones of the "perspective of development" outlined above are the ideological outcomes concerning the paths and goals of development.

- "Sustainable development" is also a warning against the risks of unsustainable development;
- "Low-carbon development" is also a warning that the high-carbon development leads to nowhere;
- "Scientific development" implies that unscientific and extensive development will harm humanity itself; and
- "Ecological civilization" indicates that human beings need to follow the "green, circular, and low carbon" path of development, bid farewell to "industrial civilization", and move toward a higher civilization.

Human innovation originates from creative ideas and thinking. The thoughts mentioned above form a philosophy about "development", and at the same time continue to enrich and deepen the philosophical view of human beings in the understanding of the society and nature.

## *Implications and Significance of Low-Carbon Development*

The direct implication of low-carbon development is to reduce greenhouse gas emissions, represented by carbon dioxide, in order to mitigate climate change. The broad understanding of low-carbon development is a new type of production mode and lifestyle, which combines the content of green development and circular development, and guides humanity to move toward a higher civilization—ecological civilization, by following the path of sustainable development. "Green" approaches commonly refer to pollution reduction and environment protection. Obviously, in the case of the energy structure in which fossil fuel takes a leading part, green and low carbon has strong synergy. The data show that particles from burning coal and oil account for more than half of the atmospheric composition of PM2.5 in China. The emissions from burning coal and oil also account for more than half of the total carbon dioxide emissions in China. This fully shows that "green" and "low carbon" are two different concepts, but they are highly consistent in the direction of work.

Climate change and air pollution have become major environmental and development issues across the globe. Low-carbon development has become a common trend of human development across countries, regions, and nations. Not only developed countries promote low-carbon transformation by proposing "low-carbon transition plan", "low-carbon society scenario", "low-carbon economic development blueprint", "low-carbon economy bill", and "climate–energy package", developing countries, including those have not yet been lifted out of poverty, have also combined "poverty alleviation" and "sustainable development" to formulate their own development strategies. Among them, energy saving, low-carbon energy, low-carbon technology innovation, low-carbon industry incubators, etc. will dominate the future low-carbon technologies and industries, and drive the sustained economic development by new areas of growth. It can be said that low-carbon development has become a global competition.

For China, low-carbon development has the aforementioned universal significance, and more importantly, special significance. According to the national conditions, China particularly needs and likely achieves low-carbon development.

National condition 1: A large population and a shortage of resources per capita. China's freshwater resource per capita is 30% of the world average, arable land resource per capita is 43% of the world average, the proven supply of oil is 1.1% of the world's reserve, and the proven supply of natural gas in China is 1.9% of the world's total. The annual volume of China's coal mining has clearly exceeded the "scientific production capacity". At the same time, the dependence on imported oil, natural gas, and coal continues to rise, among them, imported oil has reached 60% and natural gas has reached 31.6%. This basic national condition suggests that the Chinese modernization needs careful designs. We don't have the capital and reason for inefficient and blind development.

National condition 2: In the energy structure, the proportion of coal is significantly higher than the world average. Coal is the key natural endowment of the traditional energy in China. To date, coal accounts for as high as 64% of China's primary energy, twice as high as the global average. Moreover, only about half of the coal consumption in China is for power generation, while the other half is directly and separately burned (boiler, kiln, household coal stove, etc.). Moreover, due to extensive mining, washing, and transportation processes, coal releases high volume of pollutants and greenhouse gas emissions. The "low emission", a goal of clean coal utilization, should include low emissions of various pollutants as well as greenhouse gases, which is a major challenge to the low-carbon coal utilization, especially for China.

National condition 3: Limited environmental capacity. Environmental constraints are more restrictive than resources. More than 80% of China's population lives on the land east to the Aihui–Tengchong line. The land area of the eastern part of China is less than 1/30 of the total land area of the world, yet consumes about 40% of the world's coal output. In other words, in the eastern part of China, the coal consumption of per unit land area (which can be called "spatial density of coal consumption") is 12 times the global average. The statistics also show that in the eastern part of China, "fuel consumption space density" is already three times the global average; the "carbon emission space density" is six times the global average; and currently, the number of cars per 1000 persons in China is less than 1/8 of that of the United States, but the "car space density" in eastern China is approaching that of the United States! A separate statistic analysis on the Beijing–Tianjin–Hebei region shows that the "fuel consumption space density" in this region is 30 times the global average! As a result, it is easy to understand that the eastern part of China suffers the most serious smog problem in the world. If we want to have air of good quality, can we compare the energy consumption per capita and the number of cars per capita with those of the United States? Obviously not, and we should not! The population density in east China is five times the global average. Combined with the abovementioned energy consumption and energy structure, the environmental load in east China is more than five times higher than the world average! The water pollution and soil pollution are also very serious. These simple analyses paint a basic national situation in China: China's environmental capacity is significantly smaller than the world average, and China's climate capacity is also significantly smaller than the world average. China urgently needs to change its inefficient and blind growth model, promote the energy revolution, and blaze a path of development, and only in this way can it have a sustainable future.

National condition 4: The nature of the ecological environment is relatively fragile in China. The arid and semiarid areas account for 52% of the total land area of the nation. The Loess Plateau, where the soil erosion is serious, is 640,000 $km^2$; the high and cold Qinghai–Tibet Plateau covers an area of 2 million $km^2$; and the stony deserts in karst areas occupy an area of 900,000 $km^2$. These landforms are vulnerable to the adverse effects of climate change. 70% of China's annual natural disasters are related to weather. Low-carbon mitigation and orderly adaptation to address climate change are cohesive and complementary actions. The characteristics of this natural environment remind us that China needs to pay more attention to "development in a well-protected environment" and take climate change seriously.

National condition 5: Late-mover advantage. China's modernization is significantly later than that of the world's first developed countries. Therefore, we may draw on their experience and practices. Moreover, the different development paths of different types of developed countries also provide us with important references and inspirations. In addition, in the twenty-first century, compared with the time of modernization of the developed countries, there are much more advanced technologies in such aspects as information sharing, Internet, energy saving, environmental protection, low-carbon control, and new energy. China should make full use of these late-mover advantages, achieve higher efficiency and high-quality development, and embark on a path of greener and lower carbon development. Of course, if not actively seizing and making use of these advantages, the late-mover advantage will be lost.

National condition 6: The inheritance and promotion of the essence of Chinese culture. Since ancient times, the Chinese always have the concept of "the unity of heaven and man" as well as the lessons from "the stories of former sages and the history of the country, telling that the success comes from diligence and extravagance brings downfalls". Since the beginning of the twenty-first century, China has put forward the "Scientific Outlook on Development", especially the cultural concepts of "ecological civilization" and "Beautiful China" as the guiding ideology of governing the nation. These ideas are highly compatible with the green, low-carbon, and sustainable developments that are induced by and responding to climate change. When these remarkable philosophies from ancient and modern eras are upheld as guiding principles and carried out throughout the country, low-carbon development can truly be advanced.

After more than 30 years of rapid development, China is facing two competitions: domestically, the transformation of the growth model is competing with the inertia of inefficient and blind development to steer the country onto the path that is oriented toward resource-saving and environment-friendly development; while internationally, China cannot fall behind in the global green and low-carbon competition, instead of following the old path, China should obtain the leading strategic position as soon as possible and build an ecological civilization with determination.

## *The Urgency and Strategies of Low-Carbon Transition*

There has been a political consensus in the world to reduce greenhouse gas emissions so as to limit the global warming to 2 °C. With the introduction of the fifth assessment report of the United Nations Intergovernmental Panel on Climate Change (IPCC), the correlation between anthropogenic greenhouse gases and global warming has been further confirmed. As the major emitters of greenhouse gases, developed countries have the responsibility to take the lead in reducing the absolute emission of greenhouse gases. At the same time, since the total amount of remaining emissions allowed by the 2 °C temperature rise is very limited, developing countries must also make emission reduction contributions corresponding to their own development stages and capabilities. China is already the largest emitter among the emerging countries. As a responsible developing country, China takes on its international obligations and participates in global climate governance in an active and pragmatic manner. More importantly, China's international obligation in emission reduction is highly consistent with the inherent needs of domestic low-carbon development, which is the urgent need and long-term strategic demand for transforming the nation's growth model.

China's past development features high carbon emissions; the extensive development is manifested in the driving force mainly relying on five primary productivity factors: excessive resources consumption, environmental destruction, investment incentives, imported technology, and cheap labor. However, without the driving force of science and technology on top of weakening domestic demand, these five factors alone obviously won't sustain development. Another issue is the industrial structure, of which the secondary industry has prolonged dominance, especially the high proportion of high energy-consuming industries. For example, China's steel production in 2012 accounted for 45.7% of the world's total output, and cement production accounted for 57.8% of the total. The expansion of low-end industries has pushed the GDP growth, resulting in an industrial system featuring high energy consumption and high carbon emissions (including excessive backward production capacity). The 2013 data show that China accounts for 12.3% of the world GDP but consumes 21.5% of the world's energy, that is, the energy consumption per unit of GDP is nearly double that of the global average, and coal consumption accounts for 50.2% of the world's total.

This type of development has created complex and compressed environmental pollutions, and eastern China has become one of the areas with the worst smog in the world. In 2011, WHO air quality rankings of 1,082 cities in the world, Beijing ranks 1,035. According to the 2012 World Bank report, 25 out of the 112 most polluted cities in the world are in China. It is common for pollutant emissions to exceed the environmental self-purification capacity.

In parallel, carbon dioxide-based greenhouse gas emissions have increased rapidly. In 1990, China's greenhouse gas emissions accounted for only 11% of the world's total, and increased to 26% in 2012. From 1990 to 2012, China's carbon emissions increase accounted for 66% of the world's total. Among China's greenhouse gas emissions, the contribution from fossil energy combustion accounts for about 80%. China's per capita annual carbon dioxide emissions are 6 tons, approaching the level of developed countries and regions such as Europe and Japan. The per capita annual carbon dioxide emissions of some major cities in eastern China have reached more than 10 tons, exceeding the historical peaks of those of such countries and regions as Europe and Japan, and are still growing. If the central and western parts of China follow the growth model of the east China, the nation's greenhouse gas emissions will surely surge, intensifying the high-carbon energy consumption in development. Since the 16th National Congress of the Communist Party of China, the central government has repeatedly emphasized "taking a new path of industrialization". In fact, due to the lack of alternative or such new paths, the nation still navigates on the old path of "pollution followed by control and treatment", creating even more carbon emissions.

The social factors for the high carbon emission development are as follows: the easy accessibility of high-carbon energy; the low-end industries with high-carbon energy consumption can drive growth in a short, easy, and fast manner; China is objectively at the primary stage of development; international precedent cases of high-carbon energy consumption development, such as the United States, mistaken as "the inevitable law of development".

In fact, international experience has shown that high-carbon development is not the sure route to modernization. On the basis of the analyses of energy economics in developed countries, we have reached the conclusion that developed countries can be divided into two types. The data show that compared with the United States, countries and regions such as Europe and Japan have already embarked on a more energy-efficient and low-carbon path of modernization. The American way is not suitable for China's modernization in terms of temporal and spatial conditions. Although the national environmental-economic Kuznets curves give a standard appearance, the tops of the "inverted U" sit at the heights more than doubled! This is extremely instructive for China to choose its development path. In the process of modernization, China should learn from the favorable practices of countries and regions such as Europe and Japan, and give full play to its traditional cultural advantages and late-mover advantages, avoiding blindly following the "US and Canada model" or taking unnecessary efforts for "carbon reduction after high-carbon energy consumption".

China's path to modernization must be innovative, low carbon, and have Chinese characteristics. This road must be created by the Chinese people themselves.

Transitioning to low-carbon development is not easy. More than three decades of high-carbon development pose very strong inertia. Regardless of obvious adverse effects, there are still markets for the "GDP competition". At the same time, more than 30 years of economic development has also provided a material basis for the transformation. The concepts of "Scientific Outlook on Development", "ecological civilization", and "Beautiful China" have taken root among the people, providing a certain ideological basis for transformation. It is possible to transition to low-carbon development through unusual efforts, which is the significance of promoting new growth model while tackling climate change. To this end:

- It needs to explicitly stress that China is in a period of strategic opportunities, which is first of all a period of transformation of growth model, and it is crucial to transform the growth model before 2020.
- China must implement a "sound assessment and evaluation system for economic and social development" proposed by the central government. That is, a system to cover indicators that reflect the status of ecological progress, such as resource consumption, environmental damage, and ecological benefits so that the system can provide important guidance and regulations" and evaluate the performance of administration at all levels with a multidimensional "scientific development index".
- After more than 30 years of rapid development, environmental protection and ecological advancement should be placed before the goals of national development.
- For new urbanization, it is important not to compare the "urbanization rate" and "speed" but to promote the construction of smart and green cities, use low-carbon emission as an obligatory target for urbanization, and create a development path of low-carbon buildings and low-carbon transportation in China.

"The essence of the large-scale development of the western region is scientific development" instead of following the path of high-carbon energy consumption as seen in the east. It should emphasize "development within environmental capacity" and "development in a well-protected environment".

- China should strengthen adjustment of the industrial structure. With saturating high energy-consuming industries, China should continue to shutting down outdated production facilities, vigorously develop service industries that can create jobs and achieve lower carbon emissions, strive to promote strategic emerging industries, develop sorted treatment, and recycling of waste into a strategic emerging industry. In this regard, there have been successful cases and technologies that can be promoted at home and abroad.

- China should actively promote the energy revolution, reshape its energy structure, prioritize energy conservation and efficiency, limit the total consumption of coal and oil for low-carbon emissions, increase the proportion of natural gas (including unconventional gas), develop nonfossil energy, including renewable energy and nuclear energy as much as possible, and develop energy storage technologies, smart grids, and distributed energy use.
- China should aim for a synergistic effect in reducing smog and carbon emission so as to reach the peak of greenhouse gas emissions as early as possible before 2030, which can serve as the concrete goal for medium-term strategy and low-carbon development.
- China should implement the strategies to increase carbon sink of and reduce emissions to the ecosystem, involving agriculture, forestry, sea, grassland, wetlands, land use, etc.
- China should take a combination of institutional measures for transformation to low carbon emission, involving administration, economic policy, financial market, culture, public opinion, and science and technology.

The temporal and spatial conditions of China's modernization are different from that of the United States, Europe, and other countries and regions. It is impossible to rely on plundering global resources for high-carbon energy consumption and then controlling the pollutions. China relies mainly on domestic resources, and has to compact the 200-year development into decades. Naturally, the result is severe resource depletion and compressed, complex, and structural environmental pollution. Therefore, China also has to compress the enhanced green and low-carbon strategies into the rapid development so as to avoid the catastrophe of social organisms and maintain healthy and scientific development.

Low-carbon development will also bring about innovation and transformation of the society, and will have far-reaching significance in achieving the long-term strategic goals of the rejuvenation of the Chinese nation.

Low-carbon development requires developing and implementing a low-carbon production model and strengthening the support of innovative technologies. Due to the continuous heavy investments, China has a huge amount of idle excess production capacity and vacant buildings, resulting in heavy debts of resources and environments as well as imbalance between production and consumption. The production system is large but not strong, and the annual net exports of implicit energy and implicit carbon emissions account for more than 20% of domestic consumption. To maintain high economic growth, it is unsustainable to employ the traditional production methods that rely on expanding the production of various energy-intensive products as well as low value-added industries and repeated construction with high levels of investment, consumption, and pollution, and such a situation must be changed. Energy conservation, consumption curbing, and emission reduction not only promote the production transformation but also bring about a wide range of conceptual and technological innovations and advances in basic scientific research, for example, a series of new concepts and technologies distributed low-carbon energy networks, CCUS (carbon capture, utilization, and storage), and controlled nuclear fusion.

Low-carbon development requires designing and implementing low-carbon consumption patterns. A symbol of social progress is efficiency and conservation. China needs to advocate a lifestyle of "healthy material consumption, and great cultural and intellectual needs" and a low-carbon culture, so as to stop the practices of unrealistic comparison and extravagance and curb irrational consumption.

The deepening low-carbon development will create a low-carbon society. Its cells are low-carbon communities, low-carbon enterprises, low-carbon villages and towns, and even low-carbon families. This is not only conducive to the building of beautiful cities and beautiful rural areas but also will greatly enhance the quality of citizens and culture. It has a fundamental significance for the Chinese nation to become a proud and active member of the community of nations.

Looking to the future, low-carbon development represents a major opportunity for China to achieve its long-term strategic goals. If a long-term strategy for China to remain invincible in international low-carbon competitions can be formulated and pushed forward from now on, after 30 or 40 years of unremitting efforts, China will become a competitive world power with relatively strong innovation ability by the 100th anniversary of the founding of New China. Otherwise, China might lose its late-mover advantage in the international race for sustainable development and get stuck in a long-term passive situation of having no core competitiveness, which should be avoided by all means.

The firm implementation of the low-carbon development strategy will not only improve our living environment but also benefit the fundamental interests of future generations. It will have far-reaching significance in carrying out the strategic goals of the great rejuvenation of the Chinese nation as well as contributing to the progress of the mankind.

# Contents

# Chapter 1
# The History of Human Civilization

## 1.1 The Evolution of Human Civilizations

The human civilization is a history of the relationship between man and nature (Fig. 1.1). The interactions between man, economy, society, and nature have jointly promoted the progress of human civilization and historical development, of which the improvement and development of the way of production have played a decisive role. On the one hand, human gain benefits and influences nature by acquiring energy, resources, space, discharging waste, and enjoying the ecological environments; on the other hand, nature limits development due to limited supply of energy, resources, and space and the deteriorating natural environment. The relationship between man and nature evolved from primal harmony to dissonance, and to new harmony; from the low-carbon (carbon-free) development of primitive civilization and agricultural civilization, to the unsustainable high-carbon energy consumption of industrial civilization, and then to the sustainable low-carbon development of ecological civilization, forming a spiraling-upward process; therefore, low-carbon development has become an important feature of ecological civilization. Throughout the history of human civilization, human society has experienced the stages of primitive civilization, agricultural civilization, and industrial civilization, and has now entered the forming stage of ecological civilization.

(1) The primitive civilization, spanning from 2,000,000 to 10,000 BC, features fishing and hunting culture, when mankind and nature maintained a primitive harmonious relationship (Fig. 1.2). At that time, there was a small population, and human settlements were in areas with superior natural conditions. In the absence of science and technology, humans only maintained very low levels of consumption. Families and tribes were the main forms of social organization, and the nature was extremely powerful. Humans worshipped and were adapted to nature.

    The pre-civilization era was unenlightened and barbaric. In the long history, human beings were only a member of the natural ecosystem, whose survival relied entirely on the natural resources in the natural ecosystem, such as col-



X. Du et al., *Overview of Low-Carbon Development*,
https://doi.org/10.1007/978-981-13-9250-4_1

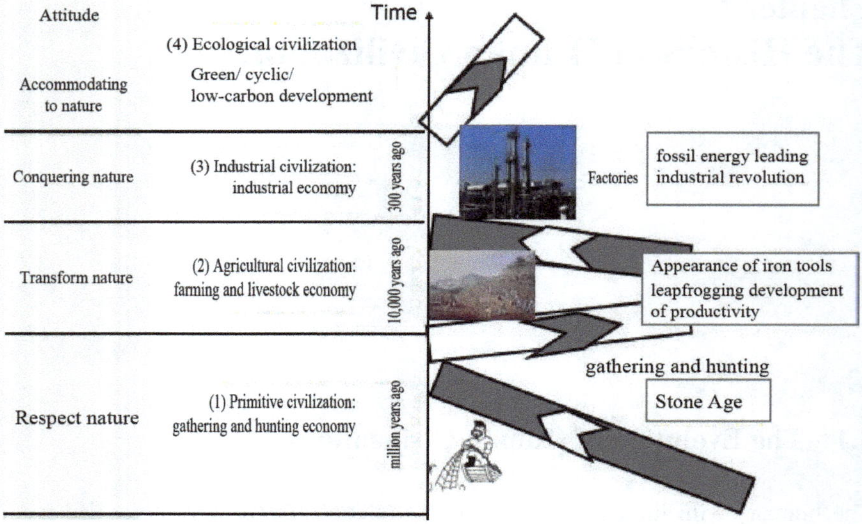


**Fig. 1.1** Degrees of coordination between socioeconomic development and nature in different civilizations

The Zhoukoudian site, located in the northern part of Longgu Mountain, Zhoukoudian Town, Fangshan District, Beijing, is a human activity site of the early Paleolithic period, with the most abundant, systematic and valuable materials in the world.

The Himba, a primitive social group that is coming to a close in Namibia, Africa, has a population of about 20,000. They are the last African people who maintain their original way of living as hunter-gathers.

**Fig. 1.2** Human primitive civilization

The transformation of farming methods: from the "slash and burn" more than 10,000 years ago, to the stone tools 8,000-9000 years ago, and to the bronze and iron plows and animal power coming down from the Warring Period of ancient China.

Longji terraced fields, located in Longji Mountain, Ping'an Village, Heping Township, Longsheng County, Guangxi, are 22 km from the county seat and 80 km from Guilin City. The Longji terraced fields were first created in the Yuan Dynasty and completed in the early Qing Dynasty. It has a history of more than 650 years.

**Fig. 1.3** Agricultural civilization

lecting wild fruits, catching insects, or hunting wild animals with simple tools made of stones. The impact of such activities on nature is insignificant when compared with the natural abundance.

(2) With the invention of tools, humans entered the stage of farming civilization (Fig. 1.3). The relationship between man and nature contains periodical and regional discords while maintaining overall harmony. With the expanded scope of living activities, mankind began to transform nature. With excessive land reclamation and felling for land, especially the frequent battles over land and water resources, the relationship between man and nature and that between people and people experienced local and temporary tension, but the nature basically retained the self-repair functions of an ecological environment.

The most obvious problem of agricultural civilization is soil erosion and land degradation caused by unreasonable use of land and the low tolerance of disasters and anti-disaster capabilities. Some data indicate that the disappearance of the Mayan civilization and the degradation of the Loess Plateau in China are caused by the contradiction between population and land, which has led to the intensified contradiction between man and nature.

(3) With the development of science and technology, the use of fossil energy and the invention of steam engines mark the entry into industrial civilization, while the relationship between man and nature becomes more intense across the world (Fig. 1.4). Human's ability to transform nature and to occupy natural resources has been unprecedentedly enhanced. With the expanding range of activity, longer life expectancy, substantially increased population, and unprecedented scale of conquest over nature with ruthless exploitation of natural resources, "Anthropocentrism" emerges. While creating a splendid modern industrial civilization, mankind abuses the natural resources to support economic development, which results in overconsumption of the earth's resources, ecological destruction, and a series of serious ecological and environmental problems.

The first industrial revolution was marked by the improvement of the steam engine. At the end of the 18th century, the steam train was invented, followed by oil locomotives and electric locomotives. Since the 1960s, many countries have developed high-speed trains.

On November 29, 2000, China unveiled its first humanoid robot. With a human appearance, a height of 1.4 meters and a weight of 20 kilograms, it can model after the basic behavior of humans. (Xinhua News Agency)

**Fig. 1.4** Industrial civilization

> The western developed countries that first enjoyed the fruits of industrial civilization, took the lead in rethinking the past and transforming their growth model after suffering the environmental degradation brought about by industrialization. After more than half a century of hard work, most developed countries have adjusted and improved their economic structure. The heavy and chemical industry, which is characterized by "high investment, high consumption, and high pollution," has been transformed into the tertiary or service industry, which is characterized by "low cost, low consumption, low pollution, and high efficiency," and the ecological environment has been significantly improved.

Carbon is an ecological factor that keeps in company with the human life. It best reflects the interaction between mankind and ecosystems as well as the impact of people on the ecosystem. High-carbon development is one of the most acute contradictions between industrial civilization and ecological environment. The unsustainable nature of high-carbon energy consumption makes low carbon a sure choice for the progress of civilization. In the ecological framework, the fundamental characteristic of the ecosystem lies in "life", while the carbon-based element is the foundation of life. The industrial civilization relies too much on the energy resources accumulated by ancient life. In the process of utilization, the organic carbon base is converted into an inorganic carbon base. For example, after the use of coal, oil, and natural gas, carbon will be converted into carbon dioxide or carbon monoxide. This "nonliving" form of energy use of industrial civilization excessively consumes a limited amount of fossil fuel that has a finite lifetime, which makes it difficult to sustain. In the future development of civilization, we have no choice but to choose "low carbon".

## 1.2 Progress and Perils of the Industrial Civilization

### *1.2.1 The Rapid Development of the Global Economy*

From 1963 to 2013, the world economy developed rapidly, and the global GDP increased from $1624.3 billion to $74909.8 billion, an increase of about 46 times. Among them, the share of global industrial value added to GDP declined from 33% in 1990 to 27% in 2011. The share of the value added of the tertiary industry to GDP increased from 60 to 70% during the period of 1990–2011, as shown in Fig. 1.5.

With the increase in productivity and the human's enhanced adaptability to nature, human population has grown rapidly. Since the beginning of human race, the world's population had grown to 1 billion in 1804, and exceeded 6 billion in 1999 (Table 1.1). In the past two centuries, the world's population has grown fivefold. Facing the fast-growing world population, the United Nations has designated October 12, 1999 as "The Day of Six Billion", reminding governments and people around the world to pay attention to population issues.

However, since the mid-1990s when population growth reached a peak of 82 million people per year, the growth rate has slowed down.

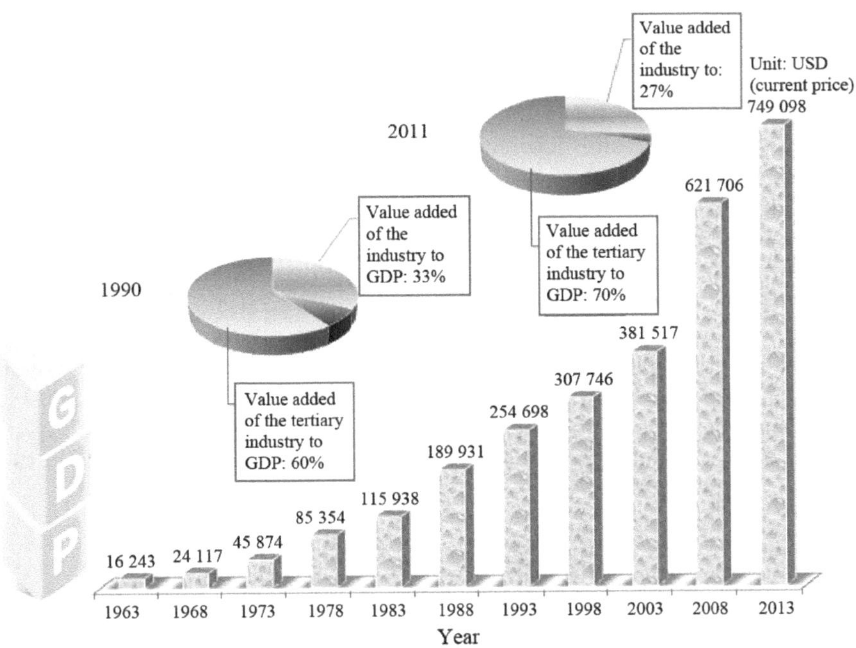


**Fig. 1.5** Rapid development of the world economy

**Table 1.1** Time for every 1 billion people increase in the world population

| | Year of each one billion people increase | Number of years required for each one billion people increase |
|---|---|---|
| The first one billion | 1804 | 200 |
| The second one billion | 1927 | 123 |
| The third one billion | 1960 | 33 |
| The fourth one billion | 1974 | 14 |
| The fifth one billion | 1987 | 13 |
| The sixth one billion | 1999 | 12 |

Since 1950, the world population growth has shown a distinct regional difference: the population growth rate of developed countries has dropped to a very low level, even to negative in some countries, while the absolute growth rate of most developing countries dropped with different degrees, but the relative level is still high.

In the twentieth century, due to the global improvement in health care, the life expectancy greatly increased. This period also saw the fastest decline of the mortality rate. The life expectancy of the world's population was only 46 years old in between 1950 and 1955, 67 years old during 2005–2010, and expected to be over 75 years old in the mid-twenty-first century.

The sustained economic development of developing countries and regions has led to a continuous declining population living in poverty in recent years. The number of people living in extreme poverty had dropped from 1.9 billion in 1981 to 1.2 billion in 2010.

The history of economic and social development in the modern world is a history of urbanization, manifested by the urban expansion, the increase in the number of cities, the growth of urban efficiency, the growth of urban population, the accelerated development of infrastructure, the advanced and standardized urban management, and the continuous improvement of civilization. World urbanization is marked by the large volume of rural populations turning into urban populations. In 1800, only 3% of the world's population lived in cities, and about 14% in 1900. By 2014, according to statistics of the World Bank, urban population had reached 53.4% of the world's population.

Globally, industrial technology innovation capabilities have improved significantly and the level of technology has continued to increase. The number of researchers per one million people in the world was 1082 in 2000, and 1284 in 2010. The total amount of high-tech export trade doubled from $987 billion in 1999 to $1933.7 billion in 2011.

### 1.2.2 Energy Resources Concerning the Global Economic Development

Natural resources are essential for human survival and development. With the global industrialization after the Second World War, the demand for natural resources is on the increase, and human pressure on the natural world is unprecedented, leading to shortages of and conflicts for natural resources.

(i) Conflicts between economic development and water consumption across the globe

The total amount of water on the earth is quite substantial, but the freshwater resources, such as rivers, lakes, and groundwater that are closely related to human production and life and easy to develop and utilize, only account for 0.3% of the total volume. The distribution of freshwater resources on land is uneven. Geographically, water resources are unevenly distributed in different continents: on the one hand, Europe and Asia have 72.19% of the world population but have only 37.61% of world's river runoff; on the other hand, South America has 5.89% of the world population, but has 25.1% of the world's river runoff. Temporally, the shortage of water resources in the dry season is a serious problem. The world's renewable inland freshwater resources per capita ($m^3$) had reduced from 13198.64 $m^3$ in 1962 to 6087.703 $m^3$ in 2013. At present, many rivers in the world are dying out and polluted to varying degrees. In addition, global climate change has caused hydrological anomalies in some areas, and we are suffering from increasing shortage of water resources.

Global water consumption has increased sixfold in the twentieth century, twice that of the population growth. The continuous growth of water demand brought about by global economic development is the core issue of water shortage. In underdeveloped regions and countries, due to outdated technology, equipment, and production processes, the waste of water for industrial production is very alarming, resulting in excessive water consumption for industrial production and further worsening the issue. In 2013, the total amount of freshwater withdrawal in the world was 3,906,738 million $m^3$, of which industrial water accounted for 17.7%.

The water pollution problem has further aggravated water scarcity. The progress of human society and economy, the process of industrialization and urbanization of human beings, especially the rapid increase in population have led to a rapid increase in water consumption, and the problem of water shortage has become increasingly prominent across the world.

(ii) The conflicts between global economic development and mineral resources consumption

"Mineral deposits" refer to the mineral resources that are formed by geological processes, stored on the earth's surface or in the crust, and can be utilized for developing national economy. More than 95% of energy and more than 80% of industrial raw materials come from mineral resources. After the 1960s, the exploitation and utilization of mineral resources have increased sharply, and mineral resources play an increasingly important role in modern industrial production and national economic development.

From 2004 to 2013, the global apparent demand for crude steel increased from 1.062 billion tons to 1.648 billion tons, at a growth rate of 55.18%. In 2013, China's apparent demand for crude steel accounted for 46.8% of the global demand, compared with only 27% in 2004. China's growth of crude steel demand has become the driving force of the growth of global crude steel consumption (Fig. 1.6).

(iii) Land resource crisis

The main problems in the development and utilization of the world's land resources are shown as follows: 1. There are more and more megacities in the world as global urbanization evolves. Urbanization accelerates the consumption of land resources, as a large number of fertile farmlands are replaced with built-up areas and ground covered with cement and asphalt. 2. Land degradation, such as soil erosion, land desertification, salinization, waterlogging, soil pollution, etc. 3. Air pollution, marine pollution, groundwater pollution, surface water pollution, and soil pollution caused by waste.

The increase in the world's population has put enormous pressure on land resources. The American agricultural economists generally believe that it is difficult to guarantee food security when the arable land per capita is less than 0.4 $hm^2$. The world's arable land per capita dropped from 0.365 $hm^2$ in 1961 to 0.199 $hm^2$ in 2012, and is still decreasing. The sharp decline in cultivated land sounded the alarm for the survival and the future of mankind.

Another manifestation of the land resource crisis is the sharp decline in forest area and the severe damages to primary forests (Figure A.1). According to statistics, in 2010, the total area of forests in the world slightly exceeded 4 billion $hm^2$, accounting for 31% of the land area and equivalent to 0.6 $hm^2$ per capita. In the past 10 years, the forest cover area has been quickly disappearing, at an average of 13 million $hm^2$

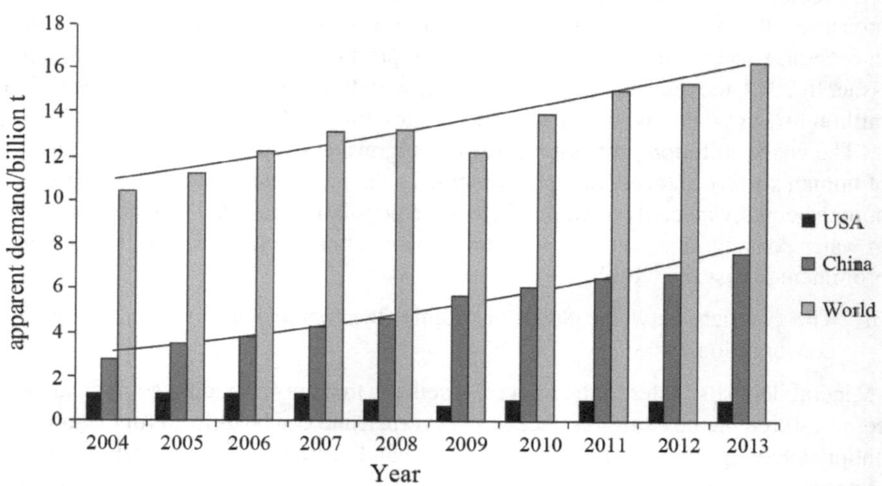


**Fig. 1.6** Comparison of apparent demands for crude steel among the world, China and the United States

every year, to serve other purposes or due to natural reasons. Although large-scale afforestation and natural expansion of forests in some countries and regions have largely reduced the net loss of global forest area, it is estimated that the average annual net loss of forest area is 5.2 million $hm^2$, slightly larger than the land area of Costa Rica and equivalent to a daily loss of more than 140 $km^2$. Regionally, South America has the largest net loss of forests, followed by Africa. Primary forests, which account for more than one-third of the world's forest area, have been severely damaged since 2000 and have shrunk by more than 40 million $hm^2$, at a rate of 0.4% annually. Primary forests, especially tropical humid forests, have the most diverse terrestrial ecosystems. The destruction of primary forests is a serious threat to terrestrial ecosystems and biodiversity.

(iv) Energy consumption and greenhouse gas emissions

Over the past 100 years, developed countries have completed industrialization and modernization through high-carbon energy consumption. The industrialization of developed countries had accelerated the massive consumption of fossil fuels on the planet. Today, many developing countries are entering the process of industrialization, which will further increase the global energy consumption. The shortage of fossil fuels has become an extremely serious constraint to the global economic development. At the same time, the greenhouse gases emitted by high fossil fuel consumption are accumulated in the atmosphere, creating greenhouse effect that causes the higher frequency of natural disasters and extreme weather and threatening the sustainable development of human society.

From 1971 to 2011, the world's energy use increased more than twice, from 7.9 billion tons of standard coal to 18.2 billion tons of standard coal (Fig. 1.7), and energy use per capita was also on the rise. Specifically, fossil fuel accounted for the biggest share of energy consumption in the 1970s, but began to decline ever since to the current level of 80%. The proportion of alternative energy and renewable resources increased and maintained at around 10%. The proportion of fossil fuel consumption is still high in China, and the proportion of nonfossil energy is rising slowly.

At present, the world energy structure is still dominated by fossil fuel, a pattern basically composed of oil, coal, and natural gas and supplemented by nuclear

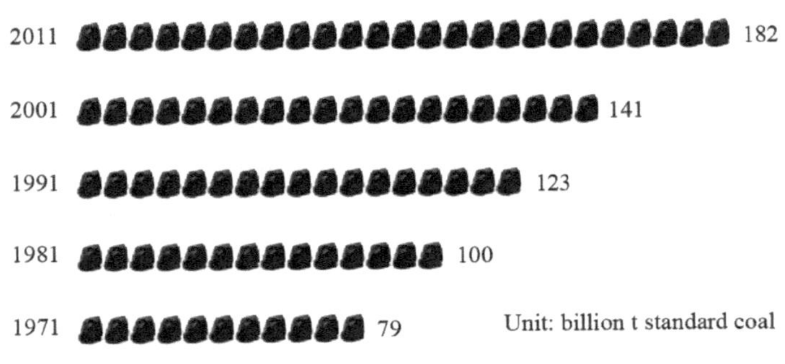


**Fig. 1.7** World energy use since 1971

**Table 1.2** Comparison of international energy consumption in 2012 unit: $10^6$t oil equivalent

| | Oil | Gas | Coal | Fossil energy consumption | Total energy consumption | Share of fossil energy consumption (%) |
|---|---|---|---|---|---|---|
| China | 483.7 | 129.5 | 1873.3 | 2486.5 | 2735.2 | 90.9 |
| USA | 819.9 | 654.0 | 437.8 | 1911.7 | 2208.8 | 86.5 |
| Europe | 879.8 | 975.0 | 516.9 | 2371.7 | 2928.5 | 81.0 |
| World | 4130.5 | 2987.1 | 3730.1 | 10,847.7 | 12,476.6 | 86.9 |

power, hydropower, and renewable energy, as shown in Table 1.2. In 2012, fossil fuel accounted for 86.9% of the world's energy consumption.

By the end of 2012, the world's proven reserves of crude oil were 1.67 trillion barrels, the proven reserves of natural gas were 187 trillion $m^3$, and the proven reserves of coal were 860.9 billion tons. According to the current energy consumption levels, current reserves can only meet the demand of 52.9, 55.7, and 109 years, respectively. Although the world's fossil energy reserve can maintain low-speed growth for a long period of time, the cost of mining and utilization will be greatly increased, and global energy demand will continue to grow in the future. The shortage of fossil fuel supply is a potential crisis for the world. Human beings must change the structure of energy consumption, reduce excessive dependence on fossil fuels, and vigorously develop new energy sources such as nuclear energy, hydropower, solar power, wind power, and bioenergy.

### *1.2.3 Global Environmental Problems Across the Globe*

Replacing natural forces with manpower to transform the natural environment is the basic feature of the industrial revolution and the root cause of environmental problems as it breaks the law of the nature. Although it embodies human's capacity to act, it fundamentally creates the imbalance of the environmental system. As humans "conquer" the natural environment all over the world, environmental problems have rapidly developed from regional issues to global issues. Population agglomeration and urbanization are the two major social outcomes of the industrial revolution as well as the basic driving forces for the outbreak of environmental problems. Industrial production requires a high population density, which results in population agglomeration. Industrial production also requires convenient transportation and a large consumer base, which is in fact urbanization. The essence of population agglomeration and urbanization is that human beings expand their territory infinitely in nature, leading to ecological destruction by reducing the habitats of other species and even endangering some species. The population on the earth has already exceeded more than 6 billion, and this number is still growing exponentially. Population growth is bound to bring about an increase in resource consumption and a tight space for living. The production and consumption of energy cause increasingly serious atmospheric pollution, and even affect climate changes. The emergence of the ozone hole and the "greenhouse effect" are all related to such activities. In addition, the constant invention of various new chemical substances and materials has caused environmental pollution that is difficult to eliminate. At the same time, in the recent decades, the pollution of the oceans, the consumption and destruction of marine living resources have reached an unparalleled level. Many rare species are on the verge of extinction due to the exploitation of wildlife. Massive deforestation causes serious consequences, such as soil erosion and land desertification. A report issued by the United Nations Development Programme shows that half of the world's wetlands have disappeared in the twentieth century; about 9% of the world's species are on the verge of extinction, and more than 130,000 $km^2$ of tropical forests are destroyed each year; two-thirds of farmlands suffer soil degradation; 30% of forest cover areas are occupied for other usage; and 20% of freshwater fish species are either extinct or endangered.

Globally, the notable environmental issues include climate change, acid rain, smog, ozone depletion, sharp decline in biodiversity, marine pollution, persistent organic pollutants, and public hazards.

(i) Climate change

It is an indisputable fact that the increasing concentrations in the atmosphere of greenhouse gases, including carbon dioxide ($CO_2$), methane ($CH_4$), ozone ($O_3$), nitrous oxide ($N_2O$), and hydrochlorofluorocarbons (CFCs), etc. have caused global warming. Carbon dioxide is the major constituent of greenhouse gas. Carbon dioxide emissions mainly come from the burning of fossil fuels. Statistics show that in 2010, the world's population reached 6.92 billion, an increase of 3.15 billion over that in 1971; the total economic output (GDP) reached 64.4 trillion US dollars, a 17-fold increase over that in 1971, while the volume of carbon dioxide emissions, for the first time, exceeded 30 billion tons, reaching 30.5 billion tons (Fig. 1.8). Carbon emissions in the process of economic development and population growth have become a global concern. In addition, other greenhouse gases such as methane, nitrous oxide, and hydrochlorofluorocarbons also increase rapidly. From 1880 to 2012, the world's average ground surface temperature increased by about 0.85 °C (Fig. 1.9).

According to the 2014 IPCC report *Climate Change 2014: Impacts, Adaptation and Vulnerability*, climate change has had an impact on all continental and marine ecosystems and human societies over the past few decades, presenting a universal risk to the safety of mankind, such as sea level rise, coastal areas threatened by high tides, urban infrastructure damages by floods and extreme weather, death and disease caused by urban heat, and food shortages caused by drought and precipitation changes.

(ii) Acid rain

During the burning of fossil fuels and processing of certain sulfur-bearing ores, a large amount of acidic substances such as sulfur dioxide ($SO_2$) and nitrogen oxides (NOx) will be released into the atmosphere; motor vehicles are also the main source

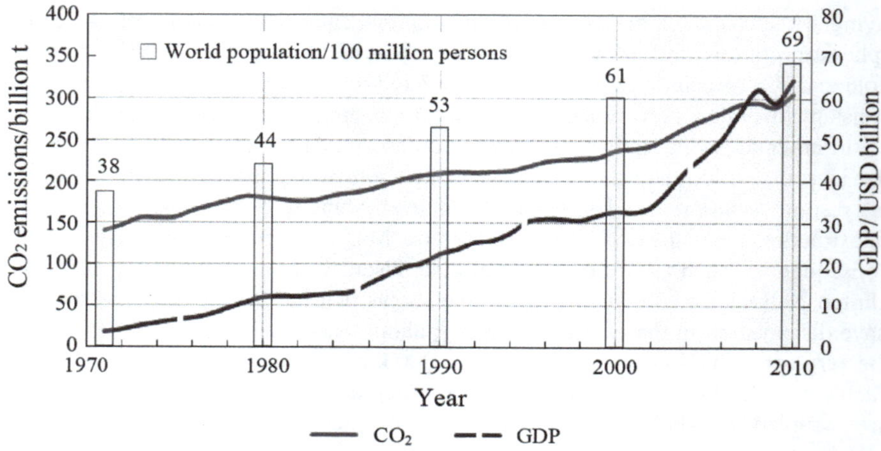


**Fig. 1.8** World GDP and $CO_2$ emissions

**Fig. 1.9** The damaged polar bear habitat as global temperatures rise and glaciers melt

of nitrogen oxide emissions. A large amount of acid gas emissions could increase the concentration of pollutants in the air, causing respiratory diseases and threatening human health. In addition to causing health concerns, when sulfur dioxide and nitrogen oxides come down to the earth through natural settling or rainfalls, etc., they pose a threat to the buildings on the ground and ecosystems. The harmful effects of acid rain mainly include: 1. Soil acidification, loss of nutrients, and disappearance of large-area forests. 2. Lower pH value of surface water body. The mineral elements lost in the soil eroded by acid rain will affect the water quality and endanger the aquatic ecosystem if they flow into the nearby water system. 3. Induce plant diseases and insect pests, resulting in a significant reduction in crop yields. 4. Strong corrosive effect on building materials, and serious damage to buildings and cultural relics.

In the 1960s, the European Air Chemistry Monitoring Network discovered acid rain areas with a pH value below 4.0 in the Netherlands, Denmark, Belgium, etc. Afterward, Swedish scientists pointed out that acid rain prevailed in Europe with increasing acidity. In the 1970s, precipitation monitoring in North America showed that acid rain covered almost the entire eastern part of North America. In some areas of the acid rain center, the pH value of precipitation dropped to 4.0. China is the third largest acid rain area in the world after, next to Europe and North America. China has the typical type of acid rain, which is caused by a large amount of sulfur dioxide emitted during coal burning. There are obvious regional differences in the distribution of acid rain in China, that is, acid rain is mainly concentrated in the region south to the Yangtze River and east to the Qinghai–Tibet Plateau. In 2012, among the 466 cities (counties) monitored, 215 cities (counties) had acid rain, accounting for 46.1%; 133 cities (counties) have the frequency of acid rain above 25%, accounting for 28.5%. The acid rain areas cover 12.2% of the country's land area.

In 1972, the United Nations held the first acid rain conference in Stockholm, inaugurating human's response to acid deposition and pollution through international cooperation. In 1979, in response to the severe cross-border transmission of atmospheric pollutants, 34 countries and the European Community signed the *Convention on Long-range Transboundary Air Pollution* (LRTAP), the first international convention with legal binding in human history to address air pollution on a regional basis. On the basis of this convention, European countries have further signed eight protocols, laying a solid foundation for a unified air pollution prevention and control plan in Europe. As coal burning is an important source of sulfur dioxide and nitrogen oxides, European countries have cut coal consumption as an important means to reduce emissions of these substances. This approach has been proven to significantly reduce emissions of sulfur dioxide and nitrogen oxides from coal consumption and contributed to the successful control of acid rain pollution in European countries.

In response to cross-regional pollution from acid rain, fine particulate matter and ozone, and to deal with pollutants capable of long-distance transmission such as sulfur dioxide and nitrogen oxides, the United States has implemented since 1994 a series of policies, such as acid rain planning (ARP), nitrogen oxide budget planning (NBP), and the Cross-State Air Pollution Rule (CSAPR), total volume control and emissions trading, and control over large and key emission sources such as power plants. Through nearly 20 years of efforts, the emissions of atmospheric pollutants such as sulfur dioxide, nitrogen oxides, inhalable particulate matter (PM10), and volatile organic compounds (VOCs) have decreased by 50–70% over the 1990 basis, and the air quality has improved significantly.

(iii) Smog

Smog (dust haze) refers to air pollution that a large number of extremely fine dust particles, soot particles, etc. (PM2.5) float evenly in the air, making the horizontal visibility of the air less than 10 km. The source of PM2.5 is very complicated and can be divided into primary source (direct discharge) and secondary source (secondary generation) according to how the particles are generated. Primary sources include coal combustion (coal burning for thermal power generation, industrial boilers, coal burning for civil use, etc.), motor vehicle emissions, industrial emissions (steel, metal smelting, building materials, chemicals, etc.), biomass burning, dust, etc. Secondary sources are gaseous pollutants (such as NOx, $SO_2$, $NH_3$, VOCs, etc.) released from sources of pollution after complex physical and chemical reactions in the atmosphere. Globally, after removing the effects of natural dust and sea salt, the areas with serious PM2.5 pollution are mainly concentrated in northern India and eastern China (Figure A.2).

The toxic and harmful particles that form the smog are suspended in the air, causing different degrees of harm to humans and ecosystems: the toxic and harmful substances in PM2.5 can directly enter the lungs through the nasal cilia in the nasal cavity, and even penetrate into the blood, which can cause a variety of diseases. Smog can absorb and scatter solar radiation, and thus affect plant photosynthesis, resulting in reduced crop yield and stunted growth of ecosystem growth. Smog can reduce visibility and tend to cause traffic congestion and traffic accidents.

(iv) Ground-level ozone pollution

Although the ozone layer in the stratosphere protects the Earth's organisms from the harm of the sun's ultraviolet rays, direct contact with ozone can be harmful to humans. The increasing concentration of ground-level ozone in the troposphere has become an environmental issue that the mankind is concerned about.

Ground-level ozone is mainly a secondary pollutant produced through a series of photochemical reactions of nitrogen oxides and volatile organic compounds. The main sources of nitrogen oxides and volatile organic compounds include human activities, such as emissions from coal burning, motor vehicle exhaust and petrochemicals, and natural emissions. Among them, motor vehicle exhaust is the major source of ozone pollution in urban environment. Ozone is a strong and irritating gas. Excessive inhalation of ozone can cause respiratory diseases or neurotoxicity, and destroy human immune function. Ozone can also damage building materials and household items as well as affect the plant growth.

Since the 1950s, European and American countries have experienced pollutions caused by soot pollutants and then motor vehicle emissions. Despite nearly 50 years of continuous efforts, ozone pollution is still the most prominent air pollution problem faced by these developed countries. China has not had the regional soot pollution under control, the regional complex air pollution characterized by fine particulate matter and ozone is increasingly prominent, which is more complicated and the situation is even more severe than those faced by the developed countries in Europe and the United States.

(v) Ozone depletion

The ozone holes are caused by ozone-depleting substances (ODS), such as hydrochlorofluorocarbons (CFCs), bromo-fluoro-alkane compounds (halon, CFCB), nitrous oxide ($N_2O$), carbon tetrachloride ($CCl_4$), and the like. Moreover, Freon and halon can stay in the atmosphere for a long period of time and are difficult to remove. Recent studies have found that nitrogen oxides also have a destructive effect on the ozone layer.

Humans already noticed the damage of the ozone layer over the Antarctic as early as in the 1970s. Rapid and massive depletion of stratospheric ozone over the Antarctic occurs in every spring, and nearly 95% of the ozone in the center of the ozone layer above the polar regions has been destroyed. Viewed from the ground, the ozone layer at high altitudes is extremely thin. Contrasted with the surrounding, it looks like a "hole" with a diameter of 1000 km, and hence the name "ozone hole". Satellite observations indicate that the area of the ozone hole is sometimes even larger than that of the United States. Figure A.3 shows the change of the ozone hole above the Antarctic from 1980 to 2014. A large area of ozone hole has also been observed over the Arctic, and even in the midlatitudes of the northern and southern hemispheres, increased ozone depletion has been observed.

The depletion of the ozone layer greatly increases the amount of ultraviolet radiation on the Earth's surface received from the sun, especially Class B ultraviolet light (UV-B) with a wavelength of 290–320 nm, which brings serious hazards to the Earth's organisms, including humans (Fig. 1.10). Increased Class B ultraviolet radiation can cause Class B burns and increase the risk of skin cancer and eye diseases such as cataracts. Prolonged intense UV radiation can also alter intracellular DNA, inhibit biological immunity, and reduce resistance to diseases including cancer and infectious diseases. Ultraviolet light also has a negative impact on many plants and may cause decline in yield and quality of certain crops. Strong UV radiation poses a hazard to aquatic ecosystems. The amount of plankton in the ozone hole area

**Fig. 1.10** The harm to the earth due to the ozone depletion

has dropped significantly. Because plankton is the basis of the marine life chain, it can lead to the decrease of fish, shrimp, shellfish, and even the extinction of certain species.

(vi) The sharp decrease of biodiversity

The diminished ecosystems, the shrinking land area and the declining health, and the change and loss of habitats of wildlife are closely related to the development of human society.

The natural extinction of species is a slow process at the speed of geological time, while the anthropogenic extinction of species is accompanied by the large-scale development of human beings. The disruptive human activities today accelerate the pace and increase the scale of species extinction. In the recent centuries, due to the extensive use of industrial technology, humans have increased the scale and intensity of the development of natural resources, and the rate of extinction and the number of endangered species have increased significantly. Studies show that in the past four centuries, human activities have caused the extinction of more than 700 species worldwide, including about 100 species of mammals and 160 species of birds. It is estimated that the number of extinct species in the last 10 years of the twentieth century is greater than the sum of species wiped out in the preceding 90 years.

## 1.3 Reflections on the Industrial Civilization

### *1.3.1 Reflections of International Communities*

Developed countries have experienced different stages in economic and social development and ecological and environmental governance. Old contradictions have been resolved, but new contradictions continue to emerge. According to the influence of ecological and environmental governance and the changes of participants, the process can be divided into two main stages: national and regional ecological and environmental protection, and global ecological and environmental governance.

(i) National and regional environmental protection

According to the different targets of resources protection and environmental governance, the national and regional environmental protection of developed countries since industrialization can be mainly divided into the following three stages.

1. The destruction and governance of natural ecological resources in developed countries (late nineteenth century—mid-1950s)

In the late eighteenth century, major western countries began their industrialization and urbanization. In order to meet the needs of infrastructure construction, human exploited the natural resources ruthlessly, causing increasingly serious deforestation and soil erosion, and higher frequency of natural disasters such as floods. Due to the destroyed natural ecosystems and recurrent disasters, humans have to think about environmental protection. Some countries introduced preliminary regulations to protect natural environment, including establishing the first national park established in 1872, and enacting the *Forest Law* in Switzerland in 1876. In addition, after the first environmental organization Sierra Club was established in the United States in 1892, the development of environmental protection organizations accelerated, and the public began to join the undertaking of environmental protection and governance. However, although the introduction of these preliminary regulations and the public efforts in environmental protection partially saved the deteriorating ecosystem in some developed countries at the early stage of capitalism, these endeavors were not able to control the overall deterioration of the environment caused by the rapid growth of traditional resource-based industrialization.

2. Worsening national and regional environmental pollution caused by industrialization and governance (mid-1950s–late 1970s)

In the mid-twentieth century, the further expansion of industrialization and urbanization and the demand for development exacerbated the destruction of the ecological environment by major developed countries, and spurred the usage of energy and resources exponentially. The lack of governance measures led to eight environmental incidents, including the 1943 Los Angeles photochemical smog and the 1952 London smog; these major incidents occurred first in the 1930s and mostly in the 1950s and 1960s, see Table 1.3 and Fig. 1.11.

The serious environmental crisis has gradually drawn the attention of governments, scholars, and the public on the importance of ecological environment. Some publications, including the *Silent Spring* by American scholar Rachel Carson in 1962, and the 1972 The Club of Rome's *Limits of Growth*, underpinned the environmental problems in the process of industrialization, illustrating that the earth's resources are limited, and that the destruction and pollution of the ecological environment beyond the capacity of the Earth will cause an uncontrollable decline in population and industrial productivity. In this context, major developed countries have established environmental protection authorities to promote clean energy production (reducing coal burning and increasing oil consumption) through policies and regulations, market means, etc. so as to increase the end-of-pipe treatment of pollutants and control major environmental pollutions. In developed countries, major environmental pollutants such as sulfur dioxide and nitrogen oxides peaked in the 1970s and began to decline steadily. The regional environmental problems, including air, water, and land pollutions, were basically handled.

3. Environmental governance on the pollution caused by excessive consumption of fossil fuels (from the 1980s to the present)

The energy strategy of "reducing coal burning and increasing oil consumption" and the end-of-pipe (EDP) treatment of pollutants have solved the problems caused by the traditional pollutants emitted from the process of industrialization and urbanization to a certain extent. However, the demand for a better quality of life since the 1980s drove the drastic increase in the use of fossil fuels by the transportation sector and

**Table 1.3** Eight major pollution incidents of the world

| | Incident | Time and place | Consequences |
|---|---|---|---|
| 1 | Meuse Valley smog event | Belgian Meuse Valley, December 1930 | The valley is 24 km long, flanked by 90 m-high mountains on both sides. With highly concentrated factories, the valley was filled with smoke. On the day of incident, the weather was foggy with temperature inversion. The air was polluted by $SO_2$, $SO_3$, and metal oxides. Thousands of people were poisoned, with many suffering symptoms of cough, shortness of breath, tearing, sore throat, nausea, vomiting, chest tightness, and suffocation. About 60 people died |
| 2 | Los Angeles photochemical smog incidents | Los Angeles, in May–October 1943 | There were about 400,000 motor vehicles of various types in the city, consuming nearly 24 million L of fuel per day and emitting over 1000 tons of exhaust into the air every day. Topographically, Los Angeles is a basin, which is not conducive to its air circulation. Therefore, under the ultraviolet light, the exhaust gas emitted by the automobile and petroleum industry generated photochemical smog that irritated people's eyes, nose, and throat, causing eye diseases and pharyngitis, resulting in many health issues among local residents and about 400 deaths of the elderly over 65-year old |
| 3 | 1948 Donora smog | Donora, USA, October 1948 | The town is located in the horseshoe-shaped river bay, a valley-shaped basin with 120 m high mountains on both sides. Due to the highly concentrated factories in town coupled with temperature inversion and foggy weather, the SO, $SO_3$ emitted from the factories and soot form sulfate aerosol. When inhaled, it causes symptoms such as cough, sore throat, vomiting, chest tightness, and diarrhea. About 6000 people were sick in 4 days, accounting for 42% of the total number of residents in the area, and 17 of them didn't survive the incident |
| 4 | London smog incident | London, the British capital, December 1952 | Due to coal the local residents used for heating during winter has high sulfur content and emits large amounts of $SO_2$ and soot when being burnt. Under the temperature inversion weather and catalyzed by metal particles, $SO_2$ generated $SO_3$, sulfuric acid, and sulfate, which were deposited on soot and inhaled into the lungs, causing chest tightness, cough, sore throat, and vomiting. In only 5 days, about 4000 people were killed. According to statistics, there were 12 incidents like this occurred in London over the years. The total number of deaths was nearly 10,000 |

(continued)

**Table 1.3** (continued)

| | Incident | Time and place | Consequences |
|---|---|---|---|
| 5 | 1953 kumamoto minamata mercury poisoning | Minamata Town, Kumamoto Prefecture, south of Kyushu, Japan, 1953–1961 | Because the local fertilizer plants used mercury chloride and mercury sulfate as catalysts in the production of nitrogen fertilizer. A large amount of converted methylmercury was discharged into wastewater and then flowed into the sea, contaminating the fish and shellfish. Local residents suffered organic mercury poisoning after consuming these methylmercury-rich fish and shellfish. The symptoms include slurred speech, unstable gait, expressionless, whole-body numbness, and finally mental disorders. By 1972, Minamata Town had more than 180 sick and 50 people died from the cause |
| 6 | Yokkaichi incident | Yokkaichi and dozens of other cities, Japan, since 1955 | It was caused by a large amount of $SO_2$ emissions from factories, coal ash, dust, and heavy metal particles such as cobalt, manganese, and titanium. When inhaled, such pollutants will cause respiratory diseases such as bronchitis, bronchial asthma, and emphysema. This incident led to more than 500 cases of sickness and 36 people died from asthma |
| 7 | Yusho disease incident | Aichi Prefecture and other 23 prefectures in Kyushu, Japan, 1968 | In the process of extracting rice bran oil, polychlorinated biphenyls were used as heat carriers. Due to poor management, PCBs leaked into the oil. Without knowing it, many people ate rice bran oil containing polychlorinated biphenyls, causing skin redness, sweating, and rash all over the body, and even symptoms such as nausea and vomiting, decline in hepatic functions, muscle pain, cough, etc. More than 5000 people were sick, 16 people died, and there were more than 10,000 victims |
| 8 | Los Angeles photochemical smog incidents | London smog incident | Yusho disease incident |
| | Fetal rickets caused by mercury pollution | | Bone pain causes severe bone deformity |

construction sector. These created the problems of urban atmospheric ozone and pollution of fine particulate matter, represented by PM10 and PM2.5, which are harmful to public health. At the same time, the excessive exploitation and consumption of fossil energy have promoted the rapid growth of greenhouse gas emissions, and the global climate change caused by rising atmospheric greenhouse gas concentrations has increasingly constrained the sustainable development of the global economy and society.

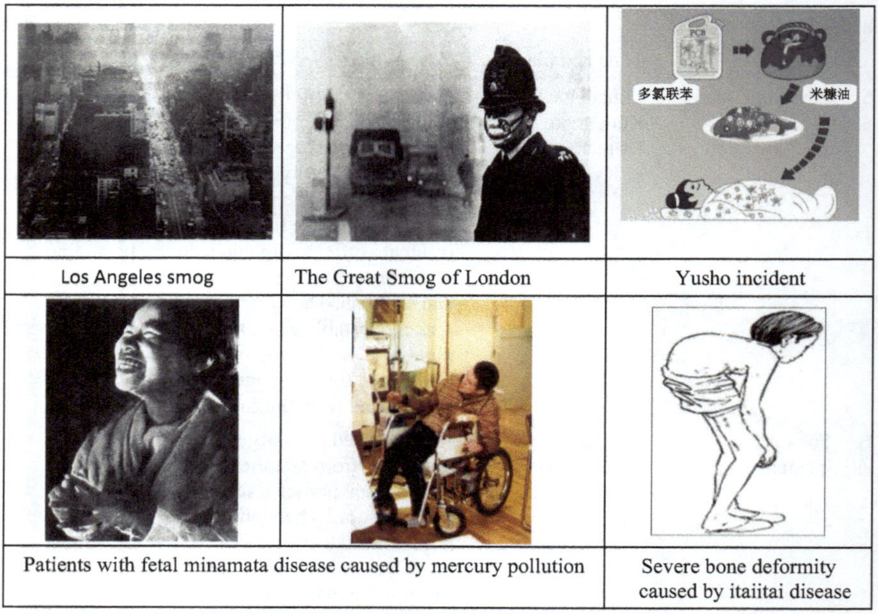


**Fig. 1.11** Some public pollution incidents in the world

In this context, major developed countries have curbed the fossil fuel demand by introducing vehicle fuel efficiency standards and establishing energy consumption and demand charges. The developed countries, such as Denmark and Germany, further increased fossil energy prices (Fig. 1.12) through far higher fossil energy environmental tax than other countries to curb high carbon emissions.

However, as the economic growth of developed countries depends on the growth of consumer demand, the increase in consumer demand will inevitably lead to the growing demand for fossil fuels, constituting a vicious circle that has not yet been fundamentally resolved.

ii Global environmental governance

As remarkable achievements were made in controlling regional environmental pollution in the developed countries in the late 1970s and early 1980s, environmental problems began to appear across the world, and developed countries began to turn their attention to international ecological governance. The international governance of the ecological environment overlaps with the regional governance of environmental issues in different countries. According to progress and achievements, global ecological governance can be divided into two stages.

1. Developed countries at the stage of active advocacy and actions (1970s–1987s)

With the increasing awareness of the global environmental protection, people are beginning to realize that the deteriorating ecological environment and the loss of

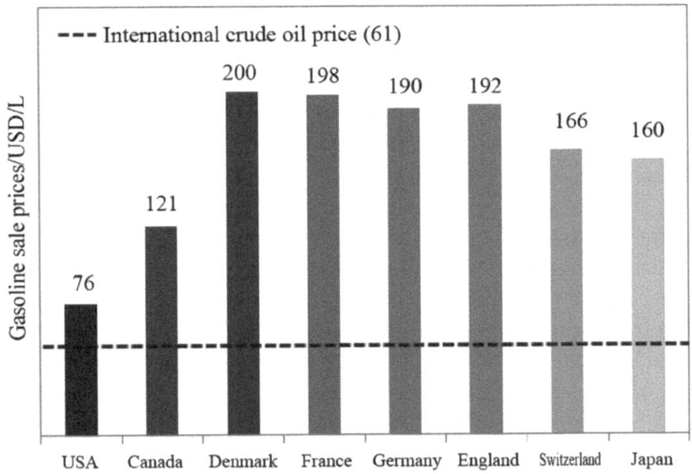


**Fig. 1.12** Comparison of gasoline sales prices in major developed countries

natural resources have significant spillover effects, and the environmental protection must rely on global joint efforts. In this context, the United Nations held the first Conference on the Human Environment in 1972 and adopted the *Declaration on the Human Environment*, which listed 26 principles to guide international and domestic activities in environmental protection, and served the foundation for the subsequent development of the *International Environmental Law*. This international consensus on environmental protection has greatly promoted the governance of environmental issues under the international framework, such as the accelerated enactment of the *Convention on the Prevention of Marine Pollution by Dumping of Wastes and Other Matters*, the *International Convention for the Prevention of Pollution from Ships*, and the *Convention on the Conservation of Migratory Wildlife Species*.

The first World Conference on Environment and Development in 1972 marks the turning point from "regional governance" to "global governance" initiated by developed countries. However, it should also be noted that the global governance during this period was limited to the advocacy and implementation of developed countries. The vast majority of developing countries did not participate in the discussion on global ecological issues and the design of governance systems. Some developing countries only dealt with the environmental and ecological issues that developed countries are concerned with after receiving their funding and technology assistance.

2. Further participation of developing countries and the polarization of the interest groups (1987–present)

On the basis of deepening the thinking and reaching a basic consensus on the relationship between economic and social development and ecological environmental protection between the 1950s and 1980s, the World Commission on Environment and Development published the report "Our Common Future" in 1987. This report provides the definition of "sustainable development", that is, "development that meets the needs of the present without compromising the ability of future generations to meet their own needs." Although the definition only points out what sustainability should be, it does not attempt to define development and the content is relatively vague. However, the concept of sustainable development has greatly promoted the international community's deliberations on the economic and social relationships with the environments and future growth models. The 2005 World Summit proposed that sustainable development consists of three interdependent yet mutually reinforcing pillars: "economic development, social progress, and environmental protection." The 2012 United Nations Conference on Sustainable Development proposes a green economy development path to coordinate the relationship between economic development, social progress, and environmental protection. The theories and practices of sustainable development have been continuously deepened and implemented.

As the concept of sustainable development deepens and evolves, the official documents on eco-conservation, including convention, provisions, and others are further enriched and created, including the major UN conventions of *1985 Vienna Convention for the Protection of the Ozone Layer*, the *1992 Convention on Biological Diversity* and the *United Nations Framework Convention on Climate Change*, the *1994 UN Convention to Combat Desertification*, and successively adopted *Agenda 21* and the *United Nations Millennium Goals*. These documents collectively illustrate the integrated action plans of governments, UN organizations, and development institutions, nongovernmental organizations, and independent groups, to address all aspects of human activities affecting the environment. They also identify the development goals that are time-bound and assessable to address prominent global issues, such as poverty, hunger, disease, illiteracy, and environmental degradation. At present, the issues of global ozone depletion and the harmless solid waste have been basically resolved, and substantial progress has been made in environmental issues such as cross-border transfer of hazardous wastes and protection of the ozone layer. However, it should also be noted that some global issues, such as desertification, biodiversity conservation and climate change, remain serious with the fundamental solution of the former two depending largely on the progress of climate change. Thus, future sustainable development must be achieved through low carbonization of energy production and consumption.

In terms of the participation in global environmental governance, the developing countries, with their rapid economic growth, have greater influence and say in the world, and their capacity and scopes of participation have expanded. At the same time, however, the traditional "two high and one resource" (high energy consumption, high pollution, and resource consumption) industries have shifted from developed countries to developing countries, where the ecological destruction and the pollutants emitted from related sectors have increased significantly. Developed countries ask the developing countries to take the responsibility for reducing pollutants and emissions, most of which are produced by the increasingly stronger economies. These developed countries, at the same time, continuously reduce their obligations to provide funding and technological assistance, where they downplay the difference with the major emerging economies but actually beyond the current development stages of these developing countries. The above reasons have led to the continuous polarized interests of the two sides in the global environmental governance. Even internally, there are confrontations that require more intense international negotiations.

Table 1.4 lists the international community's reflections on the growth model.

## *1.3.2 Reflections of the Chinese Society*

Since the founding of New China, there have been profound changes in the governance philosophy of the Party and the central government. With deepening understanding of the relationship between man and nature, the Chinese government has proposed a series of strategic ideas and practical approaches in solving resource and environmental problems since the twenty-first century (Fig. 1.13). In particular, in 2002, the 16th National Congress of the Communist Party of China put forward the idea of "taking a new path to industrialization." In 2007, the report of the 17th National Congress of the Communist Party of China clearly stated the new requirements for building an ecological civilization and an environment-friendly nation by 2020, which is regarded as one of the important criteria for building a moderately prosperous society in all respects. In 2010, the Fifth Plenary Session of the 17th CPC Central Committee stated the need to advance ecological civilization and vigorously promote green buildings, green economy, green mining, green consumerism, and government's green procurement. At the same time, "green development" was written into the "Twelfth Five-Year Plan" as an independent part, indicating China's determination and confidence in taking the green development path. In 2012, the strategic task of developing an ecological civilization was systematically, completely, and theoretically put forward in the report of the 18th CPC National Congress, highlighting "Resource consumption, environmental damage and ecological benefits should be covered by the system of standards for evaluating economic and social development, and related goals, evaluation methods and reward and punishment mechanisms should be adopted in keeping with the need of promoting ecological progress," and to integrate ecological progress into the overall plan for promoting all-round economic, political, cultural, social, and ecological progress toward socialist modernization. In 2013, the five major institutional reforms proposed in *The Decision on Major Issues Concerning Comprehensively Deepening Reforms* was adopted at the Third Plenary Session of the 18th CPC Central Committee put forward two major principles in the reform of the system for developing an ecological civilization: First, develop an ecological civilization, build a Beautiful China, work faster to establish an ecological civilization system, improve institutions and mechanisms for developing geographical space, conserving resources and protecting the ecological environment, and promote modernization featuring harmonious development between man and nature. Second, to build an ecological civilization, we must establish a systematic and complete ecological civilization system and use the system to protect the ecological environment. We must improve the property rights system for natural resources and the administration of their use, draw red lines for protecting the ecosystems,

**Table 1.4** International community's reflections on the model of development

| Years | Events | | Content and key points |
|---|---|---|---|
| 1962 | Rachel Carson "Silent Spring" | DDT advertisement in time magazine "Silent Spring" by Rachel Carson | The irrefutable facts and data point out that the abuse of DDT breaks the food chain of nature, deprives humans of a thriving spring, damages human metabolism and fertility, and induces cancer. In the face of the overwhelming siege of interest groups, the ailing Carson debated with representatives of chemical companies on television. In 1972, the United States banned the production and use of DDT. For the first time, *Silent Spring* evoked the environmental awareness of the public and became the declaration of the modern environmental movement. It exposes the conflicts between man and nature behind the capitalist industrial prosperity, raises challenges to the traditional concept of war on nature or conquering nature, sounds the alarm of environmental crisis in the industrial society, and inaugurates the progress toward ecological civilization |
| 1972 | The Club of Rome *The Limits to Growth*; United Nations Human Environment Conference *Human Environment Declaration* | The Limits to Growth<br>The standard trend of the world model | The study uses a mathematical model named the world model based on five basic variables—population, agricultural production, natural resources, industrial production, and pollution—to calculate a general decline after reaching a certain peak. The results of the continuous calculations were verified by the data from 1900 to 1970. This study shows that the earth ecosystem will collapse due to overwhelming consumption and unlimited exploitation of the resources. The countermeasure that the human society should take is to follow the path of balanced development. The results derived from the rigorous mathematical model corroborate Carson's warning, and enlighten some government officials, boosting international cooperation across cultures and nations for the purpose of environmental protection<br>That year, "Declaration on the Human Environment" was adopted at the United Nations Conference on the Human Environment, calling for more careful consideration about the consequences of any action on the environment |

(continued)

**Table 1.4** (continued)

| Years | Events | | Content and key points |
|---|---|---|---|
| 1987 | United Nations World Commission on Environment and Development, "Our Common Future" | Sustainable development Diagram of a sustainable development system | Composed of three parts of "common concern", "common challenge", and "common effort", this report with focuses on such aspects as population, food, species and genetics, resources, energy, industry, and human habitation, systematically explores a series of major economic, social, and environmental issues of the human society. It clearly puts forward three viewpoints: (1) Environmental crisis, energy crisis and development crisis are interrelated. (2) The Earth's resources and energy deposits are far from meeting the needs of human development. (3) The development model must be changed for the interests of present and future generations<br>On this basis, the report proposes the concept of "sustainable development" |
| 1992 | United Nations Conference on Environment and Development, *Rio Declaration on Environment and Development* and *Agenda 21* | United Nations Conference on Environment and Development | *Rio Declaration on Environment and Development* aims to establish new levels of cooperation between the states, key sectors of society, and the people in order to establish a new and equitable global partnership and to obtain the international agreements that respect the interests of all and safeguard the integrity of the global environment and development system. Based on the realization of our homeland, the integrity and interdependence of the planet's nature, this declaration sets out the 27 basic principles of sustainable development<br>*Agenda 21* is the "Worldwide Action Plan for Sustainable Development." It is a comprehensive action plan, but not legally binding, to address human activities that have an impact on the environment in all aspects for the governments, UN organizations, development agencies, NGOs, and independent groups around the world from 1992 to the twenty-first century |

(continued)

**Table 1.4** (continued)

| Years | Events | | Content and key points |
|---|---|---|---|
| 2002 | World Summit on Sustainable Development | – | Since international environmental development contains complex and intertwined conflicts of interest, such conventions as *Agenda 21* that aim for global sustainable development have not been effectively carried out, and the global environmental crisis has not been lifted. On the one hand, it is difficult for the developing countries to achieve economic development and environmental protection due to their own underdeveloped economy. On the other hand, developed countries have not fulfilled their obligations to provide technical and financial supports to developing countries in the implementation of the Convention. As a result, poverty is still widespread globally with ever-widening income gap. Most countries believe that it is necessary to convene more international conferences to review the guiding principle of the Rio Conference, and to discuss the new issues of the global partnership. The 2002 summit was held to prepare for this international event |
| 2012 | United Nations Commission on Sustainable Development "Our future" | Seeking consensus in disagreement Green economy indicator system | "Our future" proposes all countries in the world to re-commit to sustainability, ensuring an economically, socially. and environmentally sustainable future for our planet and the present and future generations. The major topic of the summit is the path of green economy, which drew great attention from enterprises and investors, who were concerned about the long-term forecasts regarding the shortages of energy, water, and other resources. However, in terms of capital investment in the green economy, the developed and developing countries have very conflicting views: developed countries argue that all countries, including developing countries, must transition to green economy, while developing countries claim that green economy is only an option. One of the prerequisites for transitioning to the green economy is to obtain financial assistance from developed countries |

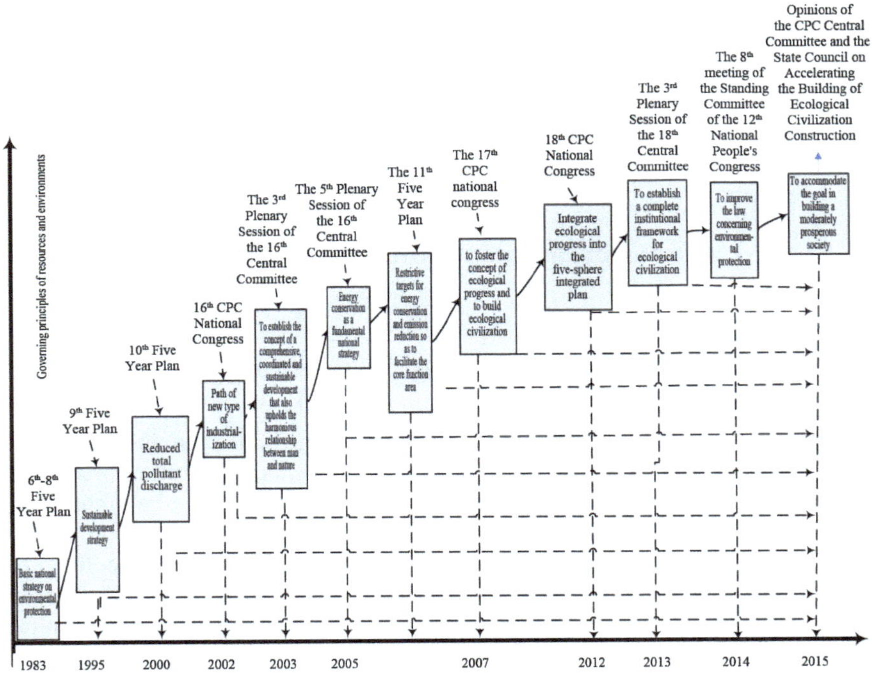


**Fig. 1.13** China's governance philosophy concerning resource and environment since 1983

establish a system for paying for resource consumption and compensating for ecological damage, and reform the ecological environment management system. The 8th meeting of the Standing Committee of the 12th National People's Congress in 2014 passed the revised draft of the *Environmental Protection Law*, proclaiming the priority of "coordinating economic and social development and environmental protection" and realizing the transition from "policy" to "implementation", marking another key milestone in the history of China's environmental legislation. The law, for the first time, has taken in the red lines for ecological protection, stipulated environmental public interest litigation, and designed a daily penalty system, and has been added with a public monitoring and early warning mechanism for environmental pollution. In April 2015, the CPC Central Committee and the State Council issued the "Opinions of the CPC Central Committee and the State Council on Accelerating the Building of Ecological Civilization," arguing that the level of ecological progress in China still lagged behind and has become a major "bottleneck" restricting sustainable economic and social development. The Opinions, composed of general requirements, nine sections, and 35 provisions, lays out guidance in building ecological civilization, breaking through the "bottleneck" and "inadequacy", and building a moderately prosperous society in all respects. The Opinions sets a goal that by 2020, significant progress will be made in the building of a resource-saving and environment-friendly society, the framework of the main functional areas will be basically formed, the quality and efficiency of economic development will be significantly improved, the ecological civilization as mainstream values will be promoted in the whole society, and the level of ecological civilization will be compatible to the goal of building a moderately prosperous society.

## 1.4 Low-Carbon Development Toward Ecological Civilization

### *1.4.1 Essential Concepts of Ecological Civilization*

Ecology refers to the state in which organisms live and grow in a certain natural environment. It also refers to the physiological and living characteristics of organisms. Civilization refers to the state in which human society advances to a higher level with a higher culture. In 1987, Ye Qianji, a Chinese ecologist and scholar, clearly defined the concept of ecological civilization from the perspective of ecology. He believed that ecological civilization is that human beings could both benefit from and benefit nature; transform and conserve the nature while keeping a harmonious relationship with nature. In 1995, the famous American writer Roy Morrison used the term "ecological civilization" in his book "Ecological Democracy" and regarded ecological civilization as a form of civilization after industrial civilization.

According to the evolution of human civilization, different scholars at home and abroad have defined ecological civilization with insights from different perspectives, including the following:

1. General perspective

Ecological civilization is a new stage in the progress of human civilization. It is a more complex, progressive, and advanced form of civilization that emerged after the industrial civilization of human society. It represents a better relationship between man and nature, between man and society, and between man and man. "Primitive civilization → agricultural civilization → industrial civilization → ecological civilization" is the inevitable development trend and destined result of human civilization, and the most stable form of the human society.

2. Specific perspective

Ecological civilization is an aspect of social civilization and an external manifestation of material, cultural and ethical, and political advancement. In the era ecological civilization that is to come, social advancement is still composed of material, cultural–ethical advancement, and political advancement. The ecological civilization is supported by ecological culture. Correspondingly, it is the ecological-economic culture that supports the material advancement, the ecological culture that supports the cultural–ethical advancement, and the ecological political culture that supports the political advancement.

3. Ecological civilization is a concept of development

On the basis of the achievements of industrial civilization, ecological civilization treats nature with a more civilized approach instead of barbaric and violent plundering, actively builds and carefully protects a good ecological environment while improving and optimizing the relationship between man and nature so as to achieve the long-term goal of sustainability of economic and social development.

In terms of development history, ecological civilization is a new form of civilization after primitive civilization, agricultural civilization, and industrial civilization. In terms of the difference between ecological civilization and other civilizations, ecological civilization is a contrast to industrial civilization that features high productivity and high carbon emissions with serious pollution and ecological damage. It emphasizes low carbon, high efficiency, high technology, low consumption, low pollution, overall coordination, recycling, renewal, healthy, and sustainability (Fig. 1.14). It is also considered as "ecologicalized industrial modernization." The essence of the concept of ecological civilization is to regard the ecological environment as the basis for the sustainable and healthy development of human beings. Any development beyond the ecological carrying capacity will bring about bad or even serious consequences.

In general, ecological civilization is the sum of the material, cultural–ethical, and institutional achievements that human beings have achieved to protect and build a beautiful ecological environment. It is a systemic project linking the entire process and all aspects of economic, political, cultural–ethical, and social advancement, reflecting the progress of a civilization and philosophy about the dialectical relationship between economic development and ecological environment. The building of an ecological civilization must be on the premise of understanding and respecting the laws of nature, with a purpose of ensuring harmony between man and nature, environment and economy, and people and society. On the basis of the carrying capacity of

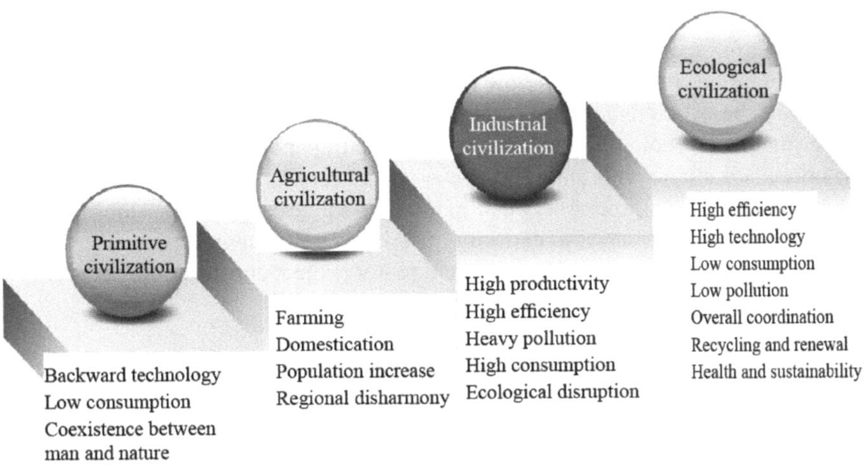


**Fig. 1.14** Comparison between ecological civilization and other civilizations

resources and environment, we shall focus on establishing a spatial pattern and industrial structure, way of production, lifestyles, and the capacity to enhance sustainable development so as to meet the essential requirement on building a resource-saving and environment-friendly society.

### *1.4.2 The Implications of Ecological Civilization*

Ecological civilization is a state that reflects the degree of harmony between human progress and nature, and is the fruit of profound deliberations on the ecological and environmental crisis brought about by traditional industrial civilization. The ecological civilization we pursue is a realm of positive interaction between and harmonious coexistence and sustainable development of human society and nature. Its essence is to establish a conservation-oriented, environment-friendly society that is based on the carrying capacity of resources and environment, follows the law of nature, and aims for sustainable development. To make in-depth analysis on gaining a clear understanding of the contents of China's ecological civilization in the current era, we must first understand the ideological concept, essential characteristics, national policies, and development approach regarding ecological civilization in China, and clarify the overall requirements and key tasks at this stage.

(i) Ideological concept of promoting ecological progress

Respecting nature, accommodating nature, and protecting nature are important ideological foundations for advancing ecological civilization as they embody new values and ecological ethics. In economic development, we pay more attention to the law of economy than the laws of nature, neglecting the carrying capacity of resources and environment in some regions, damaging natural environment, and weakening sustainability. Ecological civilization emphasizes the harmonious development of people, nature, and society. It is not only a new model of development but also a concept of value. Its essence is how to deal with the new ecological ethics and to care about the relationship between human beings and all forms of life. To build a great nation of ecological civilization, we must abandon the belief and practice that upholds human's will and power, but act in accordance with the requirements of harmonious development between man and nature and fully consider the natural conditions and the carrying capacity of resources and environment in laying out productive forces, urban development, and major project construction.

(ii) Essential characteristics of promoting ecological progress

The essential feature of promoting ecological progress is to give prominence to ecological civilization and integrate it into all aspects and processes of economic, political, cultural, and social advancement, promote fundamental changes in ways of production and lifestyle, not only in the work but also in the concept, principles, and goals in various fields regarding resources and environment. In the aspect of economic development (material civilization), we must make efforts to change the blind and inefficient economic growth featuring high carbon emissions and severe pollution to green and intensive economic growth featuring high efficiency and low carbon emissions. In the aspect of political building (political civilization), we must enhance system building to promote ecological progress, improve the legislation in the field of resources and environment, and deepen the reform of the resource and environmental management system. In the aspect of cultural building, we must establish a healthy environmental ethics and low carbon consumption concept and advocate the joint action of the whole society and promote the concept of ecological civilization. In the aspect of social building, we must promote ecological civilization education, advocate green consumption and lifestyle, and improve environmental quality and public health.

(iii) National policy for promoting ecological civilization

We must follow the fundamental national policy of conserving resources and protecting the environment, uphold the principle of prioritizing resource conservation and environmental protection, and letting nature restore itself. It is the basic guideline for resource development and utilization, low-carbon development and ecosystem protection, and serves as the fundamental provisions in formulating various economic and social policies, carrying out various tasks, and preparing various types of planning. Especially, in economic and social development, we must make analyses on resource utilization, related environmental impacts, and carrying capacity in addition to GDP growth rate.

(iv) Ways to promote ecological progress

To promote low-carbon, green, and circular development is the main path that China takes to promote its ecological civilization at the present stage, and also an important task and content of economic reform. Low-carbon development, green development, and circular development are the forms of economic development that emerged after the energy crisis, environmental crisis, and ecological crisis. Low-carbon development is the superposition and organic unification of low carbon and development. Its core mission is to improve carbon productivity and achieve low carbon consumption, low emissions, low pollution, high efficiency, high benefit, and high carbon sink in the development process. Green development focuses on the sustainable use of natural resources and improving the "green GDP" through the application of resource and energy-conserving and environmental-friendly technologies and products and an approach that is market oriented, efficient, and sound. Circular development focuses on the material recycling of the entire society, through conservation, and full utilization of resources in the whole process of production, circulation, and consumption, reducing the flux intensity of material and energy flows from natural resources into the social and economic system. To vigorously develop low-carbon economy, green economy, and circular economy is the basis and strategic way for China to advance ecological civilization at this stage.

To advance ecological civilization, we must proceed from ecological protection and governance. It is more important and urgent to carry out the ecological education in the whole society, vigorously popularize the concept of ecological civilization, strengthen moral education on ecological civilization, consolidate the foundation of ecological civilization, have the whole society's attention to the ecological environment and protection, and have the ecological civilization become the conscious and voluntary lifestyle of each of us in building a Beautiful China.

### *1.4.3 The Reality and Lasting Significance of Ecological Civilization*

Ecological civilization is a theoretical innovation of the path of socialism with Chinese characteristics. Giving prominence to ecological civilization has strategic significance for low-carbon development and building a Beautiful China.

(i) The fundamental guarantee for promoting low-carbon development

The international community has reached a political consensus on global temperature rise not to be over 2 °C, and the global response to climate change will be further strengthened. Green and low-carbon development has gradually become the direction and trend of global economic development, and many countries are accelerating in formulating relevant strategies and policies.

The next stage is a crucial period for China to build a moderately prosperous society in all respects. It is also an important period of strategic opportunity for China to vigorously promote the ecological civilization as well as green and low-carbon development. The *National Climate Change Plan (2014–2020)* proposes that active response to climate change and accelerated green and low-carbon development are the inherent requirements for achieving sustainable development and promoting ecological civilization. Integrating ecological civilization into economic, political, cultural, and social advancement will help build a green and low-carbon growth model with Chinese characteristics.

China is still in the process of industrialization and urbanization. Accelerating the green and low-carbon development and effectively controlling greenhouse gas emissions have become an inherent requirement for China to vigorously promote ecological civilization. We must solidify the concept of ecological civilization and take a sustainable development path that is in line with China's national conditions for the win-win economic development and green and low-carbon development.

(ii) The only path to build a Beautiful China

First of all, resources are the material basis for economic and social development. To finish the building of a moderately prosperous society in all respects with 1.3 billion people, the demand for resources is enormous. However, China's natural endowment is inherently insufficient. The per capita possession of strategic resources, such as oil, natural gas, coal, freshwater, and cultivated land, is only 5.8, 10, 67, 28, and 43% of the world average, respectively, indicating serious constraints on China's development. In particular, the energy resource, in 2013, China's apparent consumption of oil was 514 million tons of standard coal, and the dependence on foreign oil reached 59.5%.

At the same time, China is plagued with blind and inefficient growth model, low resource utilization efficiency, and serious wastes. In 2012, China's GDP accounted for about 11.6% of the world's total volume, but consumed 21.3% of the world's energy, 45% of steel, 43% of copper, and 54% of cement. With the rapid development of industrialization and urbanization, the imbalance between supply and demand of energy resources will become more prominent.

Second, a good ecological environment is a new demand by the times and social progress, and has become a new expectation of the people. However, the problem of environmental pollution in China is still formidable. The air quality is poor. In recent years, large-scale and prolonged smog has occurred, affecting an area of more than 1.3 million $km^2$ and 600 million people. The drinking water is unsafe as its quality is not up to standard. There is serious water pollution in some key river basins and offshore areas, and environmental problems like lake eutrophication, heavy metal pollution, grassland degradation, soil erosion, and land desertification have damaged people's health and quality of life and even led to mass incidents.

Only by carrying out ecological civilization, transforming the growth model, and promoting the transformation of resource utilization from extensive exploitation to intensive and efficient usage can we make the limited resources have the highest effect, break the "bottleneck" constraint of resources and environment, and provide material support for building a moderately prosperous society in all respects. Only by taking people-oriented approaches, exercising power in the interests of the people, accelerating the ecological civilization, and transiting the ecological approach from "pollution followed by treatment" and "destruction followed by restoration" to prioritize protection and natural restoration can we build a Beautiful China.

(iii) A requisite to promote the transformation of growth model

The ecological civilization is not only ecological restoration and reconstruction, resource conservation, and environmental governance but also involves profound changes in the entire social civilization. It is necessary to change the way of production and consumption patterns of the whole society and establish ecological civilization with green and sustainable concept of production and consumption.

1. Production

In the process of industrialization, China has completed in about 30 years the development that takes the developed countries more than 100 years to complete, and the environmental problems that developed countries have experienced in different historical stages are thus concentrated in the past 20–30 years in China. For a long time, due to factors such as geographical environment, development stage, and economic model, China's way of production is excessively extensive when compared with the international advanced ones. The increasingly limited resources are becoming a "bottleneck" that constrains the sustainable development of the economy and society.

First, with the high-speed and extensive growth of the economy, the energy consumption of production is huge and continues to rise. Reports show that China has become the world's largest consumer of iron ore, alumina, steel, copper, cement, and other resources, and increasingly dependent on imported mineral resources and energy. According to statistics, from 1990 to 2009, energy and mineral consumption increased more than twice, and the consumption of metal ores increased by 8–15 times.

Second, China's efficiency in energy utilization is still low, and emission pollution is serious. China's total energy efficiency is 33%, about 10 percentage points lower than the world average. The GDP per unit of energy use is four times that of the United States, six times that of the European Union, ten times that of Japan, and twice the world average. High consumption and inefficient way of production have brought about huge pollution hazards. In 2010, China's total wastewater discharge was 61.73 billion tons, chemical oxygen demand was 12.381 million tons, and sulfur dioxide was 2.851 million tons, all ranked first in the world. The total discharge of major pollutants exceeded the carrying capacity of the natural environment.

Finally, the industrial transformation and upgrading have always been a major problem of China's economic development and has made no significant progress so far. In terms of industrial structure, China has entered a new round of economic growth cycle since 2003. The steel, cement, automobile, and other industries, which are the basis of industrialization, have grown rapidly, and the industrial structure is dominated by heavy industry. In terms of industrial development model, China has not yet got rid of the extensive growth model characterized by input of factors of production and scale expansion. The industries of steel, cement, plate glass, coal, and chemical industry are facing the threat of overcapacity. In terms of industrial competitiveness, the basic characteristics of the industry are "big but not strong," that is, although the industry is large in scale, the product structure and technical level are low. Overall, it is still at the low end of the global industrial chain. Taking automobiles as an example, China has the world's top production and sales scale, but also faces the weak competitiveness of self-owned brands and the fact that domestic mid- to high-end markets, especially high-end markets, are almost completely occupied by foreign products.

On the whole, China's past economic growth has mainly been achieved by increasing the input of capital and labor and other factors and expanding the scale of the original production. This high-input, high-energy, and high-consumption growth model poses a serious threat to the ecological environment and causes serious environmental pollution, species extinction, and resource shortages. It will be difficult to continue the economic growth at the expense of the environment.

Therefore, in the process of building an ecological civilization, we should base ourselves on circular economy, pay attention to the ecological environment, and shift the economic growth model from extensive growth featuring low efficiency and serious pollution to intensive growth that is efficient, green, and sustainable. Meanwhile, we should energetically develop strategic emerging industries such as environmental protection, new energy, and new energy vehicles, which can promote energy conservation and emission reduction, enhance competitiveness, provide new employment opportunities, and become new economic growth areas, thereby promoting industrial restructuring.

2. Consumption

The consumption pattern is the overall state of private consumption under the guidance of certain consumption concepts. Different from the direct effect of way of production on the ecological progress, consumption patterns indirectly but significantly affect the level of ecological civilization through the downstream impacts and elastic effects of consumption. The so-called "downstream effect" means that by reducing one unit of the consumption at the bottom of the system will reduce tens of times or even hundreds or thousands of times of the resource inputs in the upstream of the system; the "elastic effect" means that the effect of improved production efficiency will be neutralized by the increase in consumption. Therefore, it is crucial to change consumption attitude. However, China's current consumption patterns have problems such as extravagance, serious waste of public expenditure, and inferior consumption.

Driven by consumerism, high consumption, excessive consumption, and one-off consumption have become the lifestyles pursued by many Chinese people. The frequent replacement of electronic and electrical products such as mobile phones, computers, televisions, etc. and the blind pursuit of luxury goods have made China a major consumer of luxury goods worldwide. According to data released by Goldman Sachs, China's luxury goods consumption in 2010 was as high as $6.5 billion, ranking first in the global growth rate for three consecutive years. In the next 3 years, China is expected to surpass Japan in terms of total consumption of luxury goods and become the world's largest consumer of luxury goods. This shows that China's luxury consumption is becoming more and more serious, and it is not compatible with China's economic and social development, traditional consumption habits, and the strategy of building ecological civilization. Luxury consumption occupies excessive social resources.

The problem of serious waste of public expenditure is another major factor restricting the upgrade of Chinese consumption pattern. Because of the undisclosed and nontransparent management, there are corruption problems derived from mixing personal interest with public property, such as the problems of "official vehicles" that draw public criticism, and using public funds to travel abroad in the name of business trip. In May 2012, the well-known consulting firm Roland Berger published the special issue of "Think: Act": according to China Luxury Market Study, the official consumption of luxury is about 20% of China's total luxury goods consumption, far more than that of developed countries, suggesting a serious waste of official expenditure.

In addition, inferior goods consumption is also a problem in China. Inferior goods, due to their poor quality or loss of functionality, can cause environmental pollution while occupying and consuming resources. Although inferior consumer goods only account for a small amount of the total social commodity in China, they can do great harm to China's economic and social development and the establishment of a socialist market economic system if the situation gets worse. For example, the current serious air pollution problems in China are affected to some extent by inferior oil products.

To build an ecological civilization in China, we must transform our consumption attitude, improve the consumption structure, guide reasonable consumption, and encourage consumption of ecological products and green products, gradually develop a healthy and resource-saving way of consumption, and to advocate consumption that is "comfortable not extravagant; to consume but not to waste."

In general, China's economy has been in the stage of extensive development for a long time. Although the total economic output ranks second in the world, the overall level of industry is low, pollution emissions are high, and the environment is greatly damaged. The essence of environmental issues is the problems of economic structure, way of production and growth model. To achieve sustainability in economic and social development and build a moderately prosperous society in all respects, we must vigorously advance of the ecological civilization, accelerate the transformation of extensive economic growth, and develop the industrial structure, growth pattern, and consumption pattern that conserve energy resources and protect the ecological environment.

(iv) The only way to upgrade ecosystem

Ecological civilization also has a profound role in promoting the improvement of China's development strategy and the enhancement of its civilization. Specifically, the strategic significance of upgrading the ecological civilization to the overall planning of social development is reflected in two aspects.

The first is the transformation and improvement of China's traditional development concept. In 1987, the 13th National Congress of the Communist Party of China proposed a "three-step" development strategy for the period before the middle of the twenty-first century, where the per capita gross national product reaches the level of moderately developed countries, people's lives are relatively rich, and modernization was basically achieved. In 1997, Jiang Zemin proposed a phased development concept for the development of China's modernization in the first half of the twenty-first century in the report of the 15th National Congress of the Communist Party of China. It was called the "new three-step" development strategy. He pointed out: "Looking into the next century, our goal is to achieve a doubling of the gross national product in the first 10 years than in 2000, so that the people's life will be more affluent and a more complete socialist market economic system will be formed. After another 10 years of hard work, the 100-year-old Party will has made the national economy more developed and the various systems better. At the 100th anniversary of the founding of the new China, it will basically realize modernization and build itself into a prosperous, strong, democratic and culturally advanced socialist country." The new strategy is to supplement and improve the original one. It is the overall goal of the Party and government and the Chinese people of all ethnic groups in the new era, and guides the development of China's social and economic development. At present, China has entered the second stage of development of the "new three-step" strategy, but the development strategy only clearly proposes the growth targets of China's per capita GDP in 2020 and 2050, and does not make clear arrangements on the social development mechanism such as the protection of resources and environment and the transformation of development methods. The proposal and development of ecological civilization will supplement China's current second and third development goals, and clarify the environmental protection objectives of ecological resources. It has been incorporated into the overall planning of China's social development, is expected to promote the improvement of China's development strategy in the new era, and guide China's material, cultural–ethical, and ecological advancement in a more comprehensive and scientific manner.

The second is to upgrade China's economic growth and social development models and promote the upgrading of China's forms of civilization. Ecological civilization is not only the innovation and improvement of China's economic growth model but also the specific operational approach and practice process of China's move toward ecological civilization, which will facilitate China to enter a period of new and higher level civilization.

(v) The duty of a major responsible country

At present, climate change and energy security have increasingly become the common challenges of human society. Green, circular, and low-carbon development has become global consensus and international trend. From a global perspective, with the rise of emerging economies such as China and India, the world's resource and environment pattern has also changed, and the global resource consumption and pollutant emissions have gradually shifted to the east.

China, with the highest and fastest growing total greenhouse gas emissions in the world and per capita emissions exceeding the world average, has become increasingly the focus of attention in international negotiations on climate change. At the same time, many developed countries have accelerated the development of "green economy" such as new energy, new materials, and energy conservation and environmental protection industries. In comparison, China is at the middle and low end of the global industrial chain, relying mainly on resource and environmental consumption and cheap labor to earn meager profits. In the face of new international development trends and competitive situations, we must only actively promote ecological civilization and take the initiative to go green. Only by doing so can we effectively control the momentum of excessive growth of greenhouse gas emissions, enhance the competitiveness of China's industrial products in the international arena, make positive contributions to coping with global climate change, fulfill the responsibility of the major responsible country, and assume the international moral high ground.

### *1.4.4 Low-Carbon Development is the Only Way to Ecological Civilization*

(i) The significance of low-carbon development to ecological progress

The current climate change is an indisputable fact. To cope with climate change, we have no choice but to take the path of low-carbon development. Climate change is a global issue, a problem challenging the survival and development of all mankind, and a problem to be solved in building ecological civilization. By means of low-carbon development, which is an effective method to tackle climate changes, we can effectively protect the ecological environment.

Ecological civilization requires the formation of a spatial pattern, industrial structure, way of production, and lifestyles that conserve resources and protect the environment. Low-carbon development is conducive to the transition of China's economic structure from the past model of "high input, high energy consumption, high pollution, low output" to "low input, low energy consumption, low pollution, high output," conducive to developing ways of environmentally beneficial production and consumption, to developing nonpolluting or low-pollution technologies, processes, and products. Also, it is conducive to creating a social and cultural atmosphere in which everyone cares for the environment, and is conducive to building a resource-saving and environment-friendly society.

Low-carbon development is a model for sustainable economic development. By improving energy efficiency and using renewable and low-carbon energy to transform high-carbon society into a low-carbon society, it is an ideal model for the development of green economy. Therefore, low-carbon development is conducive to the harmonious development between man and nature, to the principle of people-oriented, comprehensive, coordinated, and sustainable development, and building of ecological civilization and a moderately prosperous society in all respects.

(ii) Low-carbon development is the strategic choice for ecological progress

China is a big country in terms of total resources, but it is not the case in terms of per capita possession of resources. The conflicts between the total amount of resources and the growing population will exist for a long time. At present, China is experiencing rapid industrialization and urbanization, which has accelerated the consumption of various resources and made the contradictions increasingly prominent. Generally speaking, China's ecological environment is deteriorating. Although local improvement has been made, China's ability of ecological governance is far less than the rate at which it is destroyed, resulting in the expanding ecological deficit.

Low-carbon development is an important approach and means to realize ecological civilization. For a long time in the past, China mainly promoted economic development by consuming large amounts of energy resources. This process has produced a series of serious consequences. Low-carbon development will transform the society from high carbon into low carbon consumption by improving energy efficiency and promoting technologies to reduce greenhouse gas emissions. To solve the serious problems and prominent contradictions in ecological civilization, the most fundamental way is to develop low-carbon economy, improve the industrial structure, and improve production efficiency and labor productivity by developing a low-carbon economy model of "low input, low energy consumption, low pollution, and high output," so as to fundamentally reverse the deteriorating situation of China's ecological environment.

(iii) Low-carbon development is the norm for ecological civilization

Ecological civilization inevitably requires developing a low-carbon economy. The ecological progress will no longer allow industries with high energy consumption, high pollution, and high emissions to exist in the national economic system for a long time, and will no longer tolerate the huge pollution caused by these industries to the ecological environment. At the same time, low-carbon development is bound to benefit ecological civilization, and is one of the important ways to achieve ecological civilization.

Low-carbon development provides important reasons and realistic chance for ecological progress. First, low-carbon development has broad consensus and a solid foundation in modern society. While reflecting on the high-carbon setbacks, the international community has gradually reached a low-carbon consensus. Many countries regard low-carbon development as an important way to cope with global climate changes and achieve sustainable development. Furthermore, the low-carbon concept has been applied on and explored in many aspects of social development—from low-carbon materials, low-carbon construction, low-carbon transportation to low carbon consumption, low carbon production, low-carbon lifestyle, low-carbon technology, low-carbon economy, and low-carbon society. Finally, low carbon provides the most controllable standard for modern society to advance ecological civilization. Because carbon-based energy is the most widely used energy resource in modern society, people are most familiar with it and thus strong at measuring and evaluating it among many ecological factors.

Therefore, low-carbon standards are inherently rational and inevitable in advancing ecological civilization, and low carbon is a standard for ecological civilization.

## 1.5 Summary

More than 5000 years ago, the intelligent, industrious, and brave Chinese nation created a splendid Chinese civilization. More than 2000 years ago, when ancient Chinese thinkers summed up this achievement and coined the term "*wen ming*" (civilization), they had such remarks: "dragon appears in the field, all under heaven being adorned and brightened." "The attributes (of its component trigrams) are strength and vigor with elegance and brightness. (The ruling line in it) responds to (the ruling line in the symbol of) heaven, and (consequently) its action is (all) at the proper times. In this way (it is said to) indicate great progress and success". (*Book of Changes*) The ancient civilization, agricultural civilization, and flourishing industrial civilization will eventually become history. The evolution of human civilization has not yet ended, and new forms of civilization are coming to the fore, which is a low-carbon, green, and sustainable ecological civilization.

Engels pointed out in the *Dialectics of Nature*: "Let us not, however, flatter ourselves overmuch on account of our human victories over nature. For each such victory nature takes its revenge." The progress of human civilization prompts the rapid development of science and technology as well as the growing of the global resource and environment issues. In the current context, low-carbon development is an inherent requirement and entry point for ecological progress and sustainable development. Low-carbon development is an organic combination of "low carbon" and "development", requiring parallel works in reducing carbon dioxide emissions and pursuing economic and social development. To promote low-carbon development is especially significant for the improvement of source structure, protection of the environment, reform of industrial structure, and the cultivation of the ability of sustainable development. To promote low-carbon development is not only in line with China's limited national resources and environmental carrying capacity but also meets the requirements of sustainable development. It is a correct and historic choice for China, the world's largest developing country, to promote ecological civilization featuring low-carbon development.

# Chapter 2
# Modern Science of Climate Change and Proposition of Low Carbon

## 2.1 Formation of Climate Change Science

### *2.1.1 Climate Warming and Environmental Issues*

In recent years, serious air pollution hit eastern China frequently. Over a million $km^2$ of land has been subject to lasting and grave air pollution almost at the same time, especially in Beijing, Tianjin, Hebei, and the surrounding areas. Generally speaking, environmental pollution always has a source, so the main solution is controlling the source and preventing the pollutants from spreading. But some environmental problems are different. There are substances that are harmless per se and generally not called pollutants, but they can cause serious environmental problems. For instance, $CO_2$, which takes up less than 0.04% in the air, is not a kind of pollutant and will not harm human health even if its concentration is increased by tens of times to take up 1% in the air. However, $CO_2$ is an extremely important greenhouse gas (GHG), and because of its greenhouse effect, even a twofold increase of $CO_2$ concentration would cause serious global warming. Another example is CFC, which is not a natural substance but is an artificial chemical with very stable properties. Since CFC hardly reacts with other substances, we don't see it harming the organisms on earth, but exactly because of its low reactivity, its higher concentration in air would cause obvious greenhouse effect and climate warming.

We are going through the climate warming that is mainly caused by higher GHG concentration. In September 2013, the Intergovernmental Panel on Climate Change (IPCC) issued the Working Group I report of the 5th Assessment Report, which confirmed the fact of global warming in multiple aspects with a range of indicators from the atmosphere, ocean, and glacier. It was a conclusion based on comprehensive analysis of a series of observation data and has been widely accepted by the international community and scientific circle.



X. Du et al., *Overview of Low-Carbon Development*,
https://doi.org/10.1007/978-981-13-9250-4_2

Global warming has become an important environmental issue that has exerted and will continue to exert substantial impacts on the natural and social systems worldwide. If the average global temperature rose by more than 2 °C from pre-industrialized period, catastrophic consequences might befall the human society. Scientifically dealing with the challenges brought by climate change has become an international consensus and all main countries in the world have adopted positive strategies and taken active actions.

In the context of global warming, China has also undergone a warming process that is in line with the global warming over the past century, and the trend may aggravate in the future. Climate change has exerted profound impacts on China's ground surface conditions, natural ecosystem, and socioeconomic system, and the nation is facing both challenges and opportunities in dealing with this trend.

### *2.1.2 Human Understanding of Greenhouse Effect*

As mentioned earlier, GHG like $CO_2$ and CFC are not traditional "pollutants" that are environmentally harmful, and therefore the human society has emitted a tremendous amount of $CO_2$ into the air by burning fossil fuels or changing the way of land use, especially after the Industrial Revolution. In the past, people believed $CO_2$ did not have to be isolated and could be emitted into the air directly. As a matter of fact, GHGs like $CO_2$ are of very low content in the air regarding their volumetric ratio: $N_2$ takes up 78% of air, $O_2$ 21%, and Ar and others about 0.9%. In other words, over 99% of the gases in the air are not GHG. They have little reaction with incident solar radiation and basically don't react with the infrared long-wave radiation emitted from earth. This means they neither absorb nor emit thermal radiation and basically have zero effect on the climate changes on earth.

It is the many trace gases that are of extremely low content in air that have major impacts on the climatic environment on earth, such as $CO_2$, $CH_4$, $N_2O$, and ozone. These gases take up less than 0.1% of air in terms of volumetric ratio, but as they can absorb and emit radiation, they have greenhouse effect like greenhouse glass (Fig. 2.1) and function as a "blanket" in the energy balance of earth, and they also create the pleasant average temperature of 14–15 °C on the ground surface. Therefore, they are called GHGs, without whose "blanket" effect the temperature on earth would be only −18 °C, which would be unsuitable for human survival. Therefore, carbon is not just a chemical element. It is the root cause of climate change. It is thanks to its greenhouse effect that the average temperature on earth can be kept around 15 °C. But there cannot be too much carbon as excessive carbon would bring environmental disaster. Both natural and human factors may affect climate change, and human activities affect the climatic environment mainly through carbon emission, which changes the GHG concentration in air and causes greenhouse effect.

Since the mid-twentieth century, climate changes resulting from human activities have received growing attention from scientists, policy-makers, and the public in all countries. After IPCC issued the *1st Assessment Report* in 1990, five such reports

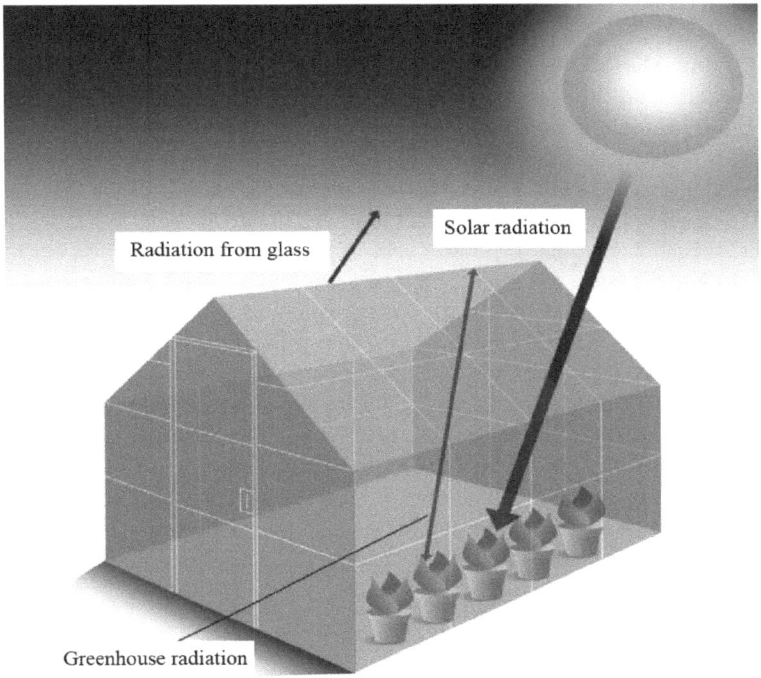


**Fig. 2.1** Greenhouse effect

had been issued by 2014, all focusing on the relation between human activity and climate change. Have human activities already caused global climate changes, are they causing and will they continue to cause such changes, and what strategies should be adopted? In the past 20 years, with the rapid development of climate science and the actual climatic evolution on earth, the mankind had a deeper scientific understanding of the climate impacts of human activities. As more evidences are gathered, the scientific community today is more positive than ever before about the human impacts on climate. The scientific contentions in this process largely boosted the progress of the scientific researches on climate change, and the results obviously changed the human perception of the nature of climate change. These new scientific findings have captured extensive attention of the governments and scientific circles in different countries, and eventually led to the international consensus on and common action of dealing with global climate change. Climate change in modern times characterized by global warming is not just a scientific issue anymore, but has evolved to be a political, diplomatic, environmental, and energy issue of global significance.

Greenhouse effect is the key physical basis for global warming. In this field, many scientists have made important contributions in the past 200-plus years. The international scientific community has passed three stages in its perception of greenhouse effect.

(I) Scientific community's early perception of greenhouse effect

In 1681, Edme Mariotle pointed out that although sunshine and its heat could easily penetrate glass and other transparent objects, heat from other sources could not. In the 1760s, Horace Benedict de Saussure conducted a simple greenhouse effect experiment with a solar-radiation thermometer (put a thermometer in a black box and cover it with a glass container), which made people realize for the first time that air itself was able to capture thermal radiation and create artificial warming.

In 1824, French scientist J. Fourier said just like the temperature on earth could rise due to air influence, the atmosphere and greenhouse glass could create similar warming effects, thus giving rise to the term greenhouse effect. Based on J. Fourier's idea, Poulliet said in 1836 that the atmospheric stratification (the higher in the troposphere, the lower the temperature) made air radiation to earth more absorptive than solar radiation. This standpoint explained for the first time the important role of temperature stratification in causing greenhouse effect. In 1839, British scientist J. Tydall measured the absorption of infrared radiation by vapor and $CO_2$, and expounded the special effect of the trace GHGs in air on the temperature change on earth. According to him, the quantitative change of any radioactive atmospheric component, such as vapor and $CO_2$, could cause the climate changes revealed by geologists' research.

(II) Measurement of $CO_2$'s greenhouse effect in late nineteenth century and early twentieth century

In 1896, Swedish scientist Arrhenius published a paper titled *The Effect of Carbonic Acid in Air on Ground Temperature* (scientists called $CO_2$ in air carbonic acid at that time). It was an extremely important paper because it made the first quantitative calculation and forecast of GHG's warming effect in the scientific history. Although Arrhenius was not the first scientist to put forth the concept of greenhouse effect, nobody before him had ever calculated the greenhouse effect of $CO_2$ in air. For the first time in human history, his paper calculated the amplitude of global temperature change caused by changed $CO_2$ concentration.

Arrhenius did not publish the paper to solve global warming resulting from the rising $CO_2$ concentration in air as this issue didn't exist at that time. Besides, given the speed of $CO_2$ emission from human activities then, it would take 3000 years for the $CO_2$ concentration to increase by 50%. Arrhenius estimated that the annual $CO_2$ emission from human activities into the air back then accounted for only 1‰ of all $CO_2$ in air and 5/6 of that was absorbed by the ocean, leaving only 1/6 in the air. Arrhenius' calculations showed that a 50% increase in $CO_2$ concentration in a span of 3000 years would increase the temperature by more than 3 °C, equivalent to 0.001 °C/a.

As mentioned earlier, British scientist Tydall pointed out that the quantitative change of any radioactive atmospheric component could cause the climate changes revealed by geologists' research. Arrhenius conducted his research with the purpose of explaining the changing mechanism of glacial period and interglacial period in history, as revealed by geologists' research. We now know that the 100,000-year-or-so glacial–interglacial cycle in the geologic age was mainly caused by changes in earth orbital parameters, which decided how much solar radiation the earth absorbed, and the solar radiation changes would, through various mechanisms, trigger the glacial–interglacial cycle of about 100,000 years. However, Arrhenius believed that instead of the change of earth orbital parameters, the change of $CO_2$ concentration in air was the main reason for the change of glacial–interglacial cycle. Another scientific viewpoint then held that the temperature difference between glacial and interglacial periods required more than 50% change in $CO_2$ concentration, which nevertheless had to be calculated and verified with relevant materials and models. Arrhenius made the calculations and finally concluded that if $CO_2$ concentration dropped by 1/3, the global temperature would drop by more than 3 °C. If $CO_2$ concentration rose by 50%, the global temperature would rise by more than 3 °C; if $CO_2$ concentration rose by 100%, the global temperature would rise by more than 5 °C. His calculations also indicated that if $CO_2$ concentration went up, the temperature difference between land and ocean, between the Equator and the temperate zone, between summer and winter, and between day and night would narrow.

His calculations denoted that if $CO_2$ concentration increased by geometrical progression, the global temperature would increase by arithmetical progression. In other words, a 50% increase in $CO_2$ concentration would lead to the average temperature rise of 3 °C and a 33% decrease in $CO_2$ concentration would lead to the average temperature drop of 3 °C. Accordingly, a 100% increase in $CO_2$ concentration would lead to the average temperature rise of more than 5 °C, and a twofold increase for more than 8 °C. Arrhenius admitted the shortcomings in his calculations. For instance, due to the lack of quantitative knowledge about carbon cycle, he could not accurately calculate the speed of the temperature rise of the earth, but the growing $CO_2$ content in air was an undeniable fact, which might affect the living environment of many generations to come.

What Arrhenius didn't expect was that $CO_2$ content in air increased much faster than he predicted. The $CO_2$ concentration in air around 1896 was less than 300 ppm,[1] and that in the early stage of the Industrial Revolution a century ago (after 1750) was about 280 ppm, indicating a 5% increase in 100 years. But a century later, the average $CO_2$ concentration worldwide reached 395 ppm in 2013, an increase of more than 30% over 100-plus years and more than 40% over a period of less than 300 years from the preindustrial period around 1750. This was nearly 10 times faster than the 50% increase over 3000 years calculated by Arrhenius.

In 1938, G. S. Callendar solved an equation set on the relation between GHG and climate change and found that the multiplication of $CO_2$ concentration would increase the global average temperature by 2 °C with more obvious warming in Polar Regions. Callendar connected the increase of fossil fuel burning, rising $CO_2$ concentration, and growing greenhouse effect, and pointed out that the human race was changing the atmospheric components at a speed drastically different from that in the geologic age, and this change would lead to notable climate changes.

[1] ppm: part per million, $10^{-6}$.

(III) Research on $CO_2$'s warming effect after the mid-twentieth century

Limited by the observation materials and models, Arrhenius, in his calculations, overestimated to varying degrees the vapor feedback and $CO_2$'s radioactive effect. The scientific circle today generally believes that the vapor feedback used by Arrhenius increased his calculations of ground temperature rise by about 30%, and his estimation of $CO_2$'s radioactive effect was 1.5 times higher. Nevertheless, his calculations based on incomplete data were surprisingly authentic. Thanks to the development of computer technology after the 1960s, it was possible to develop complicated climatic models for huge amounts of calculations, and scientists calculated the global temperature rise caused by $CO_2$ increase in air according to those complicated models. Today the global temperature rise resulting from the multiplication of atmospheric $CO_2$ concentration is referred to as equilibrium climate sensitivity (ECS), meaning the global average temperature rise after atmospheric $CO_2$ concentration is doubled and reaches the equilibrium. In 1967, scientist Syukuro Manabe at NOAA used his global atmospheric radiation convection model and came to the following conclusion for the first time: the doubling of $CO_2$ concentration in air would increase the global temperature by 2.3 °C. In the 1970s, he developed the 3D GCM to calculate climate sensitivity, which took into account the effect of changes in hydrological factors, such as the feedback of snowcap and sea ice on climate change. Syukuro Manabe's calculations based on the 3D model showed that the climate sensitivity, while considering the feedback of snowcap and sea ice on climate change, was about 3 °C, slightly higher than the result based on the radiation convection model.

(IV) Research on warming amplitude caused by $CO_2$ multiplication after 1979

In 1979, the US National Academy of Sciences (NAS) entrusted MIT's famous meteorologist Jule Charney to set up a special task force to estimate the relation between $CO_2$ and climate change. The assessment report they later published (also called the Charney Report) held that the doubling of atmospheric $CO_2$ concentration would cause the temperature rise of 3 °C (up and down 1.5 °C, thus the amplitude of 1.5–4.5 °C). During the 30-plus years afterward, scientists around the globe used all kinds of models to calculate climate sensitivity in large quantities. Ever since its first assessment report in 1990, the IPCC also assessed climate sensitivity every time, but the conclusions were basically the same. The first IPCC assessment report in 1990 concluded the global temperature rise of 3 °C (up and down 1.5 °C, thus the amplitude of 1.5–4.5 °C), and the fifth report in 2013 gave the same conclusion.

### 2.1.3 *International Scientific Research Plans and Scientific Perception*

(I) Existing international test and research plans

As people have a deeper understanding of climate change, this issue has evolved from a matter of climatic science to be a major strategic issue concerning a wide range of fields, including environment, science and technology, economy, politics, and diplomacy. Since the 1970s, the international community has taken active countermeasures and a series of actions, ranging from scientific research to scientific estimation of climate change and formulating international treaties. Typical examples include three World Climate Conferences, four scientific programs, the Future Earth program, IPCC assessment reports, and the UNFCCC conference.

The first World Climate Conference was held in 1979 in the theme of "Climate and Mankind", which led to the formation of World Climate Programme (WCP), World Climate Research Programme (WCRP), and IPCC. The second World Climate Conference held in 1990 in the theme of "global climate change and countermeasures" called for emergency international actions to stop the fast increase of GHGs in the atmosphere. This conference prompted the signing of UNFCCC in 1992 and the establishment of the Global Climate Observing System (GCOS). The third World Climate Conference was held in Geneva, Switzerland in 2009 in the theme of "climate predictions and information for decision-making," aiming to promote the development of climatic services, strengthen their applications in socioeconomic planning, and prevent and control risks of meteorological disaster. The ultimate goal of the conference was setting up a "global climate service framework" that could help global decision-makers to acquire precise and timely climate information and prediction to better tackle climate change.

The scientific community organized four scientific research programs and formed the Earth System Science Partnership (ESSP) in 2001. The ESSP consists of four scientific research programs on global environmental changes, namely, the WCRP, International Geosphere-Biosphere Programme (IGBP), International Human Dimensions Programme on Global Environmental Change (IHDP), and DIVERSITAS. It aims to boost integrated research and change research on the earth system, use these changes to study global sustainability, and consequently support political, economic, and social decision-making in the context of global climate change. Among them, the WCRP focuses on physical climate systems and their interactions and strives to obtain a quantitative understanding of the four sub-systems (global oceans, world seas, low-temperature sphere, and land surface). The IGBP closely cooperates with other programs to study the interactions between the biological, chemical and physical processes, and the human system, and offer scientific knowledge to cope with global environmental change. The IHDP studies how human activities affect (drive) global environmental changes, estimates how those changes affect human life and wealth, and discusses what countermeasures the human society should adopt to mitigate and adapt to those changes. The DIVERSITAS aims to promote the development of the biodiversity science, tighten the connection between biological, ecological, and social sciences, and provide the scientific foundation for the protection and sustainable utilization of biodiversity. Future Earth is a 10-year international scientific program initiated at the Rio + 20 in 2012. It is committed to establishing a complete scientific system of earth systems, strengthening the connection and integration of natural and social sciences, providing necessary scientific knowledge, technical methods and approaches for the world, regions and countries to cope with global environmental changes, and boosting the global and regional sustainability. Future Earth will integrate the ESSP and its four scientific programs and cross-disciplinary programs, and will gradually launch a batch of new programs and projects.

International negotiations on climate change have come to a new stage. The contracting parties' meeting held in Bali, Indonesia at the end of 2007 passed the "Bali Action Plan", formed a standing joint task force, and specified the "four wheels" for dealing with climate change—adaptation, mitigation, capital, and technology, with the emission reduction in developed countries being one of the most important contents. The Doha Conference held at the end of 2012 reached a package deal of several important issues, including the second commitment period of *Kyoto Protocol* and the long-term cooperation under UNFCCC. It wrapped up the negotiation on Bali Roadmap, pushed forward the negotiations on Durban Platform, called for efforts to establish before 2015 a post-2020 global emission reduction framework covering all main emitting countries, and urged them to step up their emission reduction actions.

(II) IPCC's latest conclusions

At present, the international community is most concerned with climate change in the past century, during which global warming is an undisputed fact. From 1880 to 2012, the global average temperature on ground surface climbed up by 0.85 °C in the northern hemisphere, and the period from 1983 to 2012 might be the warmest three decades in the past 1400 years. Since 1901, the average precipitation on midlatitude lands in the northern hemisphere increased, the global surface temperature increased, and other variables of the climate system were changing. The global average speed of sea level rise from 1901 to 2010 was 1.7 mm/a, which increased to 3.2 mm/a between 1993 and 2010. Global glaciers have subsided in general since 1971, and the size of sea ice in the North Pole shrank at the rate of 3.5–4.1% every 10 years since 1979. In 2013, the concentration of such GHGs as $CO_2$, $CH_4$, and $N_2O$ in air was 396 ppm, 1,824 ppb[2] and 325.9 ppb, respectively, the highest for nearly 800,000 years and each being 42, 153, and 21% higher than before the industrialization. The average monthly $CO_2$ concentration in the northern hemisphere kept rising and exceeded 400 ppm in April 2014, the highest since observation records ever existed (Figures A.4, A.5). Oceans were acidified because they absorbed 30% of the $CO_2$ emitted from human activities. Due to global warming, the intensity and frequency of extreme weather have changed notably since the mid-twentieth century: extreme heat increased while extreme cold decreased; heat waves were more frequent and lasted longer; strong rainfall on land increased and drought in southern Europe and western Africa was of greater intensity and longer duration; and the intensity, frequency, and duration of tropical cyclone displayed the trend of long-term increase.

The fifth IPCC assessment report further confirmed the causality between human activities and global warming and concluded that human activities contributed to over half of global warming since the 1950s. This conclusion has a credibility of 95%, higher than that released in the fourth assessment report in 2007, which is also more than 90%. The report predicted that the global surface temperature at the end of the

[2]ppb: part per billion, $10^{-9}$.

twenty-first century will be 0.3–4.8 °C higher than that at the beginning of the century, and made a rather objective estimation and statement about the emission quota against the goal of 2 °C temperature rise. It stated that if we wanted to control the global temperature rise within 2 °C by the end of the twenty-first century (compared to 1861–1880), there was room to emit 1000–1560 billion ton carbon worldwide given different probabilities. These conclusions provided the governments worldwide with options as they comprehensively weighed, negotiated, and determined their future responsibility for $CO_2$ emission reduction.

2014 marked the highest global average surface temperature since the records began in 1880, and the highest average sea surface temperature ever was observed in the tropical mid-east Pacific Ocean in the first few months of 2015, and it is possible that the global average surface temperature may hit a new high in 2015. According to changes in global average surface temperature in the past 10-plus years, the global temperature has remained high since 1998. Of the 15 warmest years in modern meteorological records, 14 appeared in the twenty-first century with the only exception of 1998. This fully proves the undisputed warming of climate system and the global warming trend is still on. Besides, although the global average surface temperature has stayed high since 1998, the speed of temperature rise is obviously lower than the average speed since 1951, while the global average concentration of GHGs like $CO_2$ continued to increase in this period. It seems that global warming is in "stagnation" since 1998, and people holding different views about climate change take it as the evidence to question global warming. In fact, the "stagnation" of global warming is a matter of decadal variability inside the climate system and it won't change the centurial trend of global warming. Figure A.6 shows that after 1850, the global average surface temperature had the decadal variability of dropping in 1870–1900 and 1940–1970, but that did not change the general trend of temperature rise. Global warming does not contradict with the "mitigation" or "temperature drop" in one or several decades. Moreover, the "mitigation" of global warming since 1998 is mainly reflected in the changes of global average surface temperature, but there is no "mitigation" as far as changes of the general climate system are concerned. Scientists discovered through research that more and more energy received by oceans penetrates to the middle (700–2,000 m under sea level) and deep level (2000 m to the bottom), while the sea surface and upper level absorbs less energy. A piece of direct evidence is that we frequently observe La Nina phenomenon but El Nino often does not reach its due intensity. Scientific studies revealed that the inter-decadal oscillation of the North Atlantic and the Pacific oceans resulted in energy redistribution inside the ocean, which might be the main reason for the "stagnation" of global warming. Many studies held that the period since 1998 was in the down curve of the 11-year cycle of solar activity, during which volcanic outburst and growing aerosol in the troposphere reduced the solar radiation reaching the earth surface, thus offsetting the temperature rise arising from greenhouse effect to some extent. In other words, the temperature drop resulting from the natural factor of climate change and the decadal variability within the climate system offset the warming effect of human activities, and "mitigated" global warming. Although the global average surface temperature has risen at a slower rate since 1998, it does not affect the long-term warming trend, and the latest studies indicated that the amplitude of global warming by 2100 will depend on human GHG emission, while the inter-decadal fluctuation of global average surface temperature has no impact on this process. The greenhouse effect of human GHG emission will offset and outweigh the effect of natural fluctuations. If global carbon emission is not considerably reduced in the next few decades, global temperature will rise to a dangerous level. Therefore, while we still have to deepen our understanding of climate change, this fact is beyond doubt and this certainty amid uncertainties is generally acknowledged.

## 2.2 Adapt to and Mitigate Climate Change

The degree of global warming in the future will mainly depend on the accumulative $CO_2$ emission worldwide. Even if emission reduction actions are taken now, the global average temperature and the sea level will continue to rise at the end of the twenty-first century. Global warming has exerted and will continue to exert substantial effects on the natural and social systems. A 2 °C increase in global average temperature may bring catastrophic consequences to the human society. Scientifically tackling climate-change-induced challenges has become a consensus in the international community and main countries in the world have all adopted positive strategies and actions. China attaches great importance to climate change. It formed a national leading group on climate change, energy conservation, and emission reduction headed by the premier of the State Council, and adopted a series of effective policies and measures for dealing with climate change according to the nation's sustainable development strategy, making positive contributions to adapting to and mitigating climate change.

### 2.2.1 *Effect of Climate Warming on the World and China*

Since the industrial revolution, the world has been going through climate changes characterized by climate warming, which has exerted substantial impacts on the natural system and human society both regionally and globally, including water resources, ecosystem, grain production, and human health. The changing precipitation and ice and snow melting in many places are changing the hydrological system and affecting water quantity and quality; the lasting glacier subsidence in many regions is affecting the downstream runoff and water resources; the permafrost in high-latitude regions and high-altitude mountains is warming up and thawing. If the average global surface temperature rises by 1 °C, an extra 7% of world population will be affected by the reduction of water resources. The geographical distribution, seasonal activities, migratory model, and natural abundance of some species have changed. Land and freshwater species are facing a bigger risk of extinction. The arctic tundra and the Amazon forest will encounter abrupt changes and irreversible high risks. The coastal system and low-lying areas have a larger chance of being flooded or suffering from coast deluge or erosion. Climate change has more cons than pros for grain output. Wheat and corn are more subject to the adverse effect of climate change than rice and soybean, and their output reduction is about 1.9 and 1.2% every 10 years on average. Climate change has increased the number of days of high temperature, heat wave, and heavy pollution, and its relation with the incidence, development, and diffusion of some epidemic diseases has been verified by scientific research. Going forward, climate change may cause more extensive influences and risks. The key risks faced by Asia mainly lie in the increase of river, sea, and city flooding, which will cause large-scale destruction of infrastructure, livelihood, and residential area; the higher risk of high-temperature-related death; and the rising risk of malnutrition caused by drought-induced water and food shortage (Fig. 2.2).

China has the same warming trend as in the world at large. According to the *China Climate Change Monitoring Bulletin* released by China Meteorological Administration, the average surface temperature in China has risen by 0.91 °C in the past century. Temperature rise in the last 60 years is especially marked, about 0.23 °C every 10 years on average, almost twice as high as the world average, particularly in the north. The first 10 years of the twenty-first century are the warmest decade in nearly 100 years.

The rainfall distribution has changed notably in China. In the past 50 years, rainfall in the western region increased by 15–50%; the phenomenon of "southern waterlogging and northern drought" was frequent in the east; rainfall in the south increased by 5–10%, while that in most northern and northeastern regions decreased by 10–30%. In the same period, the number of average annual rain days decreased, including 13%

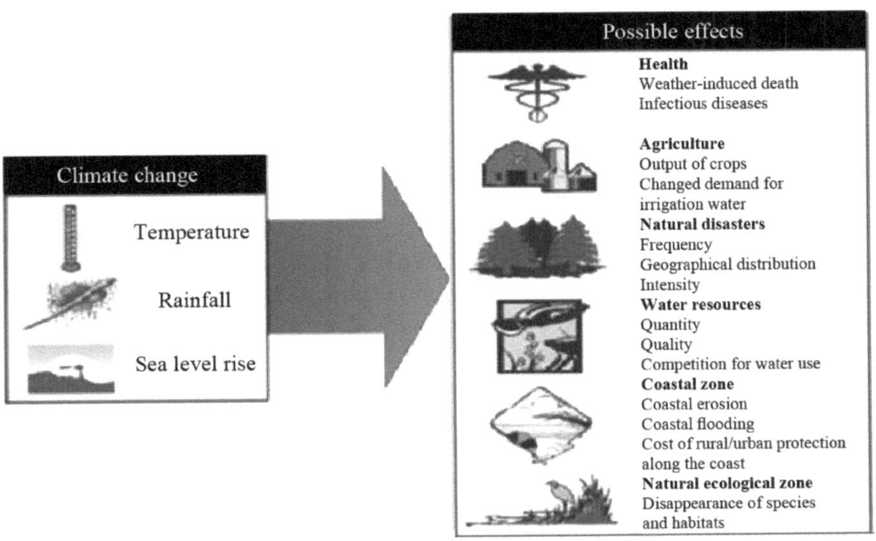


**Fig. 2.2** Effects of climate change

decrease in light rain days and 10% increase in torrential rain days. The location of main rain belts in summer changed obviously. They mainly existed in the north in the 1950s–1970s, but then they began to move southward to the Yangtze River basin and the south, and began to move to the north in the twenty-first century.

Some extreme climatic events in China have displayed obvious changes in frequency and intensity. (1) High temperature and heat wave are more frequent in summer. Especially after 1998, the number of days above 35 °C is continuously and notably larger than average. (2) Regional drought is more serious. In the previous 15 years, the number of drought days above medium intensity increased by 37% in the northeast, 16% in the north, and 10% in the southwest. (3) There are more heavy rainfalls. The past 20 years marked a high-incidence period of flooding disasters in the Yangtze River and Huaihe River basins after the 1950s. (4) Typhoons in China are obviously stronger. In the twenty-first century, eight typhoons attacked China on annual average, half of which had the maximum wind force of over the scale of 12 or 14, an increase of 14% and 1.4 times compared with the 1990s. (5) The number of smog days increased considerably while the number of fog days and sandstorm days decreased.

Climate change has had grave impacts on the natural and social systems in China. Frozen soil changes have conspicuous ecological effects and cause degradation of ecological system in the source areas of the Yangtze River and Yellow River and in mountainous areas of inland rivers. Distribution of tree species, forest line, phenological period, productivity and carbon absorption, forest fire, plant diseases, and insects have shown obvious changes. Grassland degradation is aggravated and inland wetland area shrinks with degraded functions. The diversity of animals, plants, and microorganisms; the diversity of their habitats, ecosystem, and scenery; and the degradation and extinction of some species are all somewhat related with climate change (Fig. 2.3).

Climate change has both positive and negative effects on China's agricultural development, negative effects being dominant nationwide. Climate warming expands the rice planting area in the northeast and pushes the northern planting border to near 52°N. The northern planting border of winter wheat expands to the north and west to a small extent, and wheat needs more water and is less resistant to cold in winter and spring due to temperature rise. Climate change leads to the increase of species and generations of plant diseases and pests, enlarging the scope of hazard and increasing economic damages. It also leads to the use of more fertilizers, pesticides, and herbicides, largely increasing the agricultural production cost and investment, and may raise the incidence rate of some livestock.

Climate change causes serious water problems in China. The runoff of main rivers is decreasing, and the measured runoff of main rivers in the Haihe River basin is decreased by 30–70%. The quality of national water resources is falling and water pollution is worsening; the reduced runoff undermines the diluting and self-purifying capability of water bodies, and water shortage due to quality degradation is prominent in the damp southern region.

Climate change has growing effects on the atmosphere and water environment. In 2013, the average $PM_{2.5}$ concentration in 74 cities nationwide was 72 $\mu g/m^3$, more

**Fig. 2.3** Urban waterlogging caused by heavy rainfall

than twice as high as the national standard and more than seven times the WHO standard. Rising temperature, reduced rainfall, and enhanced evaporation were bad for the diffusion and elimination of pollutants in water bodies and indirectly caused their eutrophication, leading to serious pollution like the blue algae outburst. Under the influence of temperature rise in the near sea and ocean circulation, an immense amount of enteromorpha gathered along the coast of the Shandong peninsula, which destroyed the marine ecosystem, blocked the waterway, and seriously threatened the development of coastal fishery, shipping, and tourism.

Climate change has wide-ranging and deep-going effects on China's energy security. Wind speed and direction affect wind power generation. The north of northern China and the southeastern coast, which are main areas of wind power generation in China, see wind speed reduction at the rate of 0.3 m/s every 10 years, which lowers the wind power output. Since 1961, sunshine duration in China has decreased in general, especially in winter and summer and on the northern plains, restricting solar energy exploitation and utilization. The changed river runoff also affects the safety of hydropower operation. Climate change affects energy consumption such as heating in winter and cooling in summer, and meteorological disaster affects energy production and transport. The restriction of global fossil fuel consumption imposed by UNFCCC will also pose major challenges to China's energy development.

Climate change has growing negative effects on the safety of China's major national defense and strategic projects. Rising temperature will cause the general permafrost degradation along the Qinghai–Tibet Railway and seriously threaten its safe operation. Excessive flooding in the Three Gorges reservoir area and the upstream region will increase risks in the reservoir's flood prevention, dispatching, and operation, and frequent rainstorm may cause geological disasters like landslide and debris flow, damaging the dam safety. Climate change also creates risks to the safe operation of a series of substantial projects, including the South-to-North Water Diversion project, West-to-East Gas Transmission project, China–Russia gas pipeline, and protection forest in the north, northwest, and northeast. It may even trigger grave environmental events.

Climate change and meteorological disasters pose more severe challenges to China's economic security. They have caused growing economic losses and threatened the national economic security. In the twenty-first century, the ratio of direct economic losses caused by meteorological disasters to GDP is 1.07% on annual average, more than seven times the global average (0.14%) in the same period. With the rising economic aggregate and the global and domestic economic integration, climate change and meteorological disasters are posing mounting risks to China's general economic security. They not only seriously undermine the safe economic and social operation in the country but also indirectly damage our economic security through international trade.

### 2.2.2 *Mitigating and Adapting to Climate Change*

(I) Measures to adapt to climate change

The Working Group II report of the 5th IPCC climate change assessment report issued in 2014 put forth the idea of climate resilience path. It believed that disaster risk management and strengthening the human society's resilience is an effective way of adapting to climate change and reducing fragility and exposure, and the inevitable way of proactive adaptation for sustainable development. As there are no universal risk management measures, the adaptive actions have to be customized according to local conditions. The Chinese government should set up the legal framework, protect vulnerable groups, provide information, policy, and fiscal support, and coordinate the actions of local governments on various levels. Local governments and the private sector, on the other hand, need to play a bigger role in promoting community and family risk management.

There are many ways to adapt to climate change, including institutional, technical, and engineering measures. For instance, we build flood-control infrastructure, economize on water during the dry season, and change personal behaviors. Other adaptive approaches include establishing the prewarning system of extreme weather and climate events, and intensifying climate disaster risk management, so as to lessen the impacts of climate change on human society.

In recent years, China has adopted a series of policies and taken measures to adapt to climate change in light of its economic and social development plans, and made positive results. We issued the *National Strategy for Adapting to Climate Change*, and the agricultural, forestry, water, oceanic, and health sectors formulated major policies, laws, and regulations to adapt to climate change. In the agricultural sector, for example, we developed new types of drought-resistant crops, adopted the intercropping system, kept crop stubble, treated weeds, and developed irrigation and hydroponic farming. To cope with flooding, we adopted the polder field and improved drainage approach, developed and promoted alternative crops, and adjusted the planting and harvesting time. We also developed new heat-resistant crops, changed the farming time, and monitored crop pests to deal with heat wave. In view of the observed and predicted climate change, we substantiated the adaptive measures through policy, infrastructure investment, technology, and change of behavior. The measures are all aimed to lower the risks brought about by climate change, but they come at a cost. During the 11th Five-Year period, China built new reservoirs of 38.1 billion $m^3$, increased 28.5 billion $m^3$ of water supply capacity, and added 50 million $\mu$ of net effective irrigated farm. It promoted water-saving technology in more than 400 million $\mu$ of farm and protective farming technology in over 85 million $\mu$. The area of protected wetland increased by 1.5 million $hm^2$ and the area of water and soil erosion control by 230,000 $km^2$. China stepped up efforts of ecological restoration and protection, and its ability of observing, monitoring, forecasting, and warning against extreme weather and climate events as well as their derivative disasters was significantly bettered.

(II) Measures to mitigate climate change

To make sure that climate change will not threaten the sustainable development of ecosystem, grain production, and socioeconomy for a certain period and to stabilize GHG concentrations in the atmosphere at a level that would prevent dangerous anthropogenic interference with the climate system, we must control or reduce GHG emission through policies and measures that mitigate climate change. If we want to control GHG concentration at 450 ppm $CO_2$ equivalent by 2100, we have to reduce the global emission by 40–70% in 2050 from 2010 and realize zero emission by 2100. To meet the target of 2 °C temperature rise, the energy supply sector has to be overhauled and systematic, cross-departmental emission reduction strategy has to be implemented as early as possible. Departmental and national policies and mechanisms have to be adopted to mitigate climate change, so that sectors including energy production and use, transport, construction, industry, land utilization, and human residence can take corresponding steps to achieve a stable GHG concentration in air. To mitigate climate change, we should consider the principle of sustainable development and equity, take into account a range of risks and uncertainties, face up to topics such as equality, justice, and fairness, and make allowances for the possible "symbiotic benefits" or "negative effects" when climate policies are combined with other social goals.

Main measures to mitigate climate change include the following:

1. Reducing GHG emission

Energy conservation is stressed, including technological energy conservation (e.g., intensifying technological progress and raising the efficiency of energy conversion and utilization) and structural energy conservation (e.g., changing growth model, adjusting industrial structure, promoting industrial upgrade, and increasing products' value-added rate). GHG emission during industrial production should be controlled, so is total coal and petroleum consumption. New and renewable energies like nuclear, hydro, wind, and solar power are developed to better the energy structure and reduce carbon emission per unit of energy consumption, and agricultural GHG emission must be reduced.

2. Increasing carbon sink

Technical approaches of increasing carbon sink and reducing carbon emission through forestry include carbon sequestration, storage, and substitution. Carbon sequestration means increasing land carbon storage in forest vegetation and soil by means of afforestation, reafforestation, forest operation, vegetation restoration, establishing the farming-forest composite system, and increasing forest productivity. Carbon storage means reducing interference-induced carbon emission in the forest ecosystem by logging fewer trees, improving logging approach, and controlling forest fire and diseases and pests. Carbon substitution means substituting forest products and biomass energy for fossil energy to reduce emission.

3. Earth environmental engineering—CCUS

$CO_2$ capture, utilization, and storage (CCUS) is a technology that isolates $CO_2$ from the emission source in industry or related energy industries, produces products with commercial value through physical, chemical, or biological reaction, and transmits and stores it in geological structure, thus reducing $CO_2$ emission. Despite its shortcomings like high cost, great technical difficulty, and poor certainty, CCUS, as the fundamental measure to lower atmospheric $CO_2$ concentration, is regarded by many as the inevitable choice to reduce global carbon emission. It is also an important strategic option for China and other countries to tackle climate change, and is of great significance for a nation's capability and overall competitiveness.

Other measures to mitigate climate change include the establishment of corresponding systems and mechanisms, such as establishing through administrative intervention the mechanism to encourage mitigating actions, carbon trade, and international cooperation. Market mechanism and government intervention are both indispensable.

### *2.2.3 China's Actions to Mitigate Climate Change*

GHG emission from human activities in the past 40 years accounted for about half of all such emissions since 1750, and the past decade marks the largest emission increase, with $CO_2$ from fossil fuel and industrial process being the main source of GHG increase. China's energy consumption jumped from 1.45 billion ton to 3.84 billion ton standard coal from 2000 to 2014, over 90% of which were three fossil energies—coal, petroleum, and natural gas. As a result, the $CO_2$ emission per unit of energy consumption in China is much more intense than in such developed countries as the US, Japan, and Germany. China is now the largest carbon emitter in terms of aggregate and increment. As a major energy producer and consumer, China's energy system features high carbon, and fossil energies, mainly coal, have been dominant in its energy production and consumption structure. This situation results in severe eco-environmental problems and enormous GHG emission, and seriously threatens our socioeconomic sustainability.

In recent years, China energetically develops low-carbon technologies that can mitigate climate change and stresses their promotion and demonstration through key projects. It specifically pushes low-carbon technology innovation and tries to improve core and key technologies. Through national plans for high-tech research and development and sci-tech support, we conducted R&D of energy-saving technologies in a wide range of areas, including clean and efficient energy utilization, industrial energy-saving technology and equipment development in key industries, key technology and material development for construction energy conservation, key technology and equipment development for clean production in key industries, and industrial growth model and integrated application of key technologies for low-carbon economy. These efforts have yielded a batch of invention patents and substantial achievements with independent IPR. We also advanced the construction of the national key lab of low-carbon technology and the national engineering center, established a group of state-level labs on energy conservation and emission reduction, and promoted the establishment of the industrial alliance of energy-saving, emission-reducing technologies and equipment.

China implements low-carbon technology innovations in key industries and fields, with the priority to affordable low-carbon building materials, low-carbon transport, green lighting, clean and efficient coal utilization, and other low-carbon technologies. It develops key low-carbon technologies such as solar PV battery with high PPR, solar building integration technology, large-power wind power generation, distributed natural gas, geothermal power generation, ocean power generation, smart and green grid, new energy vehicle, and power storage technology, as well as new technologies with independent IPR, such as CCUS. China has stepped up the demonstration and promotion of low-carbon technologies; formulated the policy outline for energy-saving, emission-reducing technologies; released the national catalogue of key energy-saving technologies and the national catalogue of major encouraged environmental technologies and equipment; and put in place the selection, appraisal, and promotion mechanism of energy-saving and emission-reducing technologies.

China's continued efforts lead to positive progress on our work of tackling climate change. Our ability in that aspect is strengthened constantly, the systems, mechanisms, laws, and standards are gradually improved, and the low-carbon awareness of the whole society has been raised. At the end of 2013, the $CO_2$ emission per unit of GDP in China fell by 28.56% from 2005, which was equivalent to a reduction of 2.5 billion ton $CO_2$ emission. Nonfossil energy accounted for 9.8% of primary energy in 2013. At present, China has the largest installed hydropower capacity, under-construction nuclear power capacity, solar collector area, installed wind power capacity, and artificial forestation area in the world, making positive contributions to coping with global climate change.

## 2.3 International Efforts of Tackling Climate Change

### *2.3.1 The Origin of IPCC and Its Role in UNFCCC*

The declaration of the First World Climate Conference in 1979 stated that if atmospheric $CO_2$ content continued to increase, the temperature rise would reach a measurable degree at the end of the twentieth century and become conspicuous in the mid-twenty-first century. The 1988 UN General Assembly passed the resolution to protect the climate for the current and future generations. In November 1988, WMO and UNEP jointly set up the IPCC to provide scientific advice on climate change for the international community, and the IPCC assessment reports, first issued in 1990, have been the main scientific basis for the international community to understand climate change issues, work out countermeasures and policies, and take actions. Since its founding, the IPCC has issued five assessment reports, respectively, in 1990, 1995, 2001, 2007, and 2014, which assessed the latest research results on global climate change, drew conclusions on key issues, and played an irreplaceable role in urging the international community to jointly tackle climate change. The first IPCC assessment report issued in 1990 said that emissions from human activities were considerably increasing the atmospheric GHG concentration, which prompted the signing of UNFCCC in 1992 and its coming into force in 1994. The second IPCC assessment report in 1995 said that artificial climate change was identifiable, providing solid evidences for systematically explaining the ultimate goal of the UNFCCC and pushing the adoption of the *Kyoto Protocol* in 1997. The third IPCC assessment report in 2001 further confirmed that most warming phenomena in the past 50 years might be attributed to human activities. The fourth IPCC report in 2007 expressly stated that climate change in the past 50 years was very likely caused by human activities, prompting the parties to reach a consensus at COP 15 Copenhagen. The fifth IPCC report placed more emphasis on the effects of climate change and how to adapt to and mitigate it, especially regional climate change and effect assessment, economic cost of adapting to climate change, and climate change and sustainable development. The conclusions of this report will have a great influence on the establishment of the international climate system after 2020.

### *2.3.2 Main Channels of Tackling Climate Change*

Climate change is a global issue, and international cooperation is the only way to cope with it. Since the 1990s, the international community has created some main cooperation channels represented by UNFCCC and *Kyoto Protocol* (KP), and climate change has evolved from a pure scientific issue to a political and development issue. In recent years, as the international power distribution is more balanced and political multi-polarization and economic globalization pick up speed, the international system is going through profound changes and transformation. In this process, international climate negotiations that have lasted 20 years are coming to a turning point, and non-mainstream channels other than the UNFCCC are playing a more obvious role.

UNFCCC is the world's first international convention aiming to comprehensively control the mission of GHGs like $CO_2$ and cope with the adverse effects caused by global climate change to human economy and society. It is the basic framework for international cooperation in tackling climate change. UNFCCC came into force on March 21, 1994, under which a series of deliverables were achieved through international climate negotiations over 20-plus years, represented by the *KP, Bali Roadmap, Copenhagen Accord* and *Cancun Agreement*. It put forth the basic principle of "common but differentiated responsibilities and respective capabilities" for international cooperation in tackling climate change. In provisions of Article 4 "Commitments", UNFCCC divided the contracting parties into two groups—Annex I Parties and non-Annex I Parties, and defined their "differentiated" commitments. In addition to the general obligations of non-Annex I Parties, Annex I Parties should take the initiative to change their long-term trend of artificial emission. To further differentiate the industrialized countries and countries in economic transition listed in Annex I, UNFCCC grouped industrialized developed countries as Annex II Parties and defined extra commitments for them, namely, providing financial and technical assistance for developing countries to mitigate and adapt to climate change. For Non-Annex I Parties, UNFCCC stated "The extent to which developing country Parties will effectively implement their commitments under the *Convention* will depend on the effective implementation by developed country Parties of their commitments under the *Convention* related to financial resources and transfer of technology and will take fully into account that economic and social development and poverty eradication are the first and overriding priorities of the developing country Parties."

The COP3 held in December 1997 passed the *KP* through resolution 1/CP.3, which reaffirmed the UNFCCC principle of "common but differentiated responsibilities" and provided that Annex I Parties had to cut their emission of six GHGs, including $CO_2$, in the commitment period of 2008–2012 by 5% on average from the 1990 level. In reference to the different capabilities and national conditions in different countries, *KP* also set differentiated emission reduction commitments for countries or regional integration organizations in Annex I, so as to meet the general reduction goal of 5%. Developing country Parties did not have to achieve such compulsory goals and comply with the timetable, but Annex I Parties could cooperate with developing countries by carrying out emission reduction projects in the and giving them capital and technical support through the *Clean Development Mechanism* (CDM). By then, an official GHG emission reduction mechanism was established under the UNFCCC, whereby developed countries cut emissions in a top-down manner (general reduction target was set first and then broken down among Annex I Parties) and developing countries did not bear compulsory emission reduction obligations.

On December 18, 2009, the COP15 of UNFCCC and 5th meeting of the Parties to *KP* formed the *Copenhagen Accord*, a reference document without legal binding force. The *Accord* proposed a "unified emission reduction mechanism," which meant both developed and developing countries adopted the "top-down" approach and proactively set their reduction/mitigation targets, but a verification mechanism like that under the *KP* would be carried out on developed countries while "consultation and analysis" would be conducted for developing ones. The *Cancun Agreement* adopted in December 2010 formalized part of the contents in the *Copenhagen Accord*.

The Doha World Climate Conference at the end of 2012 wrapped up the 5-year-long negotiations for the "Bali Roadmap", legally determined the second commitment period of *KP*, and rolled out the "Ad Hoc Working Group on the Durban Platform for Enhanced Action (ADP)" work plan. The international efforts against climate change realized steady transition in Doha, and international climate negotiations would shift to the "one-track negotiation" on the Durban platform in 2013. "Protocols", "other legal documents," or "agreed outcomes with legal binding force" that were applicable to all Parties would be formed at the end of 2015 and take effect in 2020.

Based on years of efforts, the Paris climate conference in December 2015 took a historic step forward.

### 2.3.3 Other International Mechanisms of Tackling Climate Change

Ever since the UNFCCC came into force in 1994, all countries have had "multilateral" negotiations for more than 20 years for its effective implementation, while several other multilateral and bilateral mechanisms related with climate change have also appeared. These mechanisms, either outside or inside the UNFCCC, constitute the international institutional arrangements for tackling climate change. As the only basic institution under the UN framework that has global engagement, the negotiation of UNFCCC and the global cooperation based on it begin to be influenced by other mechanisms. International groups and organizations, including G8, G20, APEC, IMO, and ICAO, have all taken actions regarding climate change, affecting the climate change mechanisms under the UN through collective standpoints or resolutions of international organizations.

Multilateral consultation mechanisms outside the UNFCCC related with climate change can be divided into two categories. One category derives from the topics of UNFCCC negotiations, such as the UN Secretary-General's High-level Advisory Group on Climate Change Financing. They are usually newly formed mechanisms. The other category refers to existing national groups and international organizations that are concerned about climate change for their own development, such as the IMO and ICAO. As far as the nations involved are concerned, the multilateral mechanisms can be divided into global and regional ones, which to some extent reflect their influence. Among the global multi-themed mechanisms, there are political and professional ones according to their nature, the former usually discussing all aspects of climate change, such as the G20, while the latter only focused on aspects of climate change that are related with their own expertise, such as the international aviation and shipping emission.

The mechanisms outside the UNFCCC affect it in three ways. The first way is providing technical support. The professional mechanisms, through more professional and thorough consultations, can help all parties to better understand relevant topics, urge them to seek commonly acceptable schemes, and push for relevant resolutions under the UNFCCC. The second way is providing political willingness. Political multilateral consultation mechanisms outside the UNFCCC provide the platform for various parties to reduce divergences and unify standpoint. They can reflect regional collective stance and promote progress under the UNFCCC by drawing on the power of major countries. The third way is formulating legally binding agreements. The *Conference of the Parties to Montreal Protocol*, ICAO, and other international agreements or organizations can form resolutions on climate-change-related topics within themselves that have legal binding force on their contracting parties or member states. However, in general, current multilateral consultation mechanisms involving climate change outside the UNFCCC cannot replace the *Convention* yet, nor can we negate the significance of the *Convention*'s existence.

### 2.3.4 The Future of International Climate Change Negotiations and China's Engagement

Although climate change still has uncertainties in the scientific sense, coping with it has become a global political consensus. Despite the many conflicts, international climate negotiations are essentially an earnest global effort with grave responsibilities, aiming to establish a reasonable international climate system. The essence of tackling climate change is to steer the world onto the path of low-carbon, green, and circular development and achieve sustainable development for the whole mankind. Climate negotiation should be a process that propels all parties to constructively arrive at a global climate institutional arrangement, and its ultimate approach is win-win cooperation.

China was deeply aware of the importance of climate change. Whoever takes actions obtain the leading strategic position in moral and development. As Chinese President Xi Jinping said, tackling climate change is the inherent requirement of China's sustainable development and the international obligation of a responsible major country. It is not something asked of us by others, but something we want to do out of our own will. Although China is still in the process of industrialization, IT applications, urbanization, and agricultural modernization, green development is an important feature of the whole process. China should not repeat the traditional development path that was adopted by developed countries in their industrialization stage, characterized by uninhibited GHG emission. Instead, it will explore a sustainable development path suitable for China's national conditions that will take us to the win-win outcome of economic development and climate change adaptation or mitigation.

The *China–US Joint Announcement on Climate Change* was released in November 2014, in which China announced that its $CO_2$ emission will peak around 2030, when its nonfossil energy will take up about 20% of total energy consumption. The *Joint Announcement* confirmed that the global scientific community has made clear that human activity is already changing the world's climate system, accelerating climate change has caused serious impacts, and these developments urgently require enhanced actions to tackle the challenge. It stated that smart actions on climate change now can drive innovation, strengthen economic growth, and bring broad benefits—from sustainable development to increased energy security, improved public health, and a better quality of life. Tackling climate change will also strengthen national and international security. Jointly issued by the largest developing and developed nations, the *Joint Announcement* set the principle and tone for designing the international climate institution and injected new political drive into international negotiations. It also secured the fundamental position of the "principle of common but differentiated responsibilities and respective capabilities" in the future design of international coping system, specified the emission reduction targets and paths for China and the US after 2020 in differentiated ways, and set the tone for the *Paris Agreement* reached at the Paris Climate Conference in 2015. The *Joint Announcement* established China's sound image as a positive and responsible major country and ensured that China has a greater say in the global governance system. In June 2015, China submitted the document titled *Enhanced Actions on Climate Change: China's Intended Nationally Determined Contributions* to the UNFCCC Secretariat. In addition to the targets set in the *Joint Announcement*, the new document put forth the post-2020 enhanced actions on climate change and the paths, policies, and measures to achieve the targets, including lowering $CO_2$ emissions per unit of GDP by 60–65% from the 2005 level, and increasing the forest stock volume by around 4.5 billion $m^3$ on the 2005 level. This was an active step taken by China, a contracting party of UNFCCC, as a part of the global efforts on climate change. It declared to the world the Chinese government's resolution to pursue the green, low-carbon, and circular development path characterized by growth, energy, and consumption transformation.

China has been playing a positive and constructive role in the international efforts to tackle climate change. That, first and foremost, stems from its internal needs for scientific and sustainable development, but it also reflects its fulfillment of international responsibilities as a responsible, developing major country. All in all, tackling climate change is not a trap, and low-carbon development is the future trend of the international community. China's active efforts to deal with climate change, adjust domestic economic structure, and change the growth model are the inevitable choice in order to achieve sustainable, green, and low-carbon development.

# Chapter 3
# The Concept of Low-Carbon Development

## 3.1 Proposition of the Concept of Low-Carbon Development

### *3.1.1 International Background*

(I) Transition from traditional environmental problems to climate change issues in western countries

The first Industrial Revolution that kicked off in the 1760s ushered in an era represented by the steam engine in technology and by coal in energy; the second Industrial Revolution that started in the 1870s brought the era represented by electric lamp and iron and steel in technology and by petroleum in energy. During the third Technological Revolution in the 1940s and 1950s, all kinds of electric appliances and equipment were popularized and families began to have cars. As a result, the demand for petroleum increased dramatically. A simple analysis of these three industrial and technological revolutions shows that the human world has established a technical and material system based on the consumption of fossil energies since the first revolution, and formed the industrialization paradigm, which has dominated world development for more than 200 years. Thanks to the industrialized growth model, the productivity in western developed countries was significantly improved.

With the development of industrial civilization after the 1950s, western developed countries all made tremendous economic achievements, but environmental problems appeared too, so they began to address these problems concerning water, atmosphere, and soil that occurred in the process of industrialization and urbanization. Traditional environmental problems in those countries were basically solved in the 1970s and 1980s. After that, studies on the impact of GHG emission on the human society yielded results, and western developed countries gradually came to focus on climate change issues, and set step by step the ultimate goal of tackling climate change—"to stabilize GHG concentrations in the atmosphere at a level that would prevent


X. Du et al., *Overview of Low-Carbon Development*,
https://doi.org/10.1007/978-981-13-9250-4_3

dangerous anthropogenic interference with the climate system." The *Kyoto Protocol* and other documents were signed successively afterward.

(II) Low-carbon development is the consensus of governments of all countries

While climate change reflects environmental problems on the surface, it essentially reflects issues concerning economic growth model, energy development, and sustainable development, and therefore draws close attention from all countries. In January 2004, Sir David King, Chief Scientific Advisor to the British government, published an article on the *Science* magazine, in which he said compared with power politics and terrorism, the abnormal global climate change is the biggest threat that the mankind will face. In February that year, the Pentagon, in its secret report titled *Climate Change and National Security*, pointed out that countless people will lose lives in wars and natural disasters caused by climate change in the next 20 years, which will gravely threaten the global stability. At the 2007 annual meeting of Davos World Economic Forum, global warming beats the Iraqi issue, terrorism, and Arab–Israeli conflicts to be listed as the top issue that will influence the future world. Given their close attention to climate change, all countries reached a consensus on low-carbon development and have taken active steps to practice this concept.

In 2003, the British government released the white paper *Our Energy Future—Creating a Low Carbon Economy*, which initiated the concept of "low-carbon economy". The background then was that the British government realized the decrease of domestic energy supply, and the report said Britain was going from energy self-sufficiency to an era of import dependence, and it might have to import 3/4 of its energy demand by 2020. Meanwhile, under the threat of climate warming, the global sea level rise exposed the east coast to the risk of flooding. The report set the goal that Britain will lower its emission by 60% by 2050 from the 1990 level and fundamentally change Britain into a country with one of the low-carbon economies.

In 2006, the British government issued the *Stern Report* written under the lead of Nicholas Stern, former Chief Economist of World Bank. The report pointed out that by investing 1% of the global GDP annually, we could avoid the annual loss of 5–20% of GDP in the future, and called for global efforts to shift to low-carbon economy. After issuing the *UK Climate Change Program* and *Climate Change and Sustainable Energy Act* in 2006, Britain passed the *Climate Change Act* in 2008. It was considered the world's first flagship legislation on climate change as it, for the first time, included a nation's emission reduction goal in the law and provided the long-term policy framework for tackling climate change and the long-term plan for transitioning to a low-carbon economy. Later Britain issued the white paper *Low Carbon Transition Plan* in 2009, which outlined how to transform the British economy to achieve the emission reduction goal.

The concept of low-carbon economy made all the countries confident and interested in tackling climate change with low-carbon development. The United States did not join the *Kyoto Protocol* and other international emission reduction mechanisms, but it always attached great importance to energy conservation and emission reduction. In 2007, the Senate put forth the *Low Carbon Economy Act*, which vowed to renovate the traditional high-carbon industries, accelerate low-carbon technological innovation, and urge enterprises to reduce emissions through market mechanisms. It indicated that low-carbon economy will be an important strategic choice in America's future development. In 2009, the House of Representatives passed the *American Clean Energy Security Act*, which was a comprehensive energy legislation that included the following key points: mitigating global warming through cap and trade, promoting America's economic recovery by creating millions of jobs, and enhancing America's energy security by reducing the dependence on oil import.

As the initiator of KP, Japan has been advocating and practicing "low-carbon development" since the 1990s. In 2004, it conducted a study of "Japan Low-Carbon Society Scenarios toward 2050" with the aim to provide specific measures for achieving the low-carbon society in 2050, including institutional, technological, and lifestyle changes. The study group issued the *Japan Low Carbon Society Scenarios: Feasibility Study for 70% $CO_2$ Emission Reduction by 2050 below 1990 Level* in 2007, and *A Dozen Actions towards Low-Carbon Societies* in 2008, including green architectures, convenient logistics and packaging, urban footpath design, low-carbon electricity, and low-carbon trademark. Each action has a series of technological measures, institutional reform targets, and incentive policies behind it.

The EU has been committed to pushing the world to deal with climate change together and it is taking active steps toward a low-carbon economy. In December 2008, the EU Summit reached an agreement on the energy–climate package deal to tackle climate change, enhance energy security, and strengthen low-carbon competitiveness. One of the important contents of the agreement was that the European Commission passed the *Roadmap for Moving to a Competitive Low-carbon Economy in 2050* in March 2011, which described the cost-effective path that the EU had to adopt in order to cut the GHG emission by 80–95% in 2050 from the 1990 level. It also offered a series of guiding measures on industrial policies in the economic sector, national and regional low-carbon strategies, and long-term investment.

In the meantime, EU members also actively issued all kinds of policies and rules to shift toward low-carbon economy. Germany passed the EEG in 2008, which provided a favorable environment for the development of renewable energies and remains the most important policy tool for expanding their use. Italy paid close attention to the R&D and utilization of renewable energies as it had to import over 80% of its fossil energies, and began to implement the renewable energy quota system in 1999 to promote low-carbon energy development. France has been strongly developing nonfossil and clean energies represented by nuclear power, and made notable achievements in energy conservation and carbon emission reduction in such fields as industry, construction, and transport. Sweden exercised the concept of low-carbon economy in every detail of everyday life and issued a series of policies and measures to encourage its people to use environmentally friendly cars. Denmark created a unique economy through renewable energy and clean, efficient energy technologies, including wind power generation, straw power generation, and ultra-supercritical boilers, and is acknowledged as one of the countries that have found the best solution to $CO_2$ reduction and energy issue.

Developed countries' transition toward low-carbon economy is not only an external responsibility endowed by the international community for jointly tackling climate change but also an internal requirement for economic transformation and upgrade. The *Kyoto Protocol* set absolute emission reduction targets for developed countries, which objectively accelerated their transition. As the fifth IPCC assessment report further confirmed the relation between anthropogenic GHG emission and global warming, developed countries, as main GHG emitters in history, had the obligation to reduce GHG emission and ensure the sustainable development of the earth. Besides, in face of the ever more rigorous restrictions on fossil energy, most developed countries took the initiative to shift to low-carbon energies for the sake of energy security, so as to reduce their reliance on energy import. An overview of the low-carbon development policies in those countries shows that almost all of them put energy security in a prominent position. Encouraging low-carbon technological innovation and fostering the core competitiveness of low-carbon industries are also important reasons why they transit to low-carbon economy. After the steam revolution and electric revolution, energy revolution is bound to have substantial effects on the future global landscape. Therefore, promoting economic transition and upgrade and occupying the high ground in low-carbon technology are important driving forces for developed countries to ensure the increase of domestic employment and sustained economic growth in the future.

Compared with developed countries, developing ones are subject to serious impacts of global climate change although their carbon emission only takes up a small proportion in the world. Today they have realized the consistency between poverty elimination and sustainable development and the necessity to speed up the low-carbon transition. Despite the low starting point and numerous difficulties, developing countries have made tremendous efforts for their respective low-carbon development in a wide range of aspects, including formulation of national low-carbon development strategies and policies, increasing low-carbon investment, promoting low-carbon technologies, raising energy efficiency, and encouraging low-carbon consumption. We can say that the global trend of low-carbon development has offered a rare historical opportunity for developing countries to comprehensively advance their low-carbon economic transition and sustainable development.

(III) Low-carbon development is a scientific direction, not a "trap" or "plot"

China's remarkable achievements in the past 30-plus years have drawn worldwide attention. We need to continue developing without any external interruption, but in a way, our sustained high-speed development in the past came at the price of high pollution, high emission, and high energy and resource consumption. The shift from traditional growth model to low-carbon mode naturally restricts the development of high-carbon energies and high-energy-consuming industries and consequently restricts the GDP growth driven by them. In view of China's energy endowments—abundant in coal, lacking in petroleum and gas—and the fact that it is in the stage of fast heavy industrialization and urbanization, controlling the total carbon emission means slowing the GDP growth driven by high-carbon industries. Against such a background, some people in the theoretical and practical circles have doubts about carbon control and consider it a "trap" and "plot". The book *Low-Carbon Plot* published in 2010 made it clear that pinning the responsibility for global warming on carbon and implementing "carbon tariff" and "carbon emission reduction" worldwide are huge plots. The essence of it is that developed countries and regions, including the United States and EU, try to suffocate developing countries like China with environmental issues, let them pay for GHG emission and the financial crisis, and continue to curb and exploit them, in a bid to maintain the bipolar world landscape.

The low-carbon "trap" theory and "plot" theory are the generalization for three concerns about low-carbon development. First, people are concerned that the immense amount of capital and technology invested to reduce energy consumption and control carbon emission will increase the cost and slow down economic growth. Second, people are concerned that developed countries, with their technological and capital advantages, will take low carbon as an excuse to set limitations on developing countries like China and launch a new round of economic looting on them. Third, people are concerned that carbon finance is a weapon used by developed countries to restructure the world and a financial trap for them to recontrol the international economic trend. If we accept these "carbon trap" and "carbon plot" theories indiscriminately and give up the low-carbon goal, not only our economic transformation and development will be restricted but we also may fall into the high-carbon trap.

The problem is that high-carbon development is at the expense of excessive resource consumption and environmental damage, which is against the "human-centered" growth model. It is no good for the immediate interests of the contemporaries and will harm the future society. It is an extensive and backward growth model that should be abandoned.

Now, China is in the stage of fast industrialization and urbanization, which means its huge energy demand arising from economic development will encounter the restriction imposed by nonrenewable fossil energies. Extensive growth model and high-carbon energy structure have already made China the largest $CO_2$ emitter in the world and created many signs of non-sustainability in our development. In terms of foreign trade alone, high-carbon products will draw close attention. As the carbon label of products is popularized, the change of public consumption preference will change the product supply chain. It will also prompt multinationals and dealers to take carbon content as an important assessment indicator in the purchase of raw materials, intermediate goods, and final consumer goods, so as to establish their low-carbon image and foster their low-carbon competitive advantages. If we do not speed up the transformation toward low-carbon economy, the high-carbon label on China's export goods will definitely affect the nation's position and competitiveness in the global industrial chain.

More importantly, high-carbon and high-polluting emissions in China are highly coordinated because of their basically identical roots and sources. For a long time, the Chinese economy was on the path of extensive growth featuring high input, high consumption, high emission, and low efficiency. Statistics show that capital (including resource and environmental input) contributes about 60% to China's economic growth, labor about 10%, and technical progress about 30%. It's clear that, in general, China's economic growth is based on resource and environmental consumption, whereas in developed countries, more than 60% of economic growth is contributed by sci-tech progress, with obvious superiority over China in terms of growth quality, potential, and sustainability. Although Chinese economy has realized sustained high-speed growth, the extensive high-carbon development has touched the bottom line of environmental and climatic capacity and become a main factor that restricts the scientific development of Chinese economy and society and threatens people's health and safety. Therefore, improving environmental quality, tackling climate change, and promoting low-carbon development are inherent requirements of China for its own healthy development.

While limiting high-carbon industries and eliminating backward production capacity, low-carbon development will give rise to new-type low-carbon energies, industries, and service industry; improve the industrial and employment structure; foster new economic growth areas; and improve China's innovative capacity and competitiveness. That's the only way to avoid falling into the "middle income trap" and to walk from the new normal to new-type development. Exploring the path of low-carbon development not only conforms with the world trend of "low-carbon" energy but also gives China an opportunity to change the growth model, adjust industrial structure, meet energy conservation and emission reduction goals, and achieve sustainable development.

It's safe to say that "low-carbon development" is an energy revolution that pushes for growth and consumption transformation. Neither being a "plot" nor a "trap", it is a scientific development direction, a legal, regulatory, policy, and technical standard that requires concrete actions, and a sustainable development strategy and action with definite timetable and arrangements. To make the Chinese economy sustainable, our only choice is actively promoting low-carbon development, comprehensively coordinating development with emission reduction, changing the economic development concept and behavioral pattern, and gradually establishing the development outlook that views eco-environmental protection and economic development as an integral whole rather than two parallel lines. In this way, we will explore a low-carbon, new-type modernization path that bypasses the "trap".

It must be stressed that denying the theories of low-carbon "trap" and low-carbon "plot" does not mean accepting all the low-carbon rules or systems proposed by developed countries. Due to our special national conditions, some emission reduction principles and low-carbon rules initiated by them do not apply to China. While the international community is pursing low-carbon development and building the carbon finance system, it should fully consider China's national conditions and appeals, follow the principle of differentiated responsibilities, safeguard our say for and right to development, and seek more efficient measures of energy conservation, emission, and consumption reduction. The developing China has to go through the process from carbon reduction to low-carbon development. While following the low-carbon development path, China should keep in mind the revolutionary thoughts and historical opportunities brought by the general trend and context in the world, and also respect its actual conditions and advance low-carbon development in a sound and sure way.

### 3.1.2 Concept and Connotations of Low-Carbon Development

(I) Origin of low-carbon development concept

The concept of low-carbon development originated in the 1990s against the background of global climate change, with "sustainable development" and "green economy" being its predecessors. The Rio Earth Summit in 1992 put forth the idea of "sustainable development", which emphasized the balance between economic development, social development and poverty elimination, and environmental protection. It was the first interpretation of the harmonious coexistence between man and nature and was later widely quoted by all countries. "Green economy" is a term more focused on economic development. Initiated by British environmental economist David Pearce in 1989, the core of this concept is replacing the increase of economic quantity with the improvement of economic quality. Based on the theory of rational consumer and benefit maximization in neoclassical economics, the "green economy" theory holds that we can change the consumers' economic behaviors by influencing their values and consequently protect resources and the environment while achieving economic development. Both "sustainable development" and "green economy" concepts try to integrate and coordinate economic development with environmental protection, two concepts that originally seemed contrary to each other.

The low-carbon development concept first appeared in UNFCCC and was also called low-emission development strategy (LED). It is more targeted on the impacts of climate change in comparison to "sustainable development" and "green economy". Although there is no official definition of low-carbon development yet, it usually means realizing coordinated socioeconomic development while reducing emission, or mitigating climate change and lowering the carbon emission intensity in the process of development. For developed countries, it is more often used to describe their actions of economic transition, namely, changing from the old development trajectory to low carbon and low emission. For developing countries, however, it means low-carbon economic growth in the process of development.

Foreign scholars understand low-carbon development on two dimensions—lowering carbon emission in the process of economic development and promoting sustainable economic development by controlling carbon emission. According to many international organizations that carry out low-carbon development programs, low-carbon development means realizing low carbon emission in the process of economic development with the aim of achieving sustainable economic growth. The UK Department for International Development (DFID) pointed out in its report that low-carbon development is to achieve economic growth while tackling climate change and reducing carbon emission, the latter being a means to sustainable economic growth. The Danish Institute for International Studies (DIIS) believed that low-carbon development refers to the process of minimizing atmospheric GHG emission while developing the economy. The United Nations Department of Economic and Social Affairs (UNDESA) issued *A Guidebook to the Green Economy* in 2012, which said the development strategy of low-carbon emission is an integral part of sustainable development and should be encouraged in developing countries.

The foreign academic circle has a deeper understanding of low-carbon development, believing it is the process of minimizing GHG emission in parallel with socioeconomic development and human progress, a process that requires the engagement of the whole society. Low-carbon development should not just rely on market self-adjustment or government provision of public goods, but needs vehement political intervention to solve market and systematic malfunction and ensure fairer distribution and the equal right to acquire opportunities and benefits in the process. In its report on moving toward a low-carbon society, OECD said low-carbon development is a social process of extensive social engagement. Foxon deemed that low-carbon development depends on existing mechanisms, thinking, power structure, and development paths, so it is a slow and gradual process.

The concept of low-carbon development is extensively adopted by the international community. The 4th IPCC assessment report stated that low-carbon development is more operable for developing countries. It has to achieve development goals while tackling climate change and has to make climate considerations under the current policy framework. Moreover, this concept also appeared in the ministerial statement of the Economic Power Forum 2009, in which ministers of 17 countries announced to adopt low-carbon development plans. More and more international organizations, including UNEP, UNDP, World Bank, and WWF, launched low-carbon development programs too.

(II) Domestic scholars' understanding of low-carbon development

Literally, not many domestic scholars have had definite interpretation of the concept of "low-carbon development". They tend to focus on the definition and understanding of it. Zhou Shengxian said, "low-carbon economy is an economic mode based on low energy consumption, low emission and low pollution. It is a major progress of human society after the primitive civilization, agricultural civilization and industrial civilization. Its essence is raising the energy efficiency and creating a clean energy structure, and the center is technological innovation, institutional innovation, and the change of development outlook. Developing low-carbon economy is a global revolution concerning production mode, lifestyle, values and ideas, and national rights and interests." The China Council for International Cooperation on Environment and Development stated in its report, "low-carbon economy is an economic form in the post-industrial society. It is aimed to lower GHG emission to a certain level, so that all nations and their peoples in the world will not be subject to the adverse effects of climate warming and a sustainable living environment can be eventually guaranteed on earth." He Jiankun believed, "the essential requirement of low-carbon economy is to improve the productivity of carbon, namely generating more GDP per unit of $CO_2$ emission."

Pan Jiahua's views are closer to the broad-sense understanding of low-carbon development overseas. In his opinion, low-carbon economy is a kind of economic state when carbon productivity and cultural development both reach a certain level, whose purpose is to realize the global vision of controlling GHG emission. Carbon productivity refers to GDP generated per unit of $CO_2$ emission. It can be relatively low carbon emission resulting from higher energy efficiency or absolute decrease of carbon emission thanks to clean energies and low-carbon technologies. Cultural development does not just mean sustainable economic development but also includes social progress covering health, education, eco-environmental protection, and fairness.

We can see that domestic scholars understand low-carbon economy both in the broad sense and the narrow sense. He Jiankun defined it from the narrow GDP perspective, whereas Zhou Shengxian and Pan Jiahua defined it from a broad perspective that basically covers all aspects of low-carbon development. We can say that "low-carbon economy" in the broad sense is essentially another expression of "low-carbon development".

(III) Redefinition of low-carbon development concept

A review of the views of domestic and foreign scholars indicates that low-carbon development is not a stage or continuation of the traditional industrial or high-carbon period, but a great leap forward after the industrial civilization and a profound revolution involving a wide range of aspects, including economy, political, culture, and ecology. It does not just mean low carbon emission in the process of economic development but also includes all-round social transformation at the same time in order to meet the goals of cultural development. In summary of the views of domestic and foreign scholars and experts, this book redefines low-carbon development as follows:

Low-carbon development is a new-type, high-quality, sustainable growth model that, guided by the concept of sustainable development, abandons the old growth model featured by "governance after pollution, intensive after extensive development," and uses technological and institutional innovation and industrial transformation to tackle climate change, reduce GHG emission, and coordinate the economic, social, and eco-environmental development. It is a new systematic reform concerning the production mode, lifestyle, values, and thoughts across the board, and another substantial progress from the industrial civilization characterized by fossil fuels to ecological civilization, a progress the human society experienced after the agricultural civilization.

(IV) Connotations of low-carbon development

Fundamentally, low-carbon development means finding the junction point between development and low carbon and solving their conflicts through scientific and sustainable development. It is a global revolution involving production mode, lifestyle, values, and thoughts. It's not just a development concept, but more of a growth model and politicized scientific issue, while also being a comprehensive topic involving economic, social, and environmental systems. It is a topic of rich connotations.

1. Low-carbon development is a strategy to mitigate global climate change

Effective control of the emission of GHGs represented by $CO_2$ can mitigate the trend of climate warming. While developed countries undertake their historical responsibilities by reducing the absolute volume of GHG emission in order to lead the low-carbon transformation and new-mode growth of their economy, developing ones adopt reasonable approaches consistent with their development stage to control GHG emission and achieve the win-win result of development and emission reduction as well as sustainable development.

2. Low-carbon development is an important way to practice the Scientific Outlook on Development

Scientific development is a kind of comprehensive, coordinated development, not only including material development, progress of productivity and creation of material wealth, but also the all-round improvement of people, true betterment of the human living environment, and coordinated development of the human society and natural environment. The idea of "scientific development" was initiated by western political scientists and economists in the 1970s, when they comprehensively reflected on the traditional growth model that pursued economic growth alone, seeing that enormous amounts of fossil fuels were burned after the Industrial Revolution in the west, rapid economic development caused pollution of the atmosphere, geosphere, and hydrosphere, and the excessive exploitation of natural resources resulted in energy crisis. The main measurement of low-carbon development is lowering carbon emission, and it achieves the fundamental change of economic growth mode through a series of measures, such as raising energy efficiency and adjusting energy structure. It is a specific and vivid embodiment of scientific development in the current serious situation of climate warming, and an effective way to realize the coordinated coexistence between socioeconomic development and eco-environment. In comparison with previous concepts like "sustainable development", "green development," and "circular economy", low-carbon development has a different focus, but they are all in essence different interpretations of "scientific development" in the context of various social backgrounds and requirements. These concepts were all created as people sought to realize the coordinated and harmonious development of economy and environment against the background that the natural environment was aggravated and high-speed economic development approached or even exceeded the bearing capacity of resources and environment. They are all important approaches of practicing scientific development.

3. Low-carbon development is an efficient growth model

By fundamentally changing the economic growth model, energy consumption mode, and human lifestyle, low-carbon development can comprehensively reform the modern industrial civilization based on fossil fuels (carbon-based energy), and reduce the development cost and carbon emission to the largest extent. It realizes the efficient and low-carbon economic activities and eco-friendly energy consumption in the whole process of social reproduction, covering production, exchange, distribution, and consumption, so as to ensure the clean, green, and sustainable development of the ecology, economy, and society as a whole.

4. Low-carbon development is a relative concept

Low-carbon development can be both a comparative and a categorical concept. The key lies in the development stage. Carbon emission in developed countries mainly comes from the consumer society of the post-industrial age, and developed countries of different types have growth models that are either relatively low carbon or high carbon. In comparison, carbon emission in developing countries mainly stems from capital accumulation driven by production investment and infrastructure input. Therefore, for developing countries, before their basic needs for socioeconomic development are satisfied, the relative reduction of carbon emission amid the increase of economic aggregate can be regarded as low-carbon development. But for developed countries that have reached a high level of industrialization and urbanization, the standards for low-carbon development must be higher than those for developing ones. Only the absolute reduction of total carbon emission can be regarded as low-carbon development on the premise of maintaining a high development level.

5. Low-carbon development is an orientation about international governance and order reconstruction

In the current world landscape of multi-polar development, low-carbon development has evolved from a technical and economic topic to a political issue. The UNFCCC became a new major framework in world development after the UN Charter and the *General Agreement on Tariffs and Trade* (GATT). The "fairness principle" it adopted for solving problems and the "principle of common but differentiated responsibilities" for developed and developing countries in tackling climate change have effectively included all developed and developing countries in the process of redistribution of world political interests and carbon emission rights.

6. Low-carbon development is an overall issue concerning the coordination of energy, environmental, and economic systems

Low-carbon development is not simply a technical or economic issue. It is a process of reducing the reliance on natural resources while keeping or improving the necessary speed and quality of economic development, and improving the eco-environment by changing energy structure, adjusting industrial structure, raising energy efficiency, and enhancing the capability of technical innovation and sustainable energy supply. The ultimate goal is the harmonious development of the complicated system of energy, environment, and economy.

(V) Characteristics of low-carbon development

On the basis of analyzing the concept and connotations of low-carbon development, this book summarizes the basic characteristics of low-carbon development.

1. A gradual process

To a large extent, low-carbon development is a shift of systems and an institutional innovation. From technology to economy, policy to law, industry to finance, society to environment, and business model to consumption habit, almost every link has to be adjusted accordingly. This shift of systems has to proceed gradually from a low level to high level and from part to the whole. When choosing the path of low-carbon economy, all countries started with key areas and industries and gradually moved to all-round low-carbon development integrating the five aspects of policy, energy, technology, market, and society.

2. An economical process

Maintaining economic sustainability is the central and critical content of low-carbon development. Developing and developed countries are in different development stages, differ greatly in economic situations and historical carbon emission, and have different tasks in low-carbon development. But in spite of these disparities, they all refuse to accept reduced economic growth, and the goal of sustained economic development is consistent in the whole world. Only by maintaining economic development they can understand and explore the low-carbon laws and create emission reduction models in the process of development, and they can create more material wealth and provide sufficient material guarantee for addressing the problem of high carbon emission.

3. An innovative process

The core of low-carbon development is to effectively control carbon emission; mitigate climate warming and promote and maintain global eco-balance through innovations in energy and emission reduction technologies; and the consequent adjustment of industrial structure, institutional innovation, and fundamental changes of people's ideas about consumption. In this sense, energy conservation, consumption, and emission reduction, the exploitation and utilization of renewable energies and improvement of energy consumption structure should all be based on the research, development, and popularization of low-carbon technologies. In other words, technical innovation is the fundamental way of solving environmental and energy issues and the essence of low-carbon development.

4. A strategic process

Low-carbon development is becoming an international competition for carbon emission right, capital, technology, and development space. The level of low-carbon development will determine a nation's core competitiveness and say on the international political and economic stage. Therefore, low-carbon development should be promoted as a national strategy, and energy and carbon emission should be given priority when formulating development plans.

5. A global process

The global climate system is an entirety and climate change and the impacts thereof are global phenomena concerning the common interests of the whole mankind and the future of every individual, regardless of national boundary. Therefore, low carbon emission and development require the cooperation and efforts of the whole world. Every nation, everyone living on earth has the obligation and duty to contribute to "low-carbon development".

6. A transformative process

The core of low-carbon development is promoting the improvement of energy efficiency technology, energy conservation technology, and renewable energy technology and reducing GHG emission through institutional and policy innovations. As a result, the socioeconomic growth model will shift from high energy consumption and high emission to high energy efficiency, low energy consumption, and low emission.

7. A sustainable process

Low-carbon development means promoting the harmonious and sustainable social and economic development while realizing low pollution, low emission, and high energy efficiency in life and production. This includes economic development, improvement of people's living standards, better health and education for residents, eco-environmental protection, and social equity.

## 3.2 The Relation Between Low-Carbon Development, Green Development, and Circular Development

### *3.2.1 Low-Carbon Development and Green Development*

(I) Origin and connotations of green development

Green development starts with the idea of green economy, and green economy stems from the human reflection on the relation between man and nature. Green development refers to the economic state or growth model that generates economic, social, and environmental benefits by reserving resources and protecting environment. Its basic features are low consumption, low emission, low pollution, high efficiency, and high circulation, with a special emphasis on the benign circle between man and nature, economy and environment. Green development is opposed to the "golden development" in the age of agricultural civilization and the "black development" in the age of industrial civilization. In the age of "golden development", land and labor were the most important production factors, while in the age of "black development", fossil energies like coal and petroleum were basic energies and land, labor, and capital were the most important production factors. With socioeconomic development, the scarcity of resources and environmental factors becomes more prominent, and eco-environment becomes an important production factor in the age of green development. Green development requires that economic activities should not damage the environment, which should be protected in the process of economic development.

The term "green economy" appeared in the book titled *Blueprint of a Green Economy* published by British environmental economist David Pearce and others in

1989, but it dated back to the "green revolution" that started in the 1960s, which later evolved into a global "green movement". In 1962, American biologist Rachel Carson revealed in her book *Silent Spring* the tremendous damages of the natural ecosystem caused by environmental pollution that resulted from industrial development, and advocated reducing eco-environmental pollution and destruction while pursuing industrial development. This thought was deemed as the seed of the green economy thought. 1972 was a landmark year in human reflection on green development. The research report titled *Limits to Growth* issued by the Club of Rome gave a warning to the human race: the chaotic growth of population and industry will eventually meet limits imposed by the exhaustion of resources and destruction of eco-environment on earth. In the same year, the United Nations Conference on Human Environment was held in Stockholm, which passed the resolution to form the UNEP, adopted the motto of "Only One Earth", and representatives from all countries reached the *Declaration on the Human Environment*. From then on, environmental protection was put on the agenda of human development and people gradually accepted the idea that economic development had to take into account eco-environmental protection. In 1989, British environmental economist David Pearce and his co-authors used the phrase "green economy" for the first time in their book *Blueprint of a Green Economy*, equated it with sustainable economy and discussed in depth the way to sustainable development from the perspective of environmental economy. In the 1990s, Jacobs, Postel, and others studied green economy further and especially put forth the notion of social and organization capital (SOC). In 2007, the UN Secretary-General Ban Ki-moon proposed to begin the new era of "green economy" at the UN Climate Change Conference in Bali. In October 2008, UNEP launched the Global Green New Deal and Green Economy Initiative, aiming to make global leaders and policy-makers in relevant departments realize that green economy was not a burden on growth, but an engine of growth. In the backdrop of multiple global crises, such as energy, food, and financial crisis, UNEP, for the first time, systematically rolled out the initiative of developing green economy, which received positive responses from the international community and became a new trend in the field of global environment and development.

Regarding the evolution of the thought of green economy, it is not a new concept in itself, but is a continuation of the thought of sustainable development. The core of green economy is harmony between man and nature, its purpose is sustainable development, and its connotations include the following: economic growth should respect the limitation of eco-environmental capacity and resource bearing capacity, environmental resources should be viewed as internal factors of economic development, and environmental protection be taken as an important pillar for sustainable development. The sustainable development of economy, society, and environment is taken as the goal of green economy. Green and eco-friendly economic activities, both in process and results, are taken as the contents and path of green economic development.

## (II) Comparison of low-carbon development and green development

### 1. Background

Low-carbon development and green development have different backgrounds. The former is to cope with global climate change and the serious pollutions caused by fossil-energy-based emission, while the latter derives from the green revolution and aims to cope with the black pollution caused by economic development after the Industrial Revolution.

### 2. Theoretical basis

Low-carbon development and green development have different but overlapping theoretical bases. The former's theoretical basis mainly refers to modern climate change science and economics, while the latter's includes ecology, environmental science, and ethics. Green development not only includes green production but also green consumption and low carbon consumption.

### 3. Core contents

Low-carbon development is highly targeted, primarily at the reduction of GHG emission from the use of fossil energies. Encompassing many aspects, including ethics, economy, and environment, green development is hard to assess in a quantitative way, and does not imply the restrictive conditions faced by socioeconomic development from the perspective of invested factors. The biggest difference between green development and low-carbon development is that that former does not have the rigid limitation on carbon emission.

### 4. Ultimate goal

Green development is a conceptual expression, and any economic state and growth model associated with environmental protection and sustainable development can be put in this category. Its ultimate goal is achieving the harmony between nature, society, and human race. Low-carbon development is a concept that, under the basic factors of traditional socioeconomic development (labor, land, and capital), adds more detailed factors like the consumption of energy and other natural resources and the environmental capacity for GHG emission. Under this concept, carbon emission is an invested factor and restrictive indicator of socioeconomic development.

### 5. Path of realization

The path to low-carbon development is aimed to tackle climate change, including low-carbon industry, low-carbon energy, and low-carbon cities. Green development, however, aims to solve traditional pollution, such as water pollution, air pollution, and solid waste. In other words, it seems synergy effects while realizing low-carbon development. There are many paths to green development, including low-carbon development and circular development.

### 3.2.2 Low-Carbon Development and Circular Development

(I) Origin and connotations of circular development

Circular development stemmed from the idea of circular economy, which resulted from people's reflection on the two major "environmental hazards" and the oil crisis in the twentieth century. Circular development is an economic state or growth model that generates economic, social, and environmental benefits through circular utilization of resources. Its basic features are low consumption, reuse, recycling, and high efficiency, with a special emphasis on the efficient and circular use of resources. Circular development, on the principle of "reduced consumption, reuse, and recycling", restructures the socioeconomic system according to laws governing the material cycle and energy flow of the natural ecosystem. It transforms the linear production mode of "resource—product—waste," the mainstream mode since the industrial civilization, into the feedback production process of "resource—product—regenerative resources," thus incorporating the socioeconomic system harmoniously into the material cycle of natural ecosystem. Through efficient and circular use of resources and gradient use of energy, circular development realizes low or zero discharge of pollutants and consequently the sustainable social, economic, and environmental development.

The idea of circular economy was initiated by American economist Kenneth Boulding in his essay *The Economics of the Coming Spaceship Earth* in 1966, which aimed to replace the "one-way economy" with "circular economy" in order to solve environmental pollution and resource depletion. Based on Boulding's thought of circular economy, David Pearce and Kelly Turner officially put forth the term of "circular economy" in their book *Economics of Natural Resources and the Environment* in 1990, which represented an economic growth model different from the traditional one. In their opinion, the economic system and natural ecosystem, instead of being independent from each other, were integrated and jointly formed the big ecological-economic system. The center of circular economy can be generalized as "internal–external balance, integrated circulation." "Internal–external balance" combines the internal balance that reflects the reproduction relation inside the economic system with the external balance that reflects the reproduction relation between the economic system and the ecosystem. "Integrated circulation" has twofold meanings. On the one hand, on the height of the general ecological-economic system, it governs the material cycle and flow both inside the economic system and between it and the ecosystem. On the other hand, it takes the economic system and the ecosystem as an integrated macro-system with interdependent functions, and concerns the sustainability of human economy in terms of the reproduction cycle of this macro-system. Circular economic practices began in the 1970s and 1980s, when developed countries stepped into the post-industrial stage one after another and the large volume of wastes left from industrialization and generated by the consumer society became key problems on their way to sustainable development. Against such a background, some developed countries, such as Germany and Japan, made developing circular economy and building a circular society an important approach of implementing the sustainable development strategy. They started with the disposition of wastes, reused and conducted harmless treatment of industrial wastes, and extended to the production field, with a view to establishing the sustainable production and consumption mode.

Circular development not only requires economic activities to follow the general natural, economic, and social laws but also ecological laws. It puts economic activities in the operating track of the ecosystem and strives to form a coordinated and harmonious relation between the economic system and the ecosystem. Its connotations include the following: (1) Industrial symbiosis and establishing the industrial ecosystem are technical features of circular economy, so that different enterprises can share resources and exchange by-products, and wastes generated from upstream production can be used as raw materials for downstream production. As a result, resources are well distributed among industries and regional materials and energies can be used and reused in the economic cycle. (2) The "3R" principle—reduction, reuse, and recirculation—is the core of implementing circular economy. (3) The promotion of circular economy consists of three levels—corporate level with clean production as the main content, regional level with the construction of industrial symbiotic network and ecological parks as the main content, and social level whose main content is promoting green consumption and building the network for wastes recycling and reuse.

(II) Comparison of low-carbon development and circular development

1. Background

Both low-carbon development and circular development originate in the change of economic development concept and mode in developed countries. Low-carbon development came into being against the background of global climate change caused by the use of fossil energies, and it is a growth model adopted to deal with global climate change. Emerging against the background of the oil crisis and environmental damage, circular development is a growth model adopted to raise energy efficiency and save resources.

2. Theoretical basis

The theoretical basis of low-carbon development is economics, economics of resource and environment, ecology, and sustainable development, while the theoretical basis of circular development includes ecology, environmental science, sustainable development, systematics, and thermodynamics. Their theoretical bases have much in common but not without differences. The main theoretical basis of low-carbon development is climatic economics, while that of circular development is resource economics and ecology.

3. Core contents

Circular development solves such problems as waste of resources, pollution of environment, and destruction of ecology in a broader scope, while low-carbon development focuses on dealing with the disasters and severe effects brought by the massive increase of $CO_2$ emission—the culprit for the greenhouse effect of climate change—to the earth and human society. The center of circular development is improving the use efficiency of all resources and energies in the fields of production, circulation, and consumption, so as to minimize waste discharge, including GHG emission. The center of low-carbon development is reducing the emission of $CO_2$ and other GHGs.

4. Ultimate goal

Both low-carbon development and circular development have the ultimate goal of achieving the harmonious and sustainable development between man and nature, but the latter pursues the "three-win" situation integrating economic development, resource and energy conservation, and environmental friendliness, while the former focuses on the "win-win" between economic development and climate change. The shift from high-carbon economy to low-carbon economy is not only the key of low-carbon development but also an outstanding problem that circular development has to solve, and can accelerate the deepening of circular economy. While low-carbon development can help to improve and extend the industrial chain of circular economy, the "3R" principle of circular development can totally be an important tool for low-carbon development. Low-carbon development aims to solve high energy consumption, high pollution, and high emission, while circular development is set to solve the conflict between limited resources and unlimited demand, and between economic development and environmental protection. Their goals are not completely the same. We can see that circular development and low-carbon development have similar but different ultimate goals. The former aims for full use of resources and minimal wastes, including the minimization of $CO_2$ emission, which consists of the goal of low-carbon development. But GHG emission reduction is just a small part of the requirements on circular development, but it is the central requirement of low-carbon development.

5. Path of realization

Both low-carbon development and circular development stress efficiency improvement and emission reduction, but the former is focused on reducing GHG emission by improving the energy structure and efficiency, while the latter on reducing the discharge of all kinds of wastes by raising the use efficiency of resources and energies. In light of the practices of circular development, the main difference between circular development and low-carbon development is that they correspond to different socioeconomic development stages. In other words, circular development is a mode in the industrial age when resource and energy efficiency is quite low, whereas low-carbon development is a mode developed to tackle climate change and lower the global GHG emission. Therefore, low-carbon development is the application of the circular development concept in the field of resource and energy, and circular development is the basis of it, the basis that eventually and inevitably leads to low-carbon development. Circular development is a mode every developing country in the process of industrialization has to adopt.

### 3.2.3 *Internal Logic of Green, Circular, and Low-Carbon Development*

Green, development, circular development, and low-carbon development are all sustainable modes for the development of ecological civilization. The 18th National Congress of the CPC put these three development concepts in the report in parallel. Although each of them has its own priorities, their key target is the same—coordinating the relation between man and nature and promoting the sustainable development of economy, society, and eco-environment. Compared with sustainable development, green development, circular development, and low-carbon development are more specific and targeted, and they can considerably improve the capability of the human resource–environment–socioeconomic system of sustainable development. Meanwhile, intergenerational equity and regional equity—these two dimensions advocated by sustainable development are good for implementing, green, circular, and low-carbon development in a longer term and broader sense. These concepts cannot replace each other. They jointly draw the blueprint of harmonious development between man and nature, man and man, and man and society.

(I) Philosophical consistency

1. The same systematic view

Green development, circular development, and low-carbon development share the theoretical basis of ecological economy theory and system theory. They center on the coordinated development of the ecosystem and economic system, study the big ecological system including the human race, draw on the principles of material cycle and energy conversion in ecology, consider the sustainable development of resources and environment, and explore the relation between human economic activities and natural ecology. All three development concepts make a point of making comprehensive considerations for the multiple component factors of the economic system and the ecosystem and implementing them in a coordinated way. They all pursue coordinated socioeconomic and ecological development across the board and strive for the optimal balance between ecology and economy.

2. The same development outlook

Economic development should be kept within the bearing capacity of the resources and environment. Green development, circular development, and low-carbon development all aim at protecting and improving the resources and environment, and realizing sustainable human development and an environmentally friendly society. They demand the mankind to put itself in the macro-system when considering production and consumption, and consider itself a sub-system of the macro-system in order to explore economic principles consistent with objective laws. They need the mankind to keep in mind the bearing capacity of the natural ecosystem, save natural resources as much as possible and keep improving their use efficiency, and promote the harmonious development between man and nature.

3. The same production and consumption outlook

Green, development, circular development, and low-carbon development all stress saving resource input, raising use efficiency, and clean production. For production, they require the use of recyclable and renewable resources instead of nonrenewable ones as much as possible, so that production can be reasonably based on the natural ecological cycle. They also require the use of high technologies and knowledge input instead of material input to the largest extent, so as to arrive at the harmonious unity of economy, society, and ecology, make the mankind live and work in a fine environment, and improve their living quality across the board. The new consumption concept rejects excessive waste and extravagance and advocates green consumption, namely, appropriate and layered consumption of materials.

(II) Content consistency

The relation between green development, circular development, and low-carbon development is complicated—dominant and auxiliary, dependent, and independent. Considering their different positions, the three growth models can be integrated and achieve inclusive growth. Green development is more of a kind of guiding development while circular development and low-carbon development are concrete practices, the latter being especially result oriented and strongly restrictive. The specific practices of green development, circular development, and low-carbon development depend on the actual situations in various regions, while sustainable development and green development are usually realized through low-carbon development and circular development. Therefore, their relation is as follows: sustainable development includes low-carbon development. Circular development and low-carbon development are in parallel; low-carbon development and green development are inclusive of and coordinated with each other. In practice, they are coordinated and jointly promote the harmonious development of economy, society, and environment.

Of the three growth models, low-carbon development was put forth the latest and it draws on the ideas of sustainable development, green development, and circular development in many aspects. As the mankind is seeking a way out of the global dilemma of climate change, low-carbon development suggests resolving the fundamental crux of growth model. Its core idea is controlling the emission of $CO_2$ and other GHGs by changing the high-carbon growth model and lifestyle, and realizing a new mode and lifestyle featuring low energy consumption, low emission, and low pollution. In addition to low-carbon technological innovations and low-carbon industrial structure, low-carbon development also injects new growth driver for socioeconomic development. As it focuses on energy and GHG emission, it pays limited attention to other resource and environmental issues. Although low-carbon development can indirectly reduce the consumption of some resources and help to reduce the discharge of other pollutants, some more complicated resource and environmental issues may come up in the process of low-carbon development, such as the perpetual pollution caused by the use of large amounts of polysilicon and solar panels for the development of solar energy. From this perspective, the three growth models have to be more coordinated.

In sum, green development, circular development, and low-carbon development are completely consistent in essence and the guiding thought, only with different focuses and targeted fields. These three growth models are bound to cause all-round and profound changes in modern economic development. China should attach equal importance to them and not give preference to one while neglecting the others. It should take them all into account when building the resource-saving and environmentally friendly society, and scientifically handle the relation between economy, society, and environmental system, in order to a blaze a path of sustainable development with Chinese characteristics featuring low-carbon, circular, green, and ecological development. At present, China has come to the critical period of comprehensively building a moderately prosperous society. How to break the resource and environmental "bottleneck" on economic growth and maintain sustained, fast, and healthy economic development is a major issue faced by the nation. Against such a background, we should clearly see the problems and adopt targeted measures based on our national conditions, and energetically promoted green, circular, and low-carbon development. That is an effective way for China to speed up ecological development and realize the dream of building a Beautiful China.

# Chapter 4
# High-Carbon Development is not the Only Way of Modernization

The development practices of developed countries tell us that there are more than one ways of development, and high-carbon development with high energy consumption is not the only way of modernization. Data of developed countries show that when development reaches a certain level, per capita energy consumption and emission no longer increase along with socioeconomic development, but stand still or even decrease after reaching a stable level. However, this stable level varies from one country to another, which means developed countries adopt different development paths—the relatively high-carbon path and relatively low-carbon path. The success of the low-carbon path in some developed countries implies that developing countries, when pursuing modernization, can bypass the high-carbon growth model and achieve low-carbon development. Besides, the development and application of hydropower and pumped storage power generation, modern wind power and PV, conventional natural gas and nonconventional natural gas like shale gas, low-carbon energy technologies such as renewable energy and nuclear power, and high-efficiency energy-saving technologies also offer a new path for developing countries in the process of industrialization and modernization. It is a sustainable, low-carbon growth model characterized by low energy consumption, low pollution, and low emission.

## 4.1 Different Growth Models in Developed Countries and Regions

Developed countries and regions like the United States, Canada, EU, and Japan have made tremendous socioeconomic achievements after years of development. However, although they have reached a similar development level with equivalent per capita GDP, their energy consumption and carbon emission vary greatly. The per capita energy consumption and carbon emission in developed countries represented by the United States and Canada are twice as high as that in other developed countries and regions represented by the EU and Japan.



X. Du et al., *Overview of Low-Carbon Development*,
https://doi.org/10.1007/978-981-13-9250-4_4

### *4.1.1 Study of Energy Consumption and Carbon Emission Trend in Typical Developed Countries and Regions*

In the 100 years of the twentieth century, the global GDP increased 18 times, the wealth created by the mankind exceeded the total wealth created in the past, and energy consumption increased dramatically. The humankind consumed only 720 million tons of oil equivalent in 1900, which rose to 10.33 billion in 2000 and totaled 380 billion accumulatively in the twentieth century. Having developed for several decades or over 100 years, typical developed countries and regions, represented by the US, Canada, EU, and Japan, reached a high developed level and were continuing to develop, but their per capita energy consumption and carbon emission stopped rising and even dropped after the economy reached a certain level (Figs. 4.1, 4.2, 4.3, and 4.4).

After WWII, the world economy quickly entered a fast-growing period following a short period of recovery. The rapid economic development drove the fast growth of energy consumption and massive $CO_2$ emission. The early industrialized countries and regions, including the US, Canada, EU, and Japan, came to the late stage of industrialization one after another and basically completed this process in the mid-1970s. Materials show that the middle and late stages of industrialization marked the peak of rapid economic development, concentrated and immense energy consumption, and $CO_2$ emission. The economic structure in the post-industrial developed countries went through substantial changes. The tertiary industry was highly developed and replaced the secondary industry to be the main pillar of economic development, while energy consumption and $CO_2$ emission increased at a slower pace, with zero increase in per capita energy consumption and $CO_2$ emission. In 2008, affected by the international financial crisis and reduced economic growth, the US, Canada, EU, and Japan saw all-around decrease in their total and per capita energy consumption as well as total and per capita $CO_2$ emission.

In recent years, the American government led by several presidents has changed its stance several times on the issue of climate change, barely taken any action in that field, and failed to reach a political consensus on it. But it issued *The President's Climate Action Plan* in June 2013, which proposed to make efficient use of clean energies, create the twenty-first-century transport industry, reduce energy waste in families, business, and industry, and cut the emission of GHGs like hydrofluorocarbon and methane. It asked the federal government to set an example and take the lead in using clean energies and raising energy efficiency, so as to reduce $CO_2$ emission in the country, get ready for the consequences brought by climate change, and set the direction for the international community to tackle climate change. Moreover, the US also strongly developed nonconventional natural gas, kept improving the energy structure, and rolled out a series of initiatives, including the *American Competitiveness Initiative*, *Advanced Energy Initiative* and *Climate Change Technology Program Strategic Plan*. It kept increasing the input in clean energy R&D, pushed for reforms in energy technology, and controlled and lowered energy consumption and $CO_2$ emission.

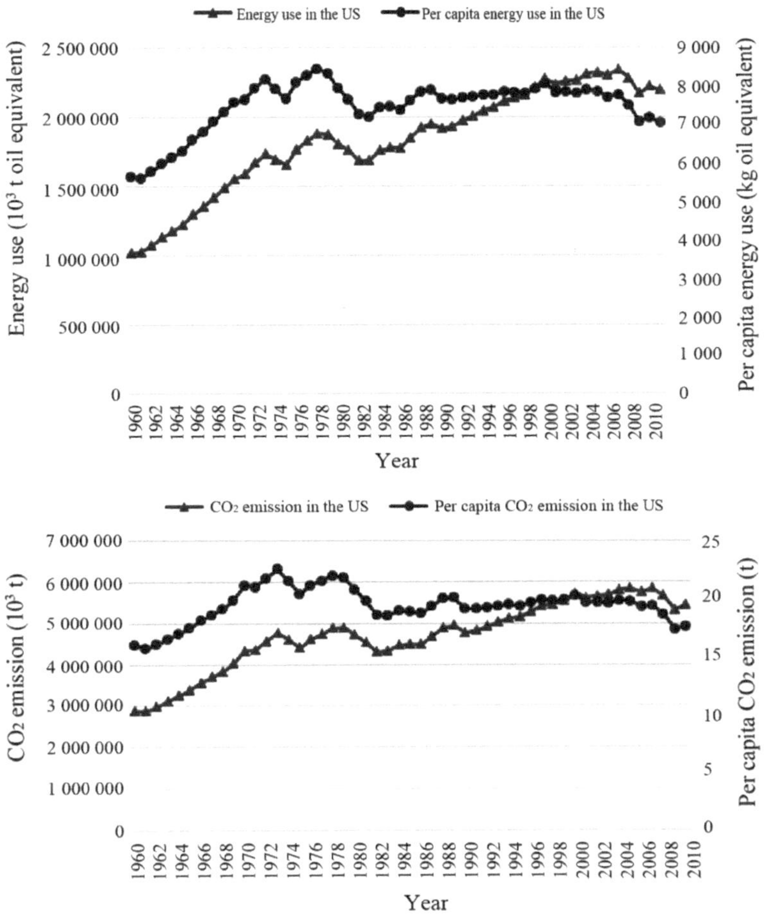


**Fig. 4.1** Energy use (upper) and $CO_2$ emission (lower) in the US

Canada had a high level of per capita energy consumption and $CO_2$ emission, but it rolled out a series of policy and measures in recent years to tackle climate change, including *Canada's Plan for Tackling Climate Change*. By making innovations and technical investments in alternative energies and biotechnology, formulating incentives, rules, standards and tax policies, and attaching importance to waste recycling, those measures aimed to improve the energy efficiency, reduce GHG emission, and promote economic development.

The EU always attached great importance to energy development strategy, actively dealt with climate change, and made green and low carbon an important strategic goal of socioeconomic development. In recent years, it has released a series of related plans and policies, including the *EU Energy Green Paper* and *A Roadmap for Moving to a Competitive Low Carbon Economy in 2050*, which established a definite and

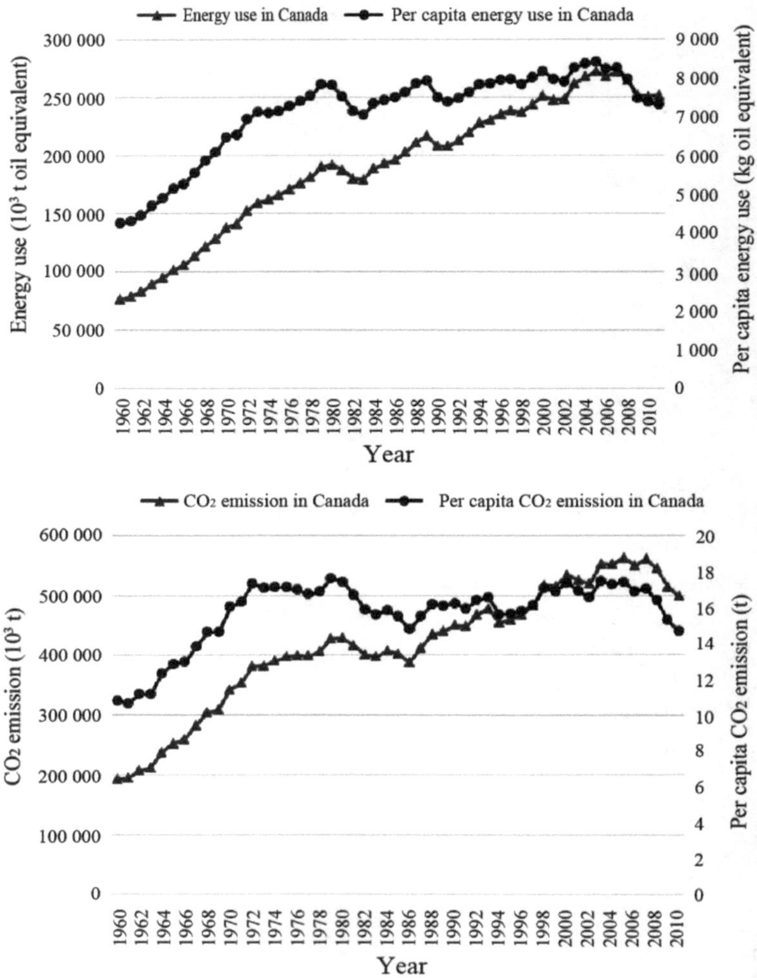


**Fig. 4.2** Energy use (upper) and $CO_2$ emission (lower) in Canada

quantitative system for achieving the goal of green and low-carbon development. On that basis, the EU created the market-based emission quota and trade mechanism, bettered the economic and energy structure, set up a long-term economic incentive and restrictive system, improved the technological competitiveness in the green and low-carbon field, improved the auxiliary system for legal standards, and created the supervisory, verifying, and reporting system. The EU tried to delink the economy and politics from energy consumption and $CO_2$ emission through policy efforts.

Japan also issued a number of strategic plans regarding energy and climate change these years, such as the "New Industrial Creation Strategy", "New National Energy Strategy", "Cool Earth 50 Initiative", "21st-century Environmental Strategy", "Cool

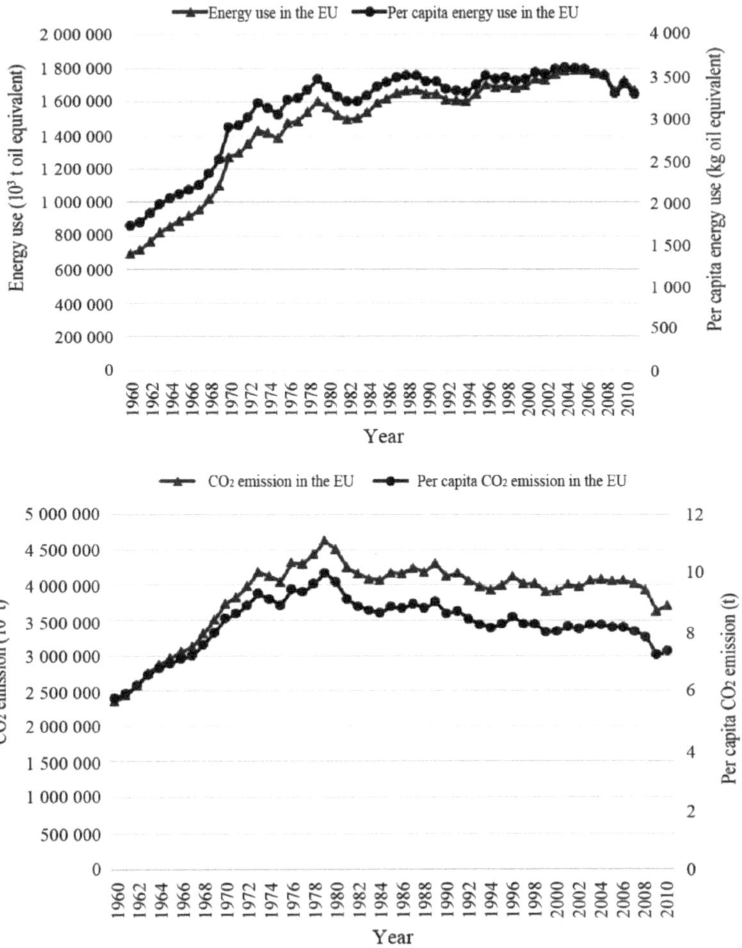


**Fig. 4.3** Energy use (upper) and $CO_2$ emission (lower) in the EU

Earth—Innovative Energy Technology Program", "Environment, Energy, Technology Innovation Strategy", "Action Plan for Building a Low-carbon Society", "Low-carbon Society Research and Development Strategy", and "Basic Energy Program". By implementing the energy conservation plan, advocating new energy technologies and independent emission reduction mechanism, mobilizing all people to join the emission reduction movement, and promoting a low-carbon life in the whole society, those initiatives aimed to effectively control energy consumption and $CO_2$ emission.

In sum, developed countries and regions including the US, Canada, EU, and Japan have issued multiple energy strategies and climate change plans in recent years. Thanks to those measures and actions, they, while maintaining economic development, achieved increasingly obvious zero growth or even reduction in total and per capita energy consumption as well as total and per capita $CO_2$ emission.

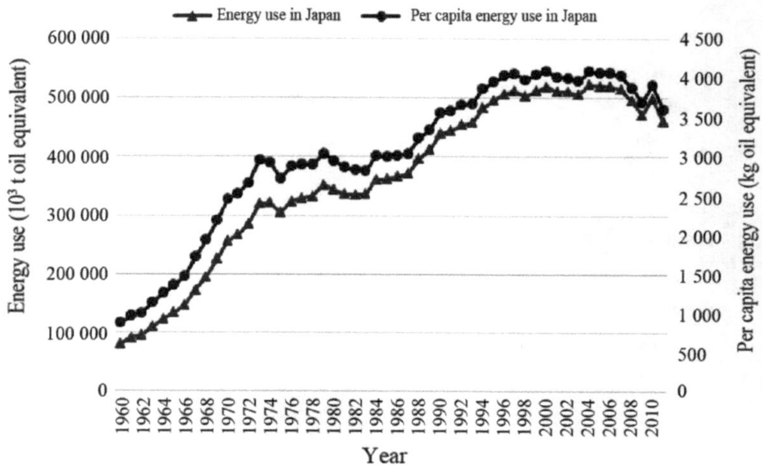


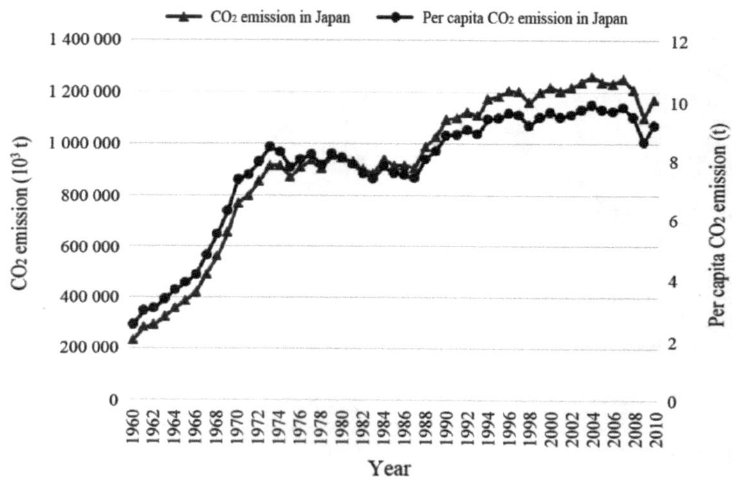


**Fig. 4.4** Energy use (upper) and $CO_2$ emission (lower) in Japan

### *4.1.2 Comparison of Typical Developed Countries and Regions in Energy Consumption and Carbon Emission*

Above is a chronological analysis of the energy use and $CO_2$ emission in developed countries, but different developed countries had different development levels in the same year. To eliminate the difference, this part will compare the energy consumption and carbon emission in those countries and regions on the basis of the same development level (per capita GDP at constant 2005 prices in US Dollars).

Figure 4.5 (upper chart) shows that with the economic growth, or the growth of per capita GDP, the per capita energy consumption in those developed countries and regions increased rapidly. But after it reached a certain point (that varied in each country and region), the per capita energy consumption no longer increased and even began to decrease while the per capita GDP continued to grow. In developed countries represented by the United States and Canada, the per capita energy consumption increased along with per capita GDP before the latter reached USD23,000, at which point the per capita energy consumption was 7–8t oil equivalent/a. After that, the

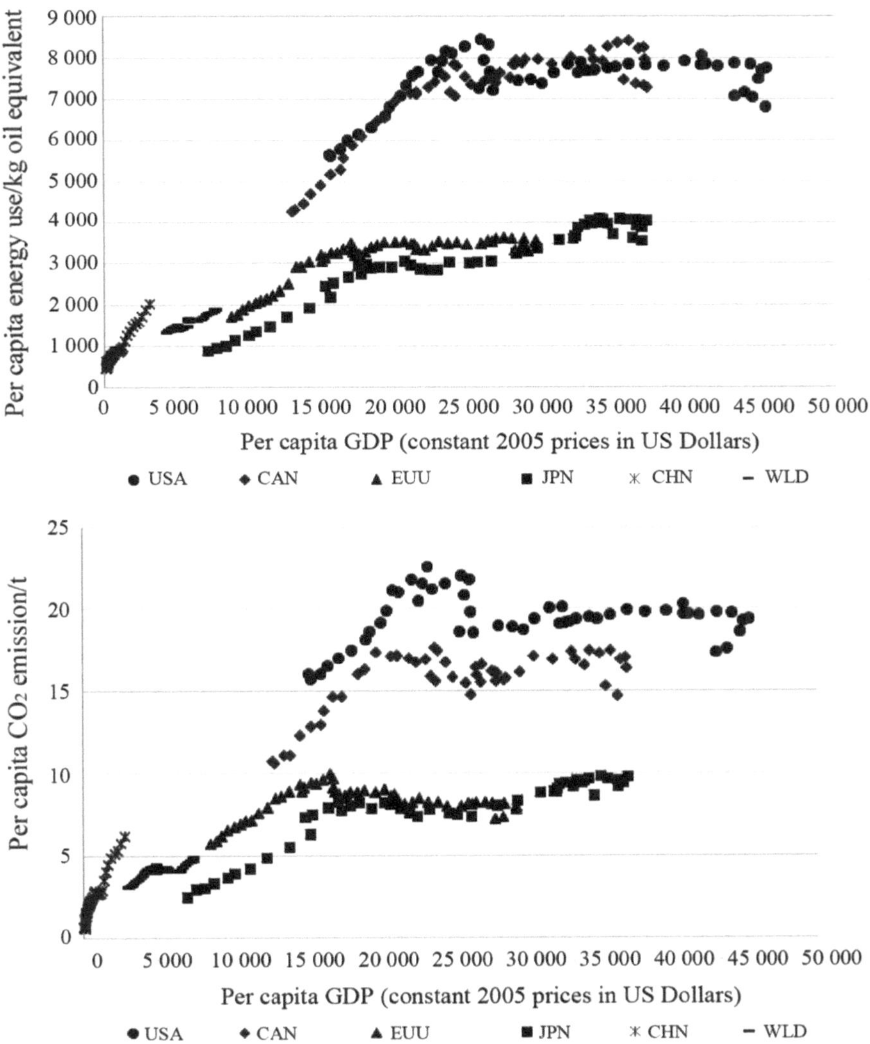


**Fig. 4.5** Comparison of energy consumption and $CO_2$ emission in typical countries and regions given the same development level

per capita energy consumption stayed at that level even though the per capita GDP continued to increase. In the EU, the per capita energy consumption increased along with per capita GDP before the latter reached USD15,000, at which point the per capita energy consumption was 3–4t oil equivalent/a. After that, the per capita energy consumption stayed at that level (about half of that in the US and Canada) even though the per capita GDP continued to increase. Japan had a similar situation as EU but with higher energy efficiency and lower energy consumption given the same economic level.

The relation between per capita $CO_2$ emission and per capita GDP follows a similar pattern, as shown in Fig. 4.5 (lower chart). With the economic growth or the growth of per capita GDP, the per capita carbon emission in those developed countries and regions increased rapidly. But after it reached a certain point (that varied in each country and region), the per capita carbon emission no longer increased and even began to decrease while the per capita GDP continued to grow. In developed countries represented by the United States and Canada, the per capita carbon emission increased along with per capita GDP before the latter reached USD23,000, at which point the per capita carbon emission was 15–20t/a. After that, the per capita carbon emission stayed at that level even though the per capita GDP continued to increase. In the EU, the per capita carbon emission increased along with per capita GDP before the latter reached USD15,000, at which point the per capita carbon emission was 7–10t/a. After that, the per capita carbon emission stayed at that level (about half of that in the US and Canada) even though the per capita GDP continued to increase. Japan had a similar situation as EU, but with lower emission given the same economic level.

Figure 4.5 also shows that compared with developed countries and regions, when per capita GDP increased by one US dollar in China, the additional energy consumption and $CO_2$ emission far exceeded that in developed countries in the same period (see the curve slope in Fig. 4.5). This means China's development in recent years has obvious high-carbon characteristics and the country faces a tremendous challenge in changing the curve slope.

Figure 4.6 compares the $CO_2$ emission intensity in the US, Canada, EU, and Japan. $CO_2$ intensity means the $CO_2$ emission per kg oil equivalent of energy consumption. It's clear that these developed countries and regions have basically the same $CO_2$ emission intensity, namely, 2–2.5 kg/kg oil equivalent of energy, because of their similar, oil-dominated fossil energy structure. According to the figure, Canada has a slightly lower $CO_2$ intensity than other developed countries because coal takes up a smaller proportion (more than ten percent) in its fossil energies than in other countries. Therefore, although $CO_2$ emission is related to other factors too, it's almost in positive correlation to energy consumption. A clear analysis of energy consumption in those countries can explain their carbon emission.

This analysis above tells us that even at the same development level (regarding per capita GDP), different developed countries and regions vary in energy consumption and carbon emission. The per capita energy consumption and carbon emission in developed countries represented by the US and Canada is twice that in the EU; Japan has a higher energy consumption and lower carbon emission than the EU given the

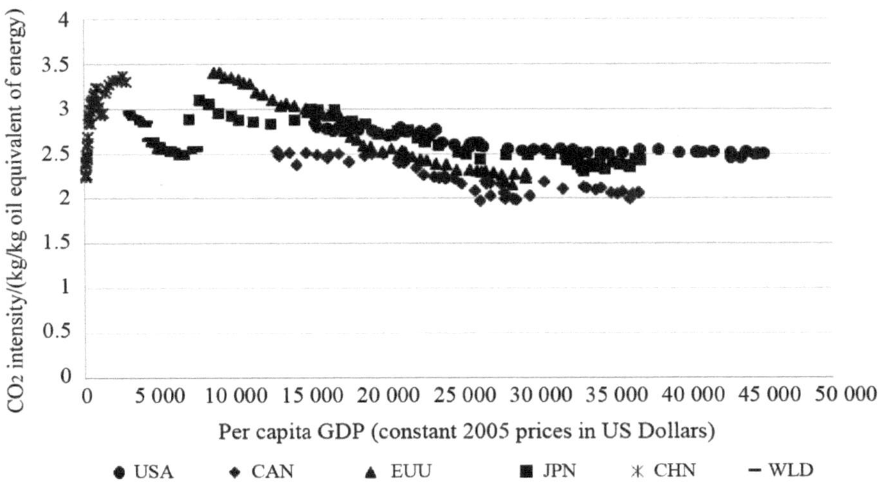


**Fig. 4.6** Comparison of $CO_2$ emission intensity in typical countries and regions

same economic level, but its per capita energy consumption and carbon emission is on a par with the EU after the stable point, about half of that in the US and Canada. We may call the energy development and carbon emission approach in developed countries represented by the US and Canada the "US, Canadian model," and that in developed countries and regions represented by the EU and Japan the "EU, Japanese model" (or subdivided into the "EU model" and "Japanese model"). We can see that the "US, Canadian model" is a path featuring high energy consumption and high emission, whereas the "EU, Japanese model" (especially the Japanese model) is a path featuring fairly low energy consumption and emission.

In sum, data of energy economics indicate that developed countries and regions do not have one common growth model. They can be divided into "two kinds of developed countries and regions" with obvious differences.

## 4.2 Analysis of the Reasons for Different Growth Models in Developed Countries and Regions

### *4.2.1 Comparison of Industrial Structure and Energy Efficiency*

Industrial structure refers to the composition of various industries and their relation and proportion. Table 4.1 shows that developed countries including the US, Canada, Britain, France, Germany, and Japan constantly adjusted their industrial structures over the years. As a result, the primary and secondary industries took an ever declining

**Table 4.1** Industrial structure in developed countries

| Nations | Type of industry | 1990 | 1995 | 2000 | 2005 | 2008 |
|---|---|---|---|---|---|---|
| US | Primary industry/% | 2.06 | 1.61 | 1.61 | 1 | 1.2 |
| | Secondary industry/% | 27.88 | 26.31 | 24.45 | 22 | 21.4 |
| | Tertiary industry/% | 70.05 | 72.08 | 73.94 | 77 | 77.4 |
| Canada | Primary industry/% | 2.91 | 2.95 | 2.3 | – | – |
| | Secondary industry/% | 31.81 | 31.23 | 33.8 | – | – |
| | Tertiary industry/% | 65.28 | 65.81 | 63.8 | – | – |
| Britain | Primary industry/% | 1.9 | 1.84 | 1.05 | 1 | 1 |
| | Secondary industry/% | 35.17 | 32 | 28.49 | 26 | 24 |
| | Tertiary industry/% | 62.92 | 66.16 | 70.46 | 73 | 76 |
| France | Primary industry/% | 3.83 | 3.36 | 2.8 | 2 | 2 |
| | Secondary industry/% | 29.7 | 27.28 | 25.48 | 21 | 20 |
| | Tertiary industry/% | 66.47 | 69.36 | 71.72 | 77 | 78 |
| Germany | Primary industry/% | 1.73 | 1.33 | 1.21 | 1 | 1 |
| | Secondary industry/% | 38.82 | 33.39 | 30.41 | 30 | 30 |
| | Tertiary industry/% | 59.45 | 65.28 | 63.38 | 69 | 69 |
| Japan | Primary industry/% | 2.74 | 1.88 | 1.38 | 2 | 1.5 |
| | Secondary industry/% | 39.19 | 34.08 | 32.1 | 30 | 28 |
| | Tertiary industry/% | 58.34 | 64.04 | 66.51 | 68 | 70.5 |

proportion (20–30% combined), while the tertiary industry was absolutely dominant (70–80%), whose proportion was close in different countries. In this sense, those developed countries did not differ much in industrial structure, which was therefore irrelevant to their difference in energy consumption and carbon emission.

Energy consumption per unit of GDP means the kg of oil equivalent energy used to produce unit of GDP in a country or region. It is the main indicator that reflects the level of energy consumption and situation of energy conservation and consumption reduction and an indicator of energy efficiency. This indicator reflects the degree of energy use in the economic activities of a country or region and the changes in economic structure and energy efficiency. $CO_2$ emission per unit of GDP refers to the $CO_2$ emission rate corresponding to GDP, or the $CO_2$ emission per unit of GDP in a certain period of time in a country or region. It measures the relation between economy and carbon emission of a country or region.

Figures 4.7 and 4.8 compare the energy consumption and $CO_2$ emission per unit of GDP in the US, Canada, the EU, Japan, and the world on average. We can see that given the same development level, the energy consumption and $CO_2$ emission per unit of GDP in the US and Canada are twice that in the EU and Japan.

This means that the industrial structure in the US, Canada, the EU, and Japan does not differ much. The disparity in energy consumption and $CO_2$ emission per unit of GDP is mainly attributed to the internal composition of their industrial structure and the distribution of added value in the international division of work.

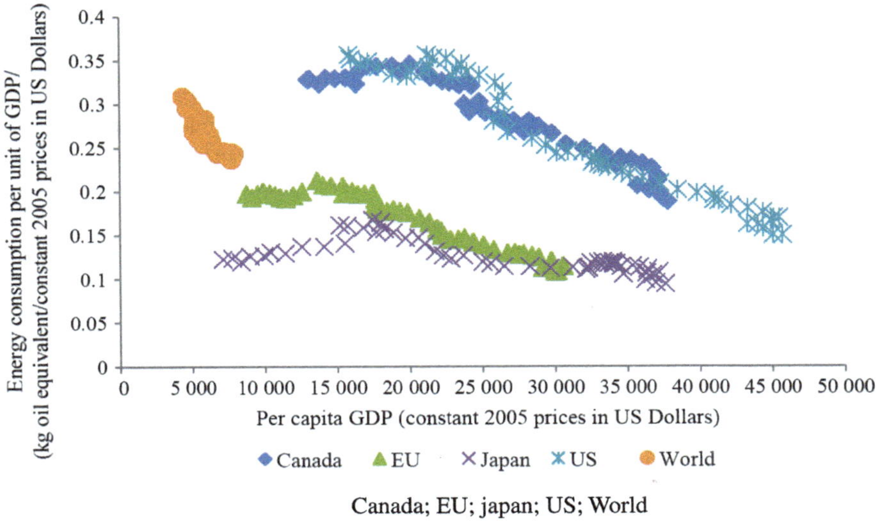


**Fig. 4.7** Comparison of energy consumption per unit of GDP

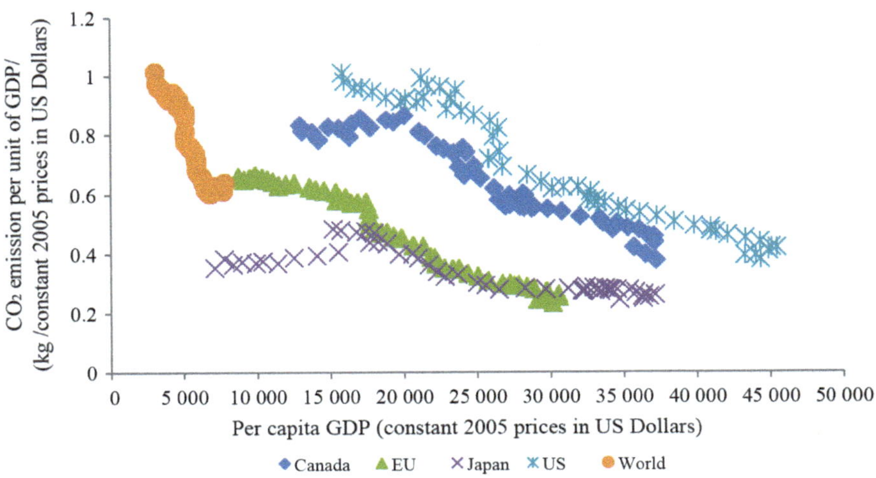


**Fig. 4.8** Comparison of $CO_2$ emission per unit of GDP

### 4.2.2 Comparison of Energy End Users

Energy end use includes energy products provided to the consumers that are not used for fuel conversion or processing conversion activities. These energy products are considered fully used and not converted to other forms. Generally speaking, there are five sectors of energy end use—industry, transport, people's living, business, and service industry. After WWII, the sector-based energy consumption structure in developed countries changed steadily. Energy consumption by the industrial sector took an ever smaller proportion, while that in the transport, living, business, and service sectors increased in both quantity and proportion.

Based on research, the reasons for the different per capita energy consumptions and carbon emissions in developed countries under their different energy growth models and the energy consumptions of each end-user sector in those countries are generalized as follows.

(1) In the industrial sector, different industries have vastly different energy consumption intensities, the main reason for which is the industrial structure in different countries, especially the structure and proportion of high-energy-consuming industries. Japan energetically implemented energy-saving measures and improved its energy efficiency, and therefore had lower energy consumption per unit product than other countries.
(2) In the transport sector, over 90% of energy consumption goes to road traffic, and the main reasons for energy consumption difference include road transport intensity, proportion of vehicle types, and vehicle population. As far as road passenger transport is concerned, the difference in consumption concept and lifestyle is most closely related to energy consumption intensity. In terms of road cargo transport, factors such as economic structure, resource endowment, and industrial layout are most closely related to energy consumption intensity.
(3) In the people's living sector, the living area per household is the main reason, so is the climatic environment in different countries.
(4) In the business and service sectors, per capita construction area is the main reason, so is the climatic environment in different countries, and the proportion of construction area and energy consumption intensity inside the sector.

### 4.2.3 Lessons Drawn from Comparative International Studies for China's Energy Development

Since we adopted the policy on reform and opening-up more than 30 years ago, we have made remarkable achievements and the Chinese economy has maintained high-speed growth. But energy consumption has also increased rapidly in this process. In 2010, China's per capita GDP reached USD2,870 (constant 2005 prices in US dollar) and energy consumption reached 2516.7 million ton oil equivalent, averaging 1.88 tons per person; its total and per capita $CO_2$ emission reached 8287 million tons and 6.2 tons, respectively. In the twenty-first century, energy supply has increased at an extremely high speed because of the huge demand, and China has the largest total energy consumption and total $CO_2$ emission in the world. To achieve the strategic goal for national economy and social development, we have to correctly understand our energy development and the reasonable room for $CO_2$ emission. We may gain enlightenments below from the comparison of energy consumption and carbon emission in developed countries.

1. The "US, Canadian model" of high energy consumption and high emission does not work. The analysis shows that under that model, the annual per capita energy consumption stays at 7–8t oil equivalent and annual per capita $CO_2$ emission at 15–20t. With this standard, even if we do not consider population increase, the energy consumption in China can increase about three times from 2010, and the average annual energy consumption will approach 10 billion ton oil equivalent, while $CO_2$ emission can increase about twice from 2010 and the average annual $CO_2$ emission will approach 25 billion tons. At that time, China's annual energy consumption and $CO_2$ emission will account for 79% of the global energy consumption and 76% of global $CO_2$ emission in 2010. If all countries in the world reach the same annual per capita energy consumption as the US, we will need four more earths and face more serious global warming. This $CO_2$ emission approach is impossible, so the "US, Canadian model" should not be imitated or promoted. But the curve in Fig. 4.5 shows that if China's energy consumption continues the current growing trend, it will lead to the high-energy-consuming "US, Canadian model," which is a dead-end.
2. Let's analyze our energy development and $CO_2$ emission limits in reference to the "EU, Japanese model," under which the annual per capita energy consumption and $CO_2$ emission stay at 3–4t oil equivalent and 7–10 tons, respectively. This takes us to an important conclusion: compared with the "US, Canadian model," the "EU, Japanese model" realized the same development level with lower energy consumption and $CO_2$ emission. Therefore, the high-carbon path is not the only way to modernization, and the "EU, Japanese model" is the best example. In reference to this model, even if we do not consider population increase, the energy consumption in China can double from 2010 and the annual energy consumption is about 5 billion ton oil equivalent, while the $CO_2$ emission can increase by about 50% from 2010 and the average annual $CO_2$ emission will reach 12.5 billion tons. The current energy efficiency in China is much lower than in the EU given the same per capita GDP. If we don't take effective measures in a timely manner to raise energy efficiency, control energy consumption, and $CO_2$ emission, it would be hard for us to achieve the same energy consumption and $CO_2$ emission level as in the "EU, Japanese model." Moreover, the "Japanese model" requires much greater efforts than the "EU model," and it is only possible if we substantially raise the energy efficiency and earnestly control energy consumption and $CO_2$ emission.

3. If we take the "average level of developed countries and regions" as the benchmark, we will go to a "quasi-US model" that's more energy-consuming than the "EU, Japanese model." If we take the average annual per capita energy consumption and $CO_2$ emission between the "US, Canadian model" and the "EU, Japanese model" as the "average level of developed countries and regions," the average annual per capita energy consumption is about 6t oil equivalent and $CO_2$ emission 15t. With this standard, even if we do not consider population increase, China's energy consumption can increase by about twice from 2010 and the annual energy consumption will be more than 7 billion ton oil equivalent, while the $CO_2$ emission can increase by less than 1.5 times from 2010, and the annual $CO_2$ emission will reach 20 billion tons. This is a "quasi-US model" that's more high carbon than the "EU, Japanese model" and also a dead end. China should be highly alert against this realistic danger.
4. Analysis of China's energy development and $CO_2$ emission limits under the model of "sustainable development." China has achieved tremendous economic achievements since the reform and opening-up was launched, but the extensive development approach caused serious resource consumption and environmental pollution, among other problems. If this situation continues, our development will not sustain. As a developing country with a large population, little per capita resources, and a heavy pressure of $CO_2$ emission, China should fully exert its latecomer's advantages, adopt advanced technologies, and choose a "new path of industrialization" that is more resource-saving and environmentally friendly than the path adopted by developed countries. Therefore, China should not take the "US, Canadian model" or follow the "average level of developed countries" in its energy development and $CO_2$ emission but should take a path with lower energy consumption, lower emission, and higher energy efficiency than the "EU, Japanese model." This means that the annual per capita energy consumption in China can increase by less than 100% from 2010 and annual $CO_2$ emission less than 50% from 2010, and we can do better than that if we make good use of our latecomer's advantages.

## 4.3 Delinking Analysis of Economic Development and Energy Consumption, Energy Consumption and $CO_2$ Emission in China

During the 30-plus years of reform and opening-up, China has maintained high-speed economic development, with the GDP rising from USD217.5 billion in 1980 to USD4.86 trillion in 2013 and per capita GDP from USD222 to USD3,583 (constant 2005 prices in US dollars)[1] in the same period. However, some problems have

[1] All data about GDP, energy consumption, and $CO_2$ emission in this paper are from the World Bank unless otherwise specified, and GDP is in constant 2005 prices in US dollars.

emerged along with the economic development and the improvement of people's living standards, such as the fast-growing energy consumption, continued increase of fossil energy consumption, and the subsequent massive emission of GHGs and polluting gases. As a result, China is under the dual pressure of energy supply and environmental protection. The 17th and the 18th National Congress of the CPC pointed out that economic development at the price of excessive resource consumption and the environment would not sustain, and China must change the growth model. To truly transform the growth model, realize win-win results between economy and environment and delink economic growth from energy consumption as soon as possible, especially from the consumption of high-carbon fossil energies like coal and oil, and delink energy consumption from $CO_2$ emission, the only choice for China is to blaze a green, low-carbon, and sustainable development path. On the basis of analyzing the delinking pattern in developed counties, exploring the delinking relation between economic growth and energy consumption, energy consumption, and $CO_2$ emission in China are of great realistic importance.

### 4.3.1 *Meaning of Delinking and Its Significance in China*

Both the World Bank and OECD defined the delinking concept. The World Bank holds that delinking is a process in which the environmental impact on economic activities is decreasing gradually, and the concept implies both dematerialization and depollution. OECD believes that delinking is breaking the connection between environmental bads and economic goods. The two definitions are essentially the same despite different expressions. Both the consumption of fossil energies and ecological degradation cause environmental damages and impacts, and they are both consequences brought by the increase of economic goods in the age of industrial civilization. Breaking the link between environmental bads (including energy consumption and eco-pollution) and economic growth and reducing the environmental impacts is called delinking.

Delinking is a process. On the premise of economic growth, when energy consumption shifts from increase to stability, we can take economic development as relatively delinked from it; when energy consumption peaks and falls continuously, we can take economic development as categorically delinked from it. On the premise of the continued increase of energy consumption, when $CO_2$ emission shifts from increase to stability, we can take energy consumption as relatively delinked from it; when $CO_2$ emission peaks and falls continuously, we can take energy consumption as categorically delinked from it.

The eco-environment has its capacity, within which the ecosystem is capable of self-recovery and environmental pollution and will not cause serious consequences. But once the environmental capacity is exceeded and eco-balance broke, environmental pressure will increase rapidly in a nonlinear fashion, and the process of reversing environmental aggravation is neither easy nor automatic. Therefore, delinking economic development from energy consumption and the emission of GHGs (and polluting gases) is the first step of improving the eco-environment. To achieve sustainable development, the mankind has to bring the atmospheric environmental pressure back to be within the environmental capacity as soon as possible.

The speed of economic growth returning to the "potential economic growth rate" is an objective requirement of scientific development. In recent years, the Chinese economy has entered a "new normal," where it shifts from high-speed growth to medium-and-high-speed growth, economic structure is improved and upgraded continuously, factor-driven development is replaced by innovation-driven development, and policies are adjusted to pay more attention to the quality and benefits of development rather than focus on GDP growth alone. This is the only way for China to achieve sustainable development that also conforms to the requirements of scientific development. From the perspective of energy economics, energy is the foundation of economic development, the same economic growth rate may be bolstered by different energy growth rates, and energy elasticity coefficient is closely related to such factors as industrial structure and energy efficiency. Developed countries' experiences tell us that when they finished industrialization and stepped into the post-industrial stage dominated by the tertiary industry, they could support economic development with a relatively low-energy elasticity coefficient, thanks to the transformation of industrial structure, technological upgrade, and improved energy efficiency. Drawing on their experiences, we should reduce the energy elasticity coefficient in China and support economic development with appropriate energy consumption, so as to build a "resource-saving" society. Regarding the science of environmental energy, the emission of polluting gases is closely related to energy consumption. Mitigated increase of energy consumption and a clean energy structure will reduce the growth rate of GHG and polluting gas emission and even decrease the total emission. Delinking energy consumption growth from polluting gas emission as soon as possible is an inevitable step on our way to building an environmentally friendly society. GHG is one of the important factors of global climate change, and it has basically the same source of emission as polluting gases. Delinking energy consumption from GHG emission is not only imperative for dealing with climate change but also for China's low-carbon transformation.

When a nation is moving toward modernization, gradually delinking economic growth from energy consumption, energy consumption from $CO_2$ emission is a general practice, but it varies from one country to another. China is still in the stage of industrialization, which coincides with IT applications, urbanization, and agricultural modernization. This is different from the situation in developed countries as their process of industrialization was independent. If China can make full use of its latecomer's advantages, speed up the transformation of economic structure, raise energy efficiency, and improve energy structure, we may be able to delink economic growth from energy consumption (especially high-carbon energy consumption), energy consumption from $CO_2$ emission as soon as possible. That will mark the transformation of China's economic growth model and push the development of its green and low-carbon "new normal." Finding out the factors that affect the delinking and exploring the ways of delinking is of great significance for developing the "ecological civilization" and building a Beautiful China, and for achieving win-win results for economy and environment.

### 4.3.2 *Analysis of Delinking Economic Development from Energy Consumption in Developed Countries and Regions*

This book takes six developed countries as examples—the US, Canada, Japan, Germany, Britain, and France. By analyzing the trend of their per capita GDP and per capita energy consumption since 1960, the book determines the delinking trend between their economic development and energy consumption and discusses the underlying reasons.

(I) Trend of economic development and energy consumption in developed countries and comparison

Per capita GDP is an indicator that gauges people's living standard in a nation. For more than half a century since 1960, developed countries represented by the US, Canada, Japan, Germany, Britain, and France saw their per capita GDP rising all the time (Fig. 4.9) with an average annual growth rate of 2–3%. By 2013, the annual per capita GDP in the US approached USD46,000, that in France USD34,000, and that in the other four nations USD37,000–38,000.

Per capita energy consumption can objectively reflect the energy consumption level in a nation. In developed countries, the per capita energy consumption did not always increase along with the per capita GDP (Fig. 4.10). It kept increasing in the late 1960s and 1970s, but came to a standstill after the 1980s and began to fall after 2005. This means that economic development and energy consumption were

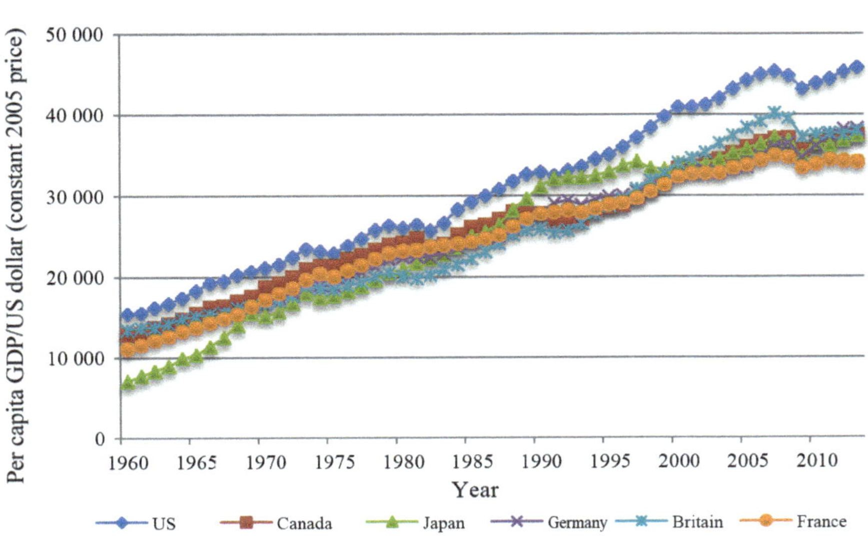


**Fig. 4.9** Comparison of annual per capita GDP in developed countries since 1960

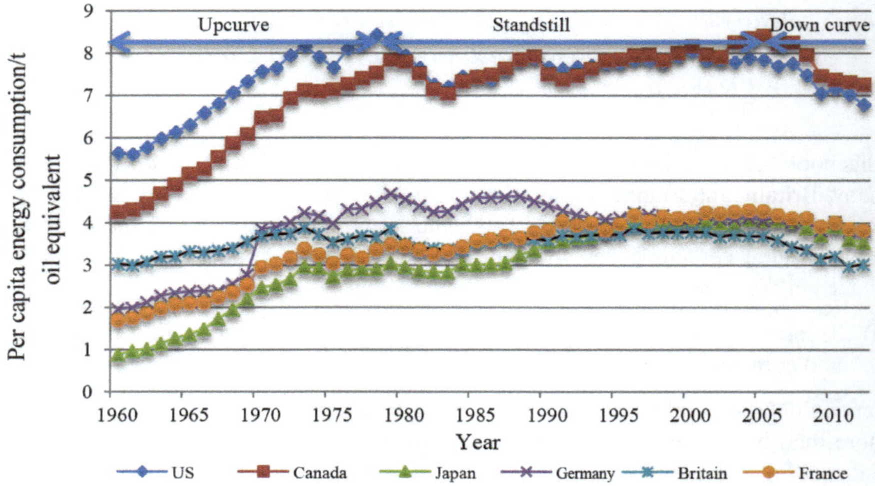


**Fig. 4.10** Comparison of annual per capita energy consumption in developed countries since 1960

relatively delinked in developed countries from the 1980s to the early twenty-first century and categorically delinked after 2005. Whether this state will continue is up to further observations.

Although economic development and energy consumption began to be delinked in developed countries after the 1980s, they did not have the same per capita energy consumption and energy intensity, but displayed the "EU, Japanese model" (represented by Europe and Japan) and the "US, Canadian model" (represented by the US and Canada). For instance, under those two models, the per capita energy consumption was about 3.5t and 7t oil equivalent/a, while the energy intensity was about 1t oil equivalent/USD10,000, 1.5t oil equivalent/USD10,000 (US) and 2t oil equivalent/USD10,000 (Canada). Such a difference in growth model is enlightening for us.

(II) Analysis of reasons for economic development–energy consumption delinking in developed countries

This paper tries to analyze the delinking trend between economic development and energy consumption in developed countries in the following two aspects.

1. Industrialization and industrial structural adjustment promote economic development–energy consumption delinking (with industrial structure in the US and Germany as an example)

In terms of the structure of the three industries: From the 1950s to the 1970s, almost all developed countries, including the US and Germany, were in the post-industrial stage, when the secondary industry took up a large proportion. In the meantime, tertiary industry began to develop quickly and eventually surpassed secondary industry, and the process of industrialization was completed. In the 1970s, tertiary industry took up 65.5% in America's output value, and the ratio in Germany was 49.1% in 1975, more than the ratio of secondary industry (48.1%) (Figs. 4.11, 4.12).

In terms of the internal structure of the secondary industry: After WWII, industrial development in the US was bolstered by productive sectors such as the capital-intensive iron and steel, automobile, construction, and mechanical and electrical products. From the late 1970s, knowledge- and technology-intensive sectors developed fast, including the aviation and space industry, equipment of automatic production of computers and software, microprocessor, robot, laser technology, optical fiber, new materials, communications technology, and bioengineering. From 1950 to 1986, automobile industry stood atop the secondary industry in Germany, while electronics and chemical industry that featured high technical content and a high level of IT application climbed up from the 9th and 5th positions to the 3rd and 4th positions based on their contribution to GDP. In contrast, the extensive iron and steel and coal mining industries dropped from the 2nd and 3rd position to the 7th and 9th.

This shows that in the period of industrialization, including the late period, the delinking between economic development and energy consumption was not obvious in developed countries. An obvious delinking trend only began to emerge after the industrialization was completed and those countries entered the post-industrial period, when the tertiary industry dominated the economic development. While the tertiary industry became the dominant industry in those countries, the secondary industry experienced structural changes internally, namely, high-energy-consuming and high-polluting industries that used to be dominant were gradually replaced by low-energy-consuming, green, and highly informatized ones. Therefore, the process

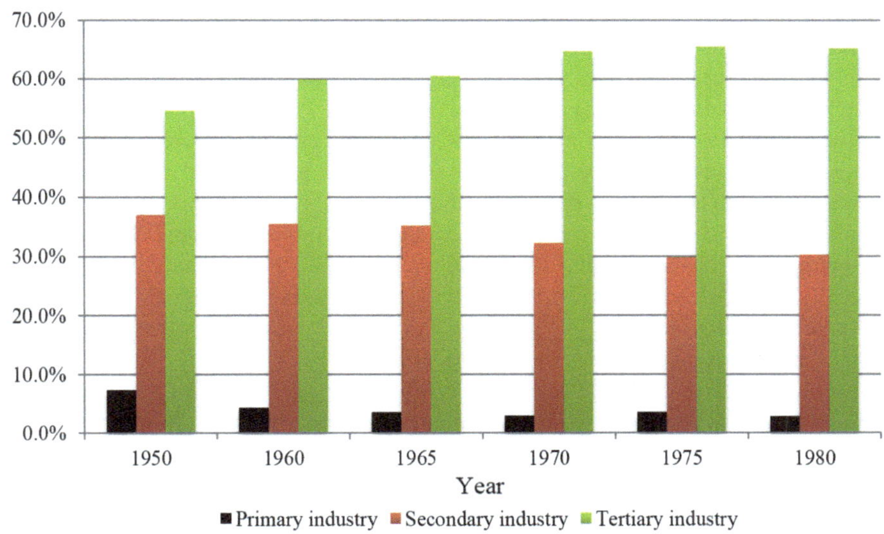


**Fig. 4.11** Structural changes of the three industries in the US

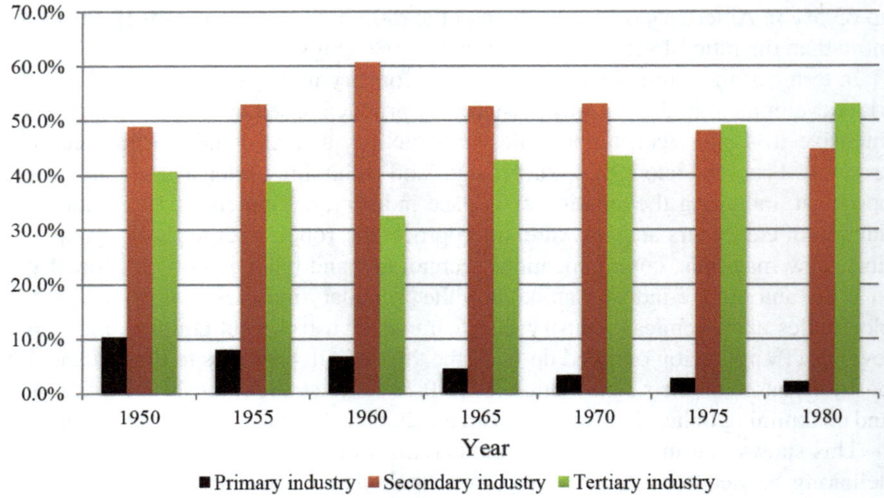


**Fig. 4.12** Structural changes of the three industries in Germany

of industrialization and industrial structural adjustment is one of the important factors for delinking economic development from energy consumption.

2. The first oil crisis helped developed countries solve the "energy security" issue and prompted them to promote and implement energy-saving and efficiency-improving measures more energetically

The first oil crisis between 1973 and 1974 sounded the alarm of energy supply in developed countries, and their per capita energy consumption began to increase at a reduced speed from then on (Fig. 4.10).

The outbreak of the oil crisis made the developed countries review their energy strategies and make adjustments. For instance, Nixon proposed in 1974 to make a new act to ensure energy self-sufficiency during energy crisis, marking its pursuit for "energy independence." Britain stepped up the exploitation and control of domestic energies (it fortunately discovered the resource-rich North Sea Oil Field) and basically realized energy self-supply in the 1980s; Germany and Japan, with a high oil import dependence of 80–90%, adopted measures and policies to strengthen oil reserve and encourage energy diversification.

In the meantime, all countries actively promoted energy-saving measures and raised energy efficiency. America implemented temporary energy-saving measures on transport, business, households, and the federal government in 1973 to effectively cope with the oil crisis. Germany made energy conservation a basic state policy of economic development from the 1970s, and adopted in the 1990s the following standard for gauging the quality of economic growth: the GDP should not grow faster than the improvement of energy efficiency. Japan formulated the *Law on Rationalization of Energy Use* (the *Energy Conservation Law*) in 1979, which supported energy conservation through tax, fiscal, and financial measures and promoted the development of energy-saving technologies and products. In early 1975, the French government made energy conservation the most pressing task of new energy policies and the "most reliable energy," and its comprehensive energy-saving actions and measures took obvious effects. It's clear that while passively coping with the energy crisis, developed countries began to proactively adopt energy-saving methods, and energy conservation and efficiency improvement was one of the important measures to delink economic development from energy consumption.

### *4.3.3 Analysis of Delinking Energy Consumption from $CO_2$ Emission in Developed Countries and Regions*

(I) Trend of delinking energy consumption from $CO_2$ emission in developed countries and regions

The per capita $CO_2$ emission in developed countries roughly passed three stages since 1960 (Fig. 4.13): increase in the 1960s–1970s, standstill in the late 1970s and early 1980s, and decrease to varying degrees after 2005. We can see that the per capita $CO_2$ emission under the "US, Canadian model" and "EU, Japanese model" is about 16t and 8t, respectively, the difference being similar to that in their per capita energy consumption.

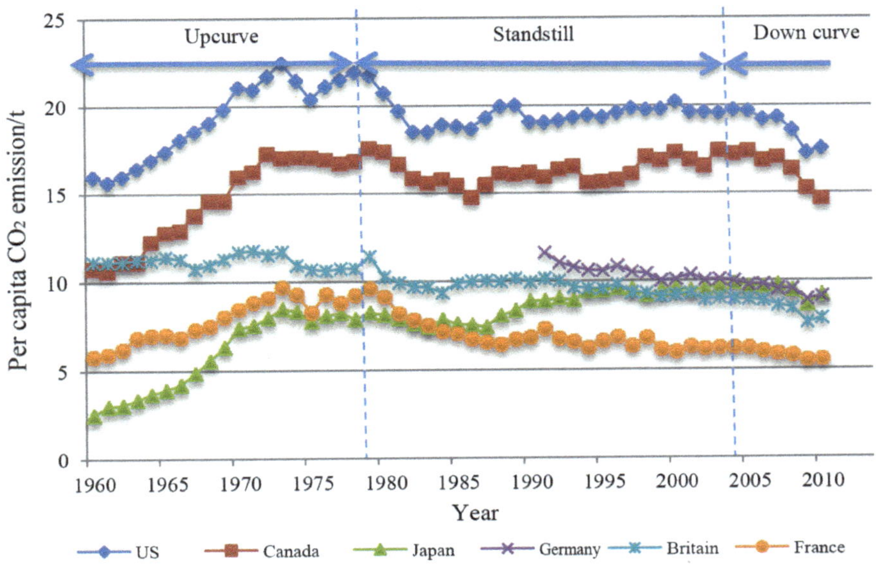


**Fig. 4.13** Per capita $CO_2$ emission in developed countries since 1960

Earlier analysis in this book shows that per capita energy consumption increased rapidly before the 1980s, came to a standstill from the 1980s to around 2005, and began to decrease afterward. The trend of energy consumption and $CO_2$ emission varied from one country to another. While the trend was similar in Britain, Germany, and France, it was more obvious in France. Canada and Japan had the similar trend to that in the US, but Japan's per capita energy consumption was only half of that in the US. This paper will analyze the trend in the US and France, two representative countries.

Further comparative analysis revealed that the per capita $CO_2$ emission and per capita energy consumption in France (Fig. 4.14) increased basically at the same pace and were not delinked before the 1970s; in the early 1970s and 1980s, energy consumption increased slowly and $CO_2$ emission peaked, and they could be considered relatively delinked; from the 1980s to 2005, energy consumption continued to increase but $CO_2$ emission decreased, so they were absolutely delinked; after 2005, $CO_2$ emission decreased faster than energy consumption, and they could be considered relatively delinked. In the US (Fig. 4.15), energy consumption and $CO_2$ emission were not delinked before the 1970s; they could be considered relatively delinked in the early 1970s–1980s when the $CO_2$ emission decreased faster than energy consumption; then they had basically the same trend and were not delinked from the 1980s to around 2005. After 2005, $CO_2$ emission decreased faster than energy consumption, and they could be considered relatively delinked. The future development of energy consumption and $CO_2$ emission and their delinking trend in various countries is up to further observations.

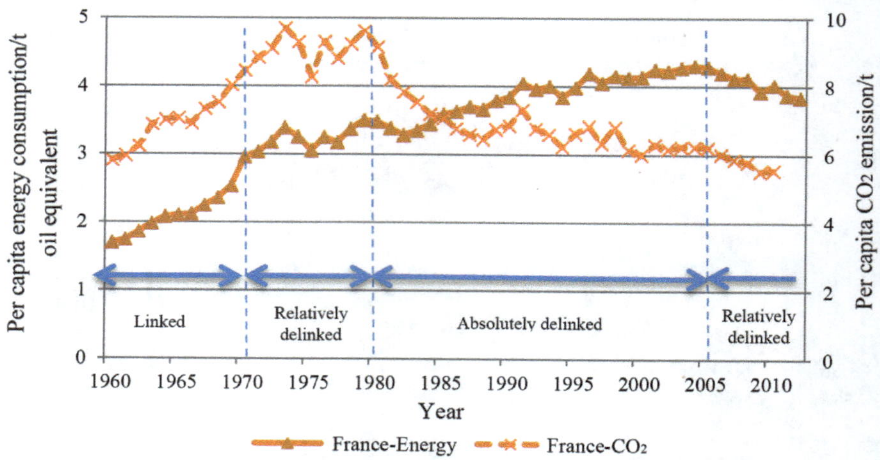


**Fig. 4.14** Trend of per capita energy consumption and $CO_2$ emission in France since 1960

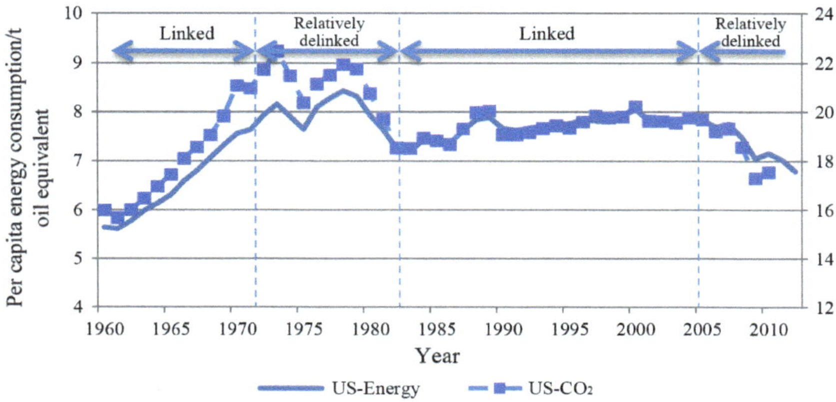


**Fig. 4.15** Trend of per capita energy consumption and $CO_2$ emission in the US since 1960

(II) Analysis of reasons for energy consumption–$CO_2$ emission delinking in developed countries

In developed countries represented by the US, energy consumption and $CO_2$ emission were relatively delinked in the 1970s–1980s. One of the reasons was that the first oil crisis and the environmental problems in the 1970s propelled those countries to generally promote energy-saving and emission-reducing measures, intensify environmental protection, and rigorously reinforce governance in the form of law. The US, for instance, set up the Environmental Protection Agency in 1970, which was in full charge of environmental management on behalf of the federal government. In the same year, the President signed the *National Environmental Policy Act* that set the tone for environmental legislation in the country, and promulgated the *Clean Air Act* that was the prototype of the principles on national air quality standards. Although the *Clean Air Act* involved the six air pollutants of $SO_2$, air-polluting particulates, $NO_x$, CO, ozone, and lead, it also caused the periodical reduction of $CO_2$ emission as those gases all had the same emission source.

A second reason was that the changed structure of energy consumption, especially the rising proportion of nonfossil energies and natural gas, which was a relatively low-carbon fossil energy, caused the relative delinking between energy consumption and $CO_2$ emission. According to America's energy consumption structure (Figs. 4.16 and 4.17), the proportion of nonfossil energies increased from 4 to 10% in the 1970s–1980s. After 2005, their proportion only increased from 11 to 14%, but the proportion of natural gas (including unconventional natural gas), which was a relatively clean fossil energy, rose from 24 to 30% in the same period, relatively delinking energy consumption from $CO_2$ emission.

The delinking between energy consumption and $CO_2$ emission was quite obvious in France mainly because that France began to energetically develop nonfossil energies in the late 1970s, whose proportion expanded from 9% in 1965 to 50% in 2014. In France, $CO_2$ emission and fossil energy consumption, especially the consumption

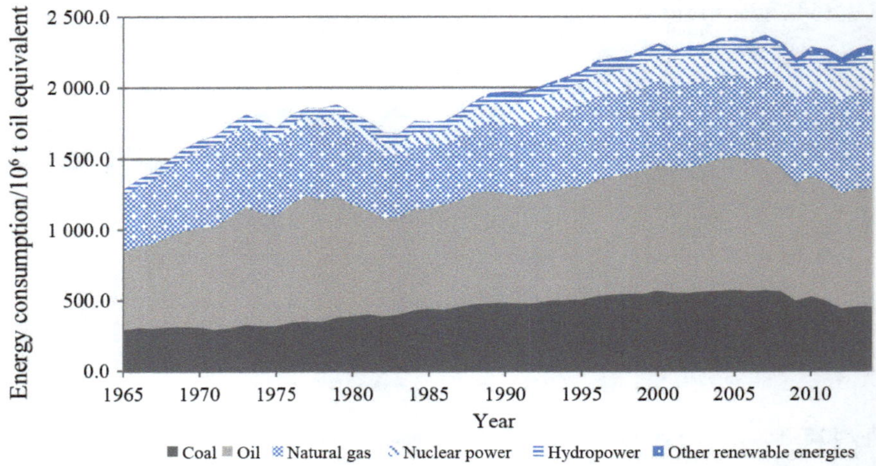


**Fig. 4.16** Energy consumption and its structure in the US between 1965 and 2014 (*Source* 2015 BP Statistical Review)

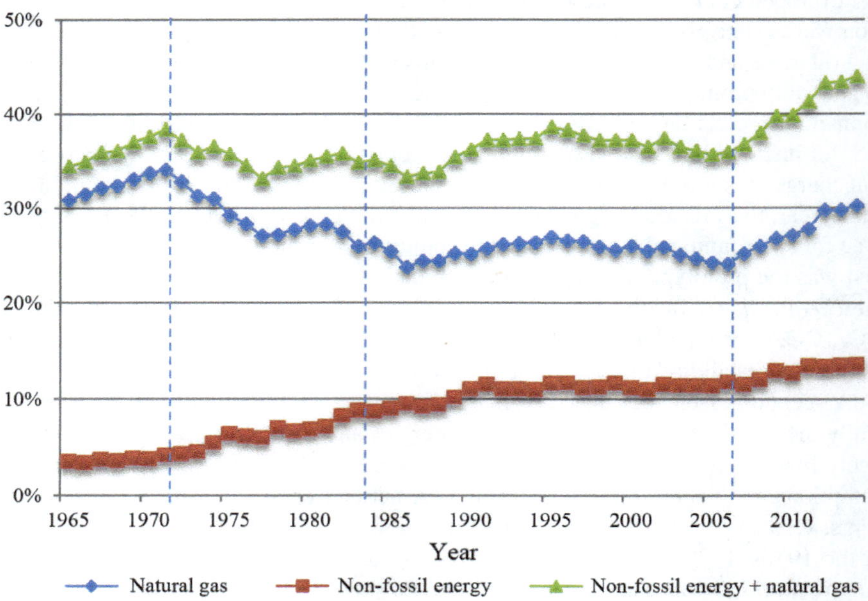


**Fig. 4.17** Proportion of nonfossil energy and natural gas consumption in the US between 1965 and 2014 (*Source* 2015 BP Statistical Review)

of high-carbon fossil energies dominated by coal and oil, had almost the identical trend (Figs. 4.14 and 4.18). For instance, between the 1960s and the 1980s, both $CO_2$ emission and fossil energy consumption increased first and then decreased; after the 1980s, total consumption of coal and oil was on the down curve in general, and while fossil energy consumption increased a little in the early twenty-first century, $CO_2$ emission did not. As a whole, France's energy consumption has kept rising for more than 50 years with a minor downswing in the recent 5 years, whereas its $CO_2$ emission has been on the decrease ever since early 1980s. Energy consumption is delinked from $CO_2$ emission.

These analyses indicate that the main factor for delinking energy consumption from $CO_2$ emission is a low-carbon energy consumption structure and increased use of nonfossil energies, which naturally reduces $CO_2$ emission, while replacing coal and oil with the relatively clean natural gas (including unconventional natural gas) can also substantially reduce $CO_2$ emission. Besides, stronger governance of air pollution, improved energy efficiency, and reduced GHG and polluting gas emission per unit of energy consumption also give a strong boost to the energy consumption–$CO_2$ emission delinking.

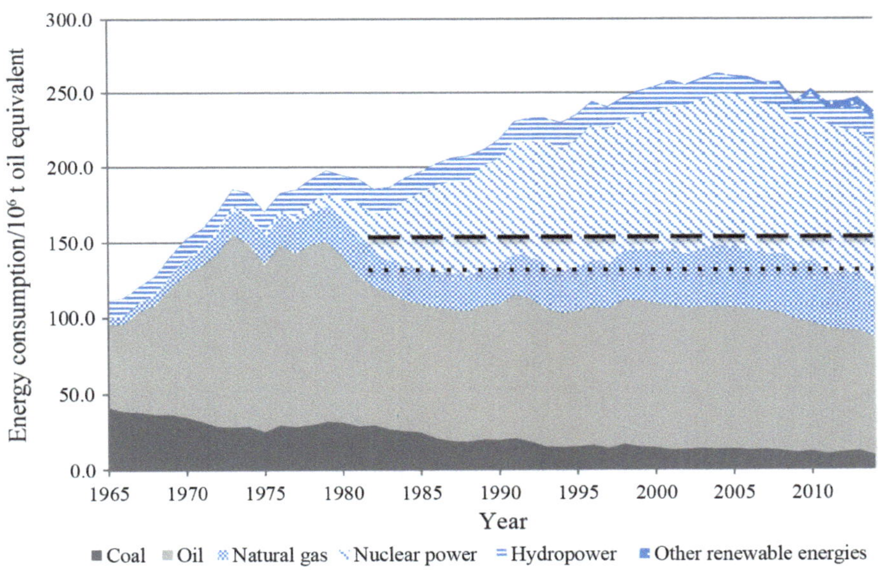


**Fig. 4.18** Energy consumption and its structure in France between 1965 and 2014 (*Source* 2015 BP Statistical Review)

### *4.3.4 Analysis of Delinking of Economic Development, Energy Consumption, and $CO_2$ Emission in China*

(I) Development trend in the past 30-plus years

As far as economic development and energy consumption are concerned, China's energy consumption has increased along with economic growth in the 30-plus years of reform and opening-up (Fig. 4.19) and the energy intensity has kept falling. From 1980 to 2013, China's GDP grew by 22.4 times from USD217.5 billion to USD4,864 billion; energy consumption increased by 6.9 times from 422 million ton to 2.92 billion ton oil equivalent[2]; per capita GDP increased by 16 times from USD222 to USD3,583; per capita energy consumption increased by 5 times from 0.43 ton to 2.15 ton oil equivalent, and energy intensity fell by 69% from 19.4 ton to 6 ton oil equivalent/USD10,000.

Although China has made great achievements in economic development, the fast growth of energy consumption should not be ignored. Especially in the twenty-first century, China's energy consumption has increased rapidly owing to the accelerated industrialization and urbanization, and its global percentage has kept rising, exceeding 20% in 2010, when its GDP was only 7.4% of the world's total. This indicated the low energy efficiency in China. Per capita energy consumption also increased

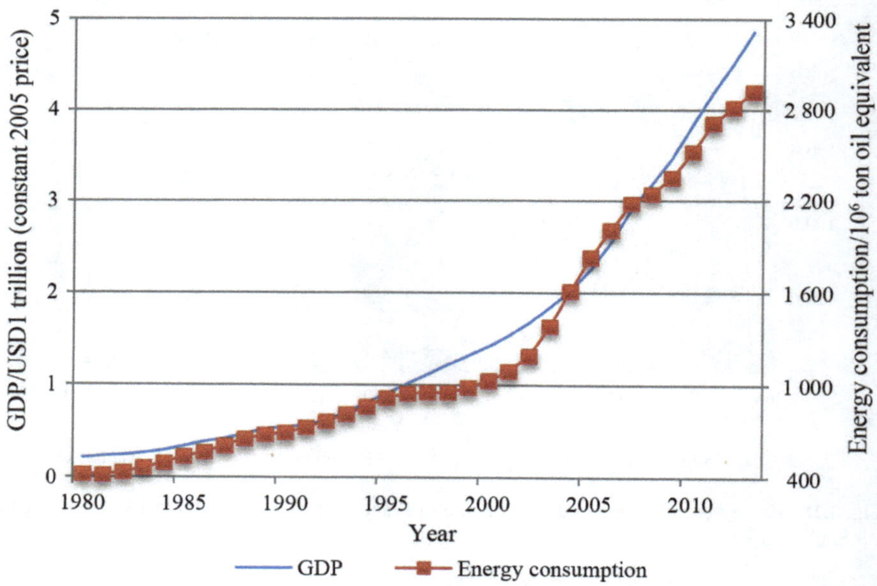


**Fig. 4.19** China's GDP and energy consumption between 1980 and 2013

[2]In this section, data about energy consumption come from the website of China National Bureau of Statistics and 2015 BP Statistical Review.

fast, reaching 2.15 ton oil equivalent in 2013, more than 60% of that under the "EU, Japanese model," while our per capita GDP was less than 10% of that in developed countries.

In regard to energy consumption and $CO_2$ emission, China's $CO_2$ emission has increased rapidly along with the growing energy consumption over the past 30-plus years (Fig. 4.20). In 2014, China emitted 9.76 billion ton $CO_2$, accounting for 27.5% of the world's total, and its energy consumption accounted for 23.0% of global consumption. In that year, our annual per capita $CO_2$ emission exceeded 7 tons, approaching the average level in the "EU, Japanese model."

China's energy consumption is dominated by coal and oil (Fig. 4.21). Although the proportion of nonfossil energy consumption increased year by year (from 3% in 1980 to 11% in 2014), the sustained and increase of coal and oil consumption resulted in rapid increase of $CO_2$ emission. Meanwhile, the extensive energy production and consumption model, which failed to effectively control $CO_2$ emission in the process, gave China a high $CO_2$ emission per unit of energy consumption.

In recent years, smog has been frequent in central and eastern China, and the emission of $SO_2$, $NO_x$, and other polluting gases, which had much to do with smog, was not optimistic. According to relevant research results, the $SO_2$ and $NO_x$ emissions in China exceeded the environmental capacity by 63% and 91%, respectively, in 2011. In the current stage, China is undergoing industrialization and urbanization in parallel, and polluting gases and GHGs like $CO_2$ basically have the same emission sources, so controlling their emission can achieve the same goal. When energy con-

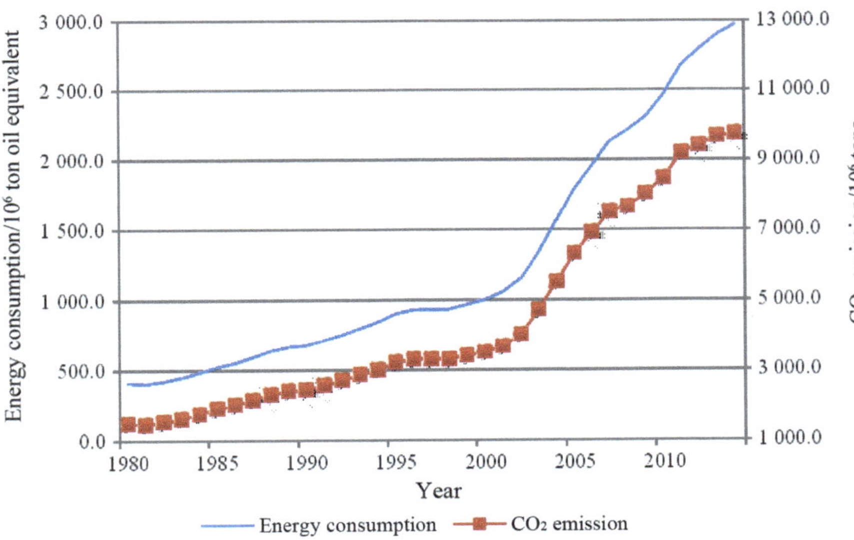


**Fig. 4.20** Energy consumption and $CO_2$ emission in China between 1980 and 2014 (*Source* 2015 BP Statistical Review)

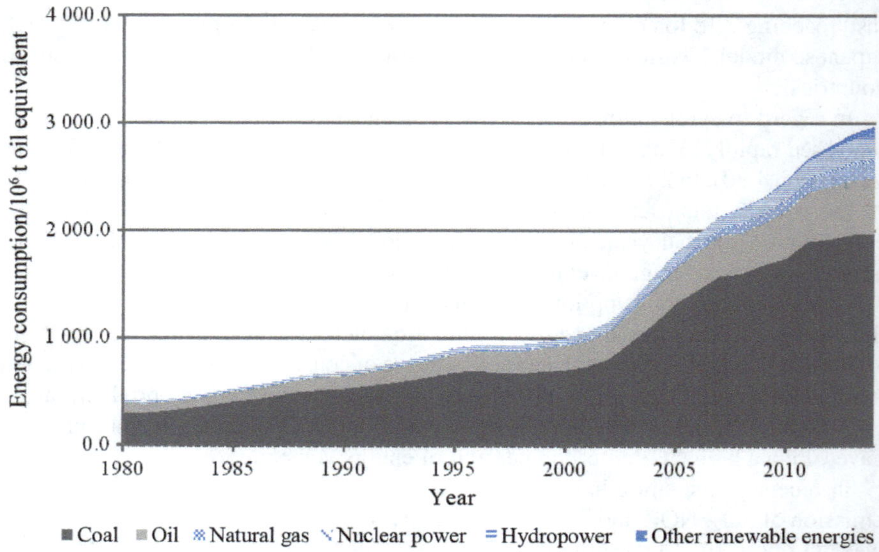


**Fig. 4.21** Energy consumption and its structure in China 1980–2014 (*Source* 2015 BP Statistical Review)

sumption is delinked from $CO_2$ emission to a certain level, it will also be delinked from polluting gas emission to some extent.

(II) Delinking analysis in low-carbon scenarios

To study the future scenarios and based on the judgment of China's future development trend, this paper makes the following assumptions about GDP growth and population, as shown in Table 4.2. According to calculations, China's GDP will reach USD24.59 trillion in 2050 (constant 2005 price in US dollar) and per capita GDP USD18,000 assuming a population of 1353 million people, basically reaching the level of a moderately developed country.

Assume the low-carbon scenario is one where the scientific production capacity of coal is realized as soon as possible. The "research on medium- and long-term energy development strategy (2030, 2050) of China," a major advisory project of Chinese Academy of Engineering, proposed the target of controlling the total energy consumption in view of China's scientific energy production and use. The State Council issued in 2014 the *Energy Development Strategy Action Plan (2014, 2020)*, which

**Table 4.2** Main parameters in China's future development scenarios

| Years | 2010 | 2020 | 2030 | 2040 | 2050 |
|---|---|---|---|---|---|
| Population/100 million | 13.38 | 14.00 | 14.20 | 14.02 | 13.53 |
| GDP growth (annual average) | – | 7.2% | 5.2% | 3.8% | 2.8% |

stated that the consumption of primary energy will be controlled around 4.8 billion ton standard coal (3.36 billion ton oil equivalent) in 2020. Considering China's actual energy consumption in recent years and its hope to restore the scenario of scientific production capacity of coal as soon as possible, this paper assumes the following parameters of a low-carbon scenario as shown in Table 4.3 and Fig. 4.22, including the annual use of primary energy, fossil energy consumption, and its proportion.

Given the changed structure of fossil energy consumption and technological progress, and based on the assumptions above, fossil energy consumption, especially coal and oil consumption, will peak before 2030, while primary energy consumption will continue to grow slowly. If CCUS is not adopted, China's $CO_2$ emission will

**Table 4.3** Primary energy consumption, fossil energy consumption, and its proportion in China in a low-carbon scenario

| Years | 2020 | 2030 | 2040 | 2050 |
|---|---|---|---|---|
| Energy consumption/$10^8$t oil equivalent | 33.6 | 36.8 | 39.3 | 41.8 |
| Proportion of coal and oil consumption | 75% | 72% | 58% | 48% |
| Proportion of natural gas consumption | 10% | 13% | 14% | 15% |
| Fossil energy consumption/$10^8$t oil equivalent | 28.6 | 31.3 | 28.3 | 26.3 |
| Coal & oil consumption/$10^8$t oil equivalent | 25.2 | 26.5 | 22.8 | 20.0 |

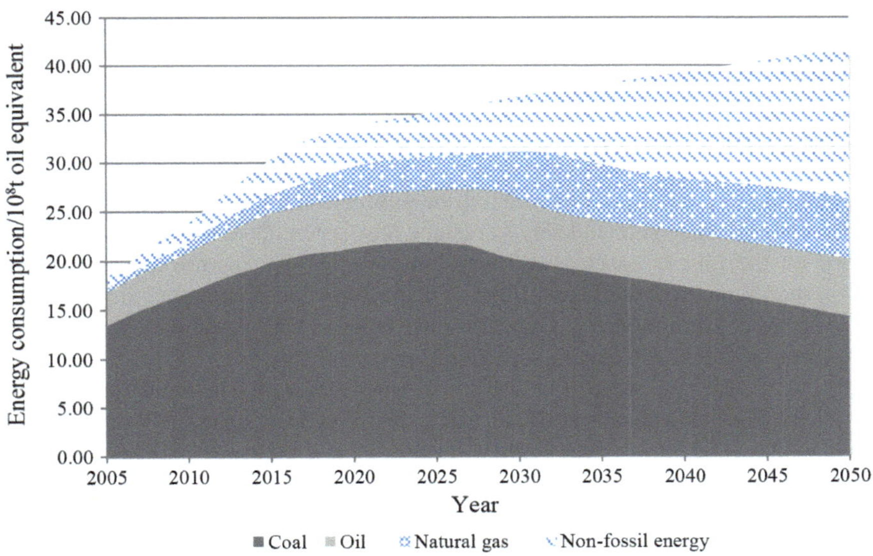


**Fig. 4.22** Trend of primary energy consumption in China in a low-carbon scenario

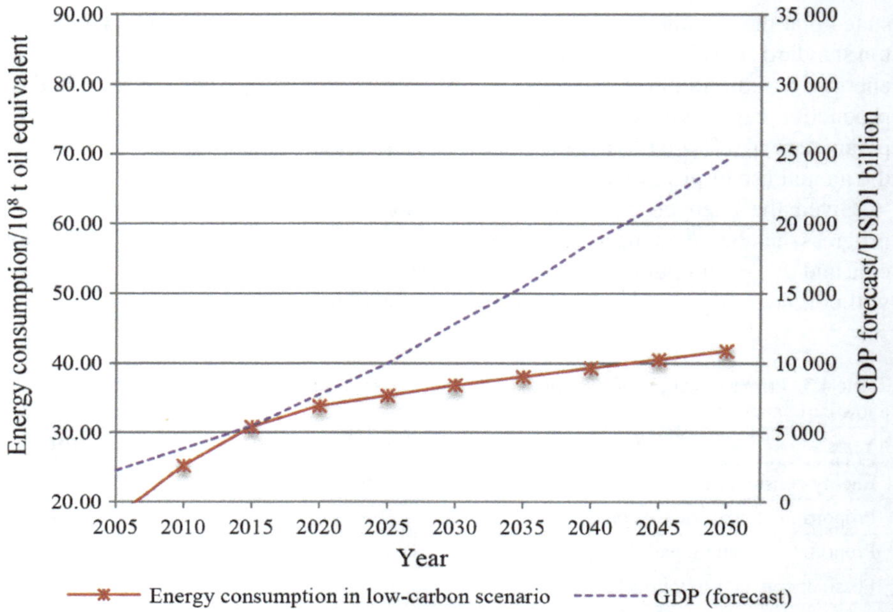


**Fig. 4.23** Trend of economic development and energy consumption in China 2005–2050

peak (about 11 billion t[3]) in or before 2030 and then gradually decrease, and will decrease by a large margin in 2050.

In a low-carbon scenario, the trend of economic development and energy consumption, energy consumption, and $CO_2$ emission in China in 2005–2050 is shown in Fig. 4.23 and Fig. 4.24, respectively.

After the Chinese economy entered the new normal, the GDP growth forecast is reduced and there is a tendency to resume the "potential economic growth rate." Based on the primary energy consumption and energy structure assumed in the low-carbon scenario, a delinking trend will appear between economic development and energy consumption. As $CO_2$ emission will peak in or before 2030, energy consumption will be absolutely delinked from it after that.

China is in the later stage of industrialization, during which its industrial structure will be transformed and upgraded more quickly, the proportion of GDP generated by the secondary and tertiary industries will continue to grow, and the service industry will take a larger percentage. At the same time, the internal structure of the secondary industry will be adjusted too, whereby high-energy-consuming and high-emission industries will be gradually transformed and upgraded into clean ones with high energy efficiency, and the output value of green, low-carbon industries will take an

[3]$CO_2$ emission is calculated based on the amount of fossil energy use. $CO_2$ emission from other sources is not considered. The emission coefficient is in reference to BP Statistical Review.

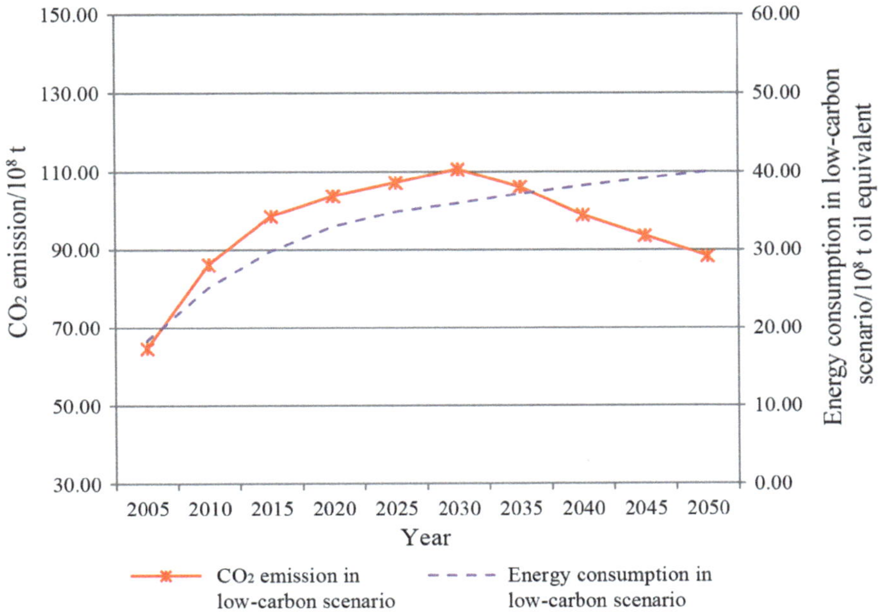


**Fig. 4.24** Trend of energy consumption and $CO_2$ emission in China between 2005 and 2050

ever rising proportion. Therefore, China's energy consumption may peak during its future economic development and be delinked from it.

In the background that the global efforts for tackling climate change, the emission of GHGs including $CO_2$ in China will not follow the "US, Canadian model." As a responsible major country, China, on the one hand, will win the necessary space for its own development. On the other hand, in view of its actual air quality, GHGs and polluting gases have basically the same emission sources, so there is a common direction and goal for the control of their emissions. By doing this, China is not just doing its bit for tackling global climate change but is also working for the health of Chinese people of this generation and for a green environment for generations to come. The fundamental solution is changing the energy structure dominated by coal and oil as soon as possible. In the future energy system dominated by nonfossil energies, there will not be any risks even if the energy consumption increases, and people's living standard will be improved and the living environment will be clean and beautiful.

At present, China should tighten the control of energy consumption, especially the use of high-carbon fossil energies such as coal and oil. While changing the energy consumption structure, we should also intensify the control and utilization of GHG and polluting gas emissions during energy production and consumption. Energy production should be efficient and clean and energy consumption energy-saving, green, and environmentally friendly. We must lower the energy consumption per unit of GDP and control the increase of per capita energy consumption, reduce the GHG and polluting gas generated per unit of energy consumption, and control per capita $CO_2$ emission.

As for China's future path of energy development, we, on the premise of meeting the reasonable demand for supply in a scientific way, should implement domestic energy production based on a scientific production capacity, save energy and raise energy efficiency, and cut the quantity and intensity of $CO_2$ and polluting gas emission. If we follow this low-carbon scenario, coal and oil consumption will peak before 2030, so will the energy consumption in 2050 with almost zero increment, but the economy will continue to grow, meaning that China's economic development is almost absolutely delinked from energy consumption. When $CO_2$ emission peaks in or before 2030, it will be absolutely delinked from energy consumption.

# Chapter 5
# The Necessity to Transform Growth Model

## 5.1 Achievements and Crisis Brought by China's High-Carbon Growth

The economic growth model refers to the development strategy for national economy and the specific mechanisms and principles driving and governing the increase and operation of productivity factors in a certain period of time. It includes the goals, approaches, priorities, and steps of economic growth. Under given technological conditions, a country is restricted by available natural resources in carrying out the strategy of economic growth. The approach to economic growth includes means, methods, and model in pursuing economic development. The notion is encompassing, involving not only the growth model but also structural issues (economic structure, industrial structure, urban–rural structural and regional structure, among others), growth quality, economic efficacy, income distribution, environmental protection, urbanization, industrialization, modernization, and others. In the past 30 years, China has chosen a growth model focusing on quantitative increase and expansion in scale. This model has the following features. First, the main target is high-speed growth as a way to catch up the developed countries; second, it is imbalanced in terms of economic structure as it sacrifices sectors like agriculture and light industry; third, as an extensive growth model, it features excessive pursuit of expansion in scale and increase of products with more inputs of workforce and capital. This traditional growth model used to boost the economy under certain historical conditions, but its flaws become increasingly prominent over time.

In this chapter, the book looks back on the economic achievements brought by the growth model over the last 30 years after the adoption of reform and opening-up policy. It continues with analysis of environmental issues in this process to show the adverse consequences of the model. Finally, it outlines the characteristics of this growth model and argues that it is imperative to transform it.


X. Du et al., *Overview of Low-Carbon Development*,
https://doi.org/10.1007/978-981-13-9250-4_5

### *5.1.1 Remarkable Achievements Brought by Previous Growth Model*

i. Rapid economic growth and greatly enhanced overall national strength

Since the adoption of reform and opening-up policy, China has witnessed a booming economy and the tremendous improvement in terms of overall national strength and international influence. The GDP swelled to 56.8845 trillion yuan in 2013 from 364.5 billion yuan in 1978, an average annual growth rate of 15.5%. The figure outnumbers the world average of 6.4% during the same period, and the highest level of Japan and South Korea, which stood at 9.2% and 8.5%, respectively (Fig. 5.1). The country became the second largest economy after the United States from the 10th ranking in 1978. The GDP per capita rocketed to 41,805 yuan in 2013 from 381 yuan in 1978, an annual increase rate of 14.4% on average (Fig. 5.2). The national

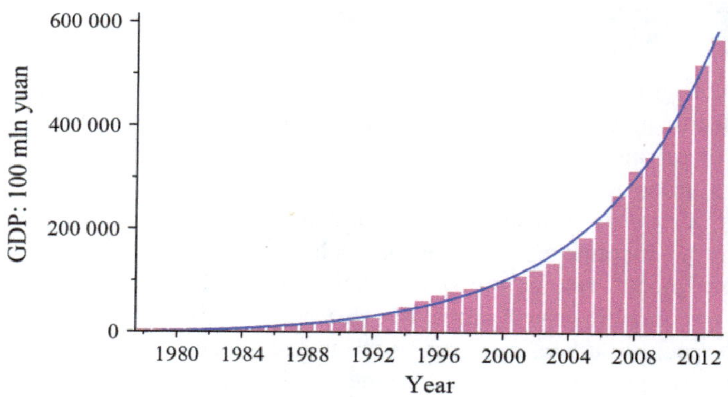


**Fig. 5.1** China's GDP in 1978–2013

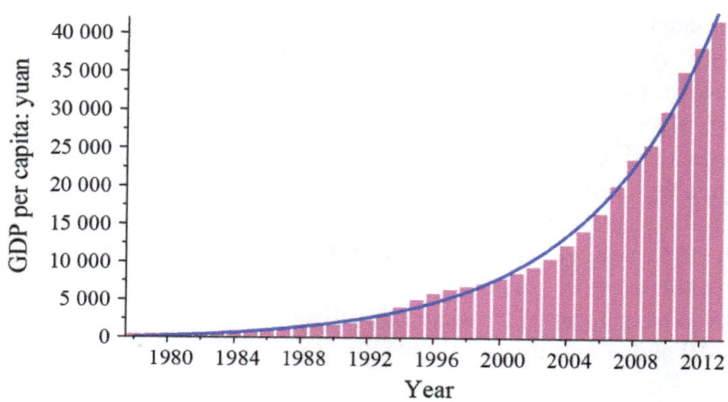


**Fig. 5.2** China's GDP per capita in 1978–2013

fiscal revenue increased at an annual rate of 14.5% from 113.2 billion yuan in 1978 to 12.9143 trillion yuan in 2013 (Fig. 5.3). From a country strapped for foreign exchanges, China is able to maintain abundant foreign exchange reserves, which grew from $167 million in 1978 to $3.82 trillion in 2013. The annual growth rate was 33.2% (Fig. 5.4), ranking first in the world. Chinese enterprises grew from a few weak ones to currently a large number of competitive big companies. In 2004, 100 out of the Fortune 500 companies were from the Chinese mainland, seven of which were included for the first time. Sinopec Group jumped to the third place, and China National Petroleum Corporation stayed next to it.

China's economic structure has been continuously improved and upgraded since the reform and opening-up. The three sectors have maintained sustainable development. The agriculture has been strengthened as the foundation of the national economy. Both the manufacturing and service sectors have grown rapidly. In 1979–2013, the added values of the primary, secondary, and tertiary sectors increased at an annual

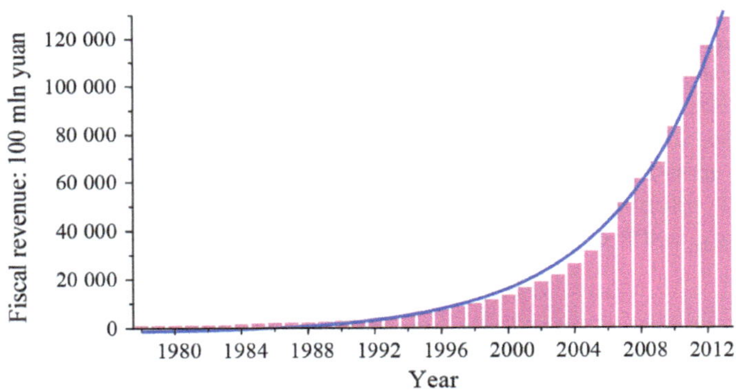


**Fig. 5.3** China's fiscal revenue in 1978–2013

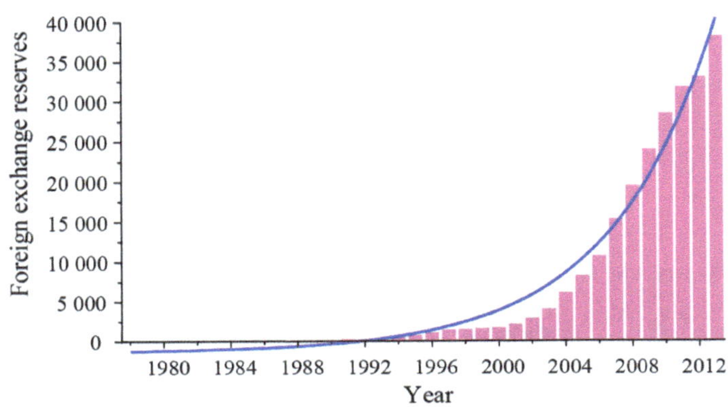


**Fig. 5.4** China's foreign exchange reserves in 1978–2013

rate of 12.2%, 15.2%, and 17.7%, respectively. The proportion of them to the economy changed from 28.2:47.9:23.9 in 1978 to 10.0:43.9:46.1 in 2013. The proportion of the primary and secondary sectors to the economy dropped by 18.2 and 4.0% points, respectively, whereas the proportion of the tertiary sector increased sharply by 22.2% points.

The supply capability of agricultural products has been steadily improved. The added value of the primary sector increased from 102.8 billion yuan in 1978 to 5.6957 trillion yuan in 2013, an increase of 9.5-fold after adjusting for inflation. The annual average growth rate is 6.7%. The outputs of primary agricultural products increased exponentially (Table 5.1) and moved up in the world ranking (Table 5.2).

The manufacturing capacity has been rapidly enhanced. In 2013, the added value of the manufacturing sector was 24.9684 trillion yuan. Calculated at comparable prices, it expanded by 25-fold from 1978 with an average annual growth rate of 9.6%. The output of major industrial products has mushroomed (Table 5.3). The per capita output of industrial products in major years has increased significantly (Table 5.4). In terms of industrial structure, there is a significant shift from low-technology, labor-intensive, and single-category industries to capital- and technology-intensive and complete-categories ones. Over the past three decades, many Chinese products have ranked among the world's top in terms of output. By the end of 2010, the outputs of 220 Chinese industrial products have ranked first across the world, and the rankings of many primary industrial product outputs have moved up (Table 5.5).

Since the reform and opening-up, traditional manufacturing sectors, like metallurgy, energy, textiles, machinery, and shipping among others, have been continu-

**Table 5.1** Comparison of the yields of primary agricultural product in 1978 and 2013 (10,000 tons)

| | 1978 | 2013 | Folds of increase |
|---|---|---|---|
| Grain | 30,477 | 60,194 | 2.0 |
| Cotton | 217 | 631 | 2.9 |
| Oil plants | 522 | 3531 | 6.8 |
| Bast fiber plants | 135 | 23 | 0.2 |
| Sugarcane | 2112 | 12,820 | 6.1 |
| Beet | 270 | 926 | 3.4 |
| Tea | 27 | 193 | 7.1 |
| Fruit | 657 | 25,093 | 38.2 |

**Table 5.2** The world ranking of China's primary agricultural product yields

| | Grain | Meat | Cotton | Peanut | Rapeseed | Tea | Fruit | Sugarcane | Soybean |
|---|---|---|---|---|---|---|---|---|---|
| Ranking in 1978 | 2 | 3 | 3 | 2 | 2 | 2 | 8 | 6 | 3 |
| Ranking in 2013 | 1 | 1 | 1 | 1 | 1 | 1 | 1 | 2 | 4 |

Table 5.3 The comparison of the output of major industrial products in 1978 and 2013

| | 1978 | 2013 | Folds of increase |
|---|---|---|---|
| Automobile/10,000 units | 14.91 | 2211.7 | 148 |
| Cement/10,000 tons | 6524 | 242,000 | 37 |
| Crude steel/10,000 tons | 3178 | 77,904.1 | 25 |
| Electricity generation/trillion kW per h | 256.6 | 53,975.9 | 210 |
| Natural gas/100 million $m^3$ | 137 | 1170.5 | 8.5 |
| Raw coal/100 million tons | 6.18 | 36.8 | 6 |
| Domestic refrigerator/10,000 | 2.8 | 9261.0 | 3308 |
| Color TV set/10,000 | 0.4 | 12,776.1 | 31,940 |
| Mobile phone/10,000 | – | 145,561.0 | – |
| Computer/10,000 | – | 33,661.0 | – |

Table 5.4 The comparison of the per capita output of major industrial products in 1978 and 2013

| | 1978 | 2013 | Folds of increase |
|---|---|---|---|
| Cement/kg | 68.2 | 1778.5 | 26.1 |
| Crude steel/kg | 33.2 | 572.5 | 17.2 |
| Raw coal/t | 0.65 | 2.70 | 4.2 |
| Crude oil/kg | 108.8 | 153.6 | 1.4 |
| Cloth/m | 11.5 | 64.9 | 5.6 |

ously upgraded. Taking the steel-making sector as an example, the continuous casting ratio was improved to 99.64% in 2013, about 90% points higher than that in 1980, suggesting it reached international advanced level. Meanwhile, high-tech sectors, including electronic information, bioengineering, aerospace, pharmaceuticals, new energy, and new materials, have grown from scratch and continued to flourish. They have become the robust driving forces for China's leapfrog development in manufacturing sector.

The tertiary sector has grown rapidly. With deepening understanding of and increasing input in the tertiary sector, the sector has sustained rapid growth. In 2013,

Table 5.5 The world rankings of China's major industrial products output

| | Steel | Coal | Crude oil | Electricity generation | Cement | Chemical fertilizer |
|---|---|---|---|---|---|---|
| Ranking in 1978 | 5 | 3 | 8 | 7 | 4 | 3 |
| Ranking in 2013 | 1 | 1 | 4 | 1 | 1 | 1 |
| The year ranking first place | 1996 | 1990 | – | 2012 | 1985 | 1996 |

**Table 5.6** The comparison of the outputs of the major service products in 1978 and 2013

| Item | The scale in 1978 | The scale in 2013 | Folds of increase |
|---|---|---|---|
| Outstanding loans/trillion yuan | 0.2 | 76.6 | 401 |
| Insurance revenues/100 million yuan | 400.5 | 17,222 | 43 |
| Passenger volume/100 million per km | 1896.6 | 36,036 | 19 |
| Goods volume/100 million tons per km | 10969.3 | 186,478 | 17 |
| Postal and telecommunication services/100 million yuan | 34.2 | 16,679 | 488 |
| Added value of the real estate sector/100 million yuan | 80.2 | 33,277 | 415 |

the added value of the sector totaled 26.2204 trillion yuan. Calculated at comparable prices, it expanded by 51.8-fold from 1978 with an average annual growth rate of 11.9%. All services sectors have witnessed rapid growth (Table 5.6).

ii. Remarkable improvement in people's livelihood

Statistics by the World Bank show that China's gross national income per capita rose from USD154.97 in 1978 to USD 4,026.02 in 2013. By World Bank standards, China has emerged from a low-income country to one of the upper middle-income economies in the world. It is indeed a remarkable achievement for China, which is a country with weak economic foundation but large population.

Urbanization is speeding up (Fig. 5.5). The urbanization rate increased to 53.73% in 2013 from 17.9% in 1978, up by 35.73% points and an annual average rise of 1.02% points. A large quantity of rural residents migrated to urban areas, boosting balanced economic development between the urban and rural areas. It is predictable that the trend will maintain for a long period of time.

People's income is significantly improved and people are getting affluent. The disposable income per capita in urban areas rocketed from RMB 343 in 1978 to RMB 26,955 in 2013, an increase of 13.5-fold in inflation-adjusted terms. The average annual growth rate is 7.7%. Rural residents' disposable income per capita jumped from RMB 134–RMB 8,896, an increase of 11.4-fold in real terms. The average annual growth rate is 7.2%. Family wealth in both rural and urban areas is on the rise. The RMB savings account balance of rural and urban residents amounted to RMB 40.0 trillion at the end of 2012. In real terms, it is an increase of 327-fold from RMB 21.06 billion in late 1978. Per capita figure jumped from RMB 21.9–RMB 29,396, an increase of 231-fold in real term, the highest in the world. In addition, China's saving rate has exceeded 50%, far higher than the world average. The share of financial assets like stocks and bonds is expanding in household asset portfolio.

Residents in both rural and urban areas witness a significantly improved living standard and life quality (Fig. 5.6). In 1978, people spent an average of RMB 184; however, in 2013, the figure jumped to RMB 18,311, an increase of 17.1-fold in real

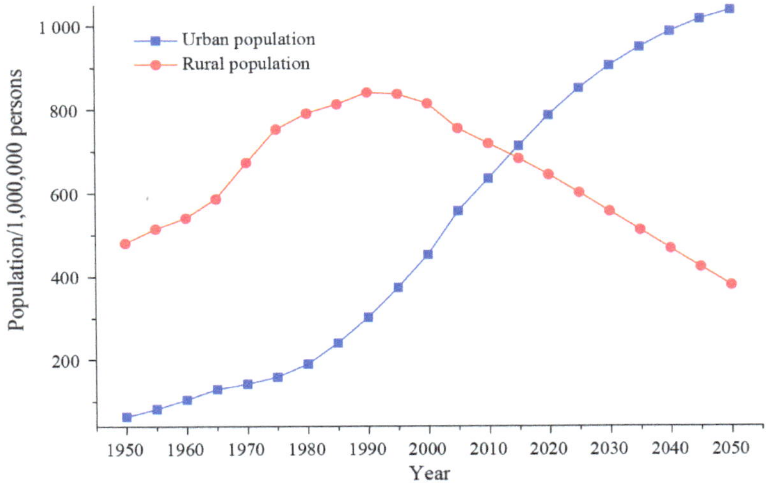


**Fig. 5.5** The urban and rural population in China in 1950–2050 (*Source* The Department of Economic and Social Affairs, United Nations)

**Fig. 5.6** People's living conditions before and after the reform and opening-up

terms. The average annual growth rate is 8.5%. Household expenditure on food in urban areas fell from 57.5 to 35.0%. In rural areas, the figure is down from 67.7 to 37.7%. By UN standards, China is a relatively prosperous country. However, the Gini coefficient, which is the most commonly used to measure equality in income or wealth distribution, rose from 0.24 in 1978 to 0.473 in 2013, exceeding the international warning line of 0.4 as shown in Fig. 5.7. Its maximum value is one, which expresses maximal income disparity, meaning that only one person has all the incomes, and all others have none. A Gini coefficient of zero expresses perfect equality, where everyone has the same income. In terms of consumer durable goods, color TV sets, washing machines, refrigerators, air conditioners, telephones, and others have been widely used in urban areas. The ownership of high-end durable goods such as automobiles and personal computers has increased significantly as shown in Fig. 5.8. The penetration of color TV sets, electric fans, washing machines, and motorcycles among others is also on the rise in rural areas as shown in Fig. 5.9.

As people's living standards have been improved, China's population in absolute poverty has reduced dramatically. In 1981, the number was 835 million, and in 2013, it fell to 68 million. In 1981, China's absolute poor people accounted for 43.1% of the world's total, and in 2010, the figure fell to 13%. In 2013, the incidence of poverty in China further decreased to 5.0%, which is far below the world average level of 20.58% in 2010. Absolute poverty has been fundamentally eradicated in China. UN and the World Bank said that the achievements in poverty reduction of mankind should be largely attributed to China.

China's rapid economic growth cannot be achieved without its national policy that advances with the times and the people's resolution to get rid of poverty. However, due to the national conditions at that time, it was not the best development path relying on technological advances and industrial innovation. The main assets China

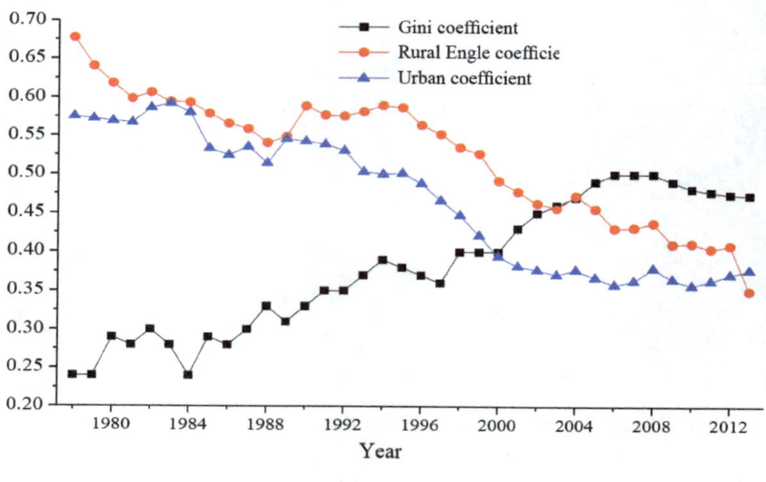


**Fig. 5.7** Chinese people's consumption in rural and urban areas in 1978–2013 (*Source* China Statistical Yearbook)

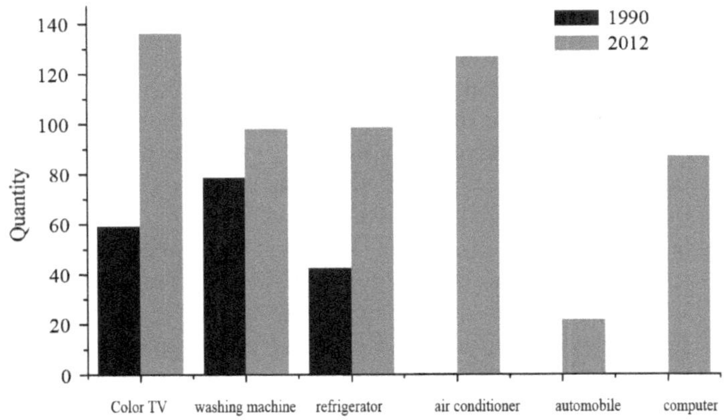


**Fig. 5.8** Ownership of durable goods in every 100 urban households

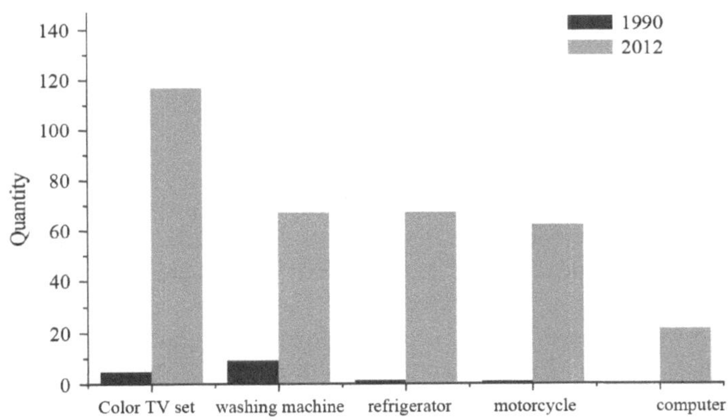


**Fig. 5.9** Ownership of durable goods in every 100 rural households (unit)

had were abundant natural resources and vast labor forces at the primary level. As a result, it could only depend on the expansion of low-end industries like the real estate, automaking, and others to boost the economy. In the past 30 years, the momentums of China's fast-growing economy include the following:

(1) Excessive consumption of natural resources, which imposes pressure on the environment. Preference was given to high-resource regions over those less rich regions and to shallow resources over the deep ones. This drastically lowered the cost of resource acquisition. After being dug out, mineral resources would be directly put into use or extensively processed. At the end of resource utilization, industrial wastewater, exhausts, and residues were directly discharged into nonindustrial areas like the farmland and wasteland without treatment. In this extensive utilization model, inputs before and after using the resources were

low, in turn reducing the production cost of natural resources, but leading to a series of environmental problems.

(2) Scientific and technological advances. Economic growth is inseparable from technological advances and innovation. New products and technologies improve productivity and facilitate people's life and work. In the past, technological innovation and improvement depended on high inputs of resources. Low-carbon innovation has not yet been widely carried out.

(3) Vast labor forces at the primary level. The rapid economic growth in the coastal areas, southeast China, was inseparable from the cheap vast labor forces, which enabled Chinese manufacturing sector to compete with businesses of developed countries regardless of workload and failures. However, the quality of such labor forces was relatively low. Most of them created value by trading their physical strengths but only a few would think about sustainable, low-carbon development in their work. As a result, the characteristic of high-carbon growth was more evident in the first line of production.

### *5.1.2 Previous Economic Growth Exacerbating Resource and Energy Issues*

Resources are the basic elements of a country's development. Resource is the support of economic development, and its reserve, distribution, structure, availability, utilization efficiency, and degree of self-sufficiency are all closely related to national security and development. Insufficient supply of resources will seriously affect people's life and work, impede economic development, and undermine social stability. As China has attained remarkable economic achievements, the country faces increasingly endemic shortage in resources supply.

i. Abundant resource reserve, but low per capita quantity

With a vast territory, China is rich in natural resources. The reserves of some important resources rank among the highest of the world. But from the perspective of per capita quantity, China is a "small country" in terms of natural resources as the quantity is lower than world average level.

China has a wide range of land resources, including large areas of arable land, forest land, grassland, desert, and mudflats, among others. Figure 5.10 shows the proportion of each category. The absolute reserve of land resources in China is large, ranking third in the world. China's arable land accounts for 7.7% of the world's total, ranking fourth across the world, grassland for 10%, ranking third, and forest land for 4.1%, ranking eighth. But the amounts per capita are low. China's arable land per capita is less than 40% of the world average, grassland per capita lower than 50%, and forest area per capita less than 20%.

According to statistics by *World Factbook 2012* released by the Central Intelligence Agency of the United States, China's recyclable water resources totaled 2,840

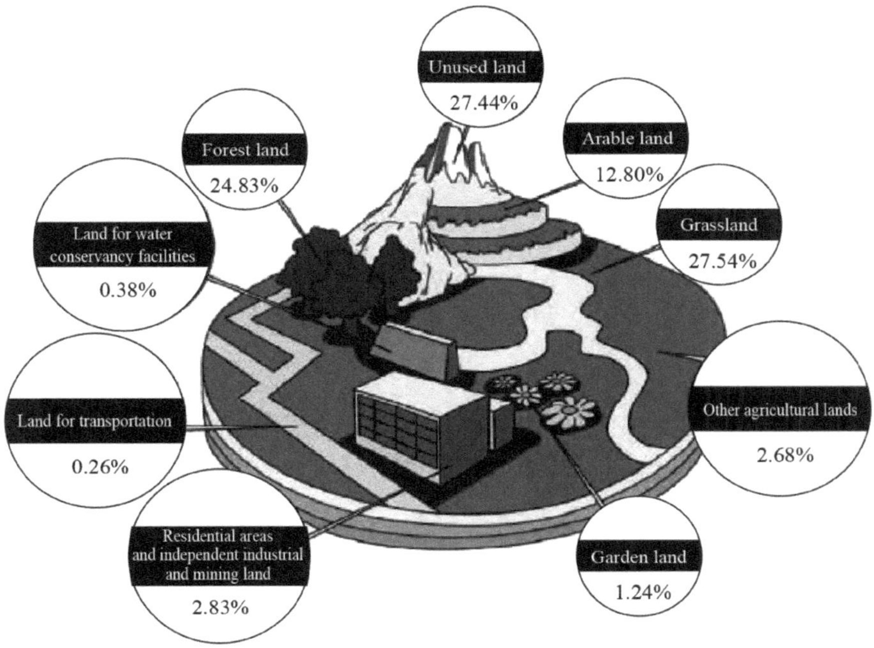


**Fig. 5.10** The distribution and utilization of land resources in China. *Source* (Peng Keshan, Sustainable Use of Land Resources in China [J]. Journal of Capital Norman University (Natural Science Edition), P61–65, 35(4), 2014)

cubic km, ranking fifth after Brazil, Russia, Canada, and the United States. But its water resource per capita was only about 2100 $m^3$, 28% of the world average, 12% points than its arable land per capita. Nationwide, only Tibet Autonomous Region and Qinghai Province have water resources per capita higher than the world average as shown in Fig. 5.11. The annual water shortage nationwide exceeded 50 billion $m^3$. Two-thirds of cities faced the problem of water shortage. Nearly 300 million rural residents had no access to safe clean water. The Ministry of Water Resources has predicted that as China's population reaches 1.6 billion in 2030, water resources per capita will be only 1750 $m^3$. Given measures for water conservancy are fully carried out, the total water resources are estimated to be 700–800 billion $m^3$. Additional water supplies of 130–230 billion $m^3$ are required to be realized. The actual amount of water resources available in the country is close to the ceiling of reasonable use. It is extremely difficult for the development of water resources.

Mineral resources are important natural resources. They are formed over tens of millions even hundreds of millions of years and are important material basis for social production and development. People in modern society cannot live or work without mineral resources. As nonrenewable resources, mineral resources have limited reserves. There are more than 160 types of minerals in the world, over 80 of which are widely used in production. By characteristic and use, they are classified into three

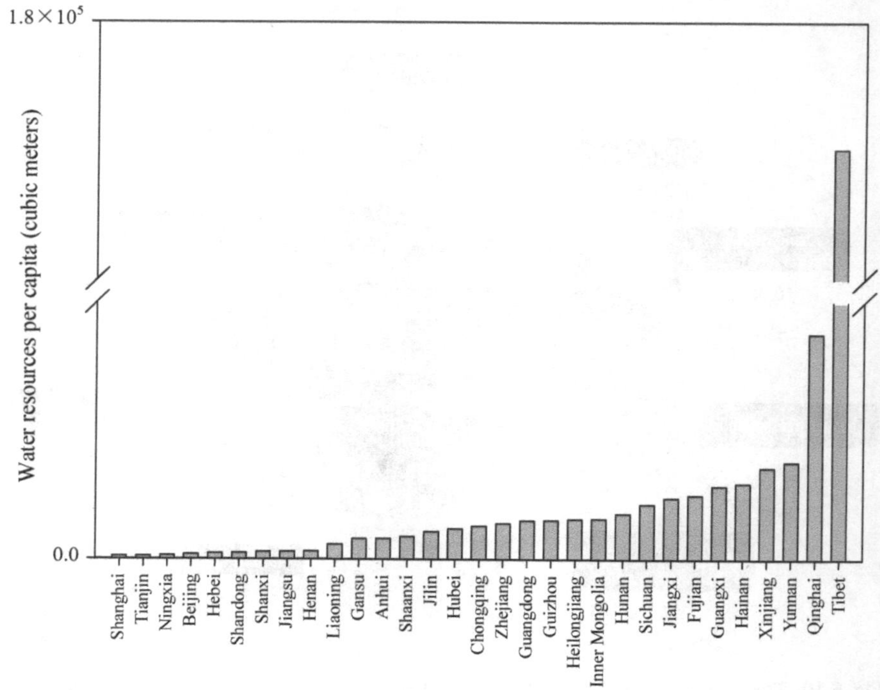


**Fig. 5.11** Water resources per capita of each province, autonomous region, and municipality (excluding Hong Kong, Macao, and Taiwan) in China

major types of metal minerals, nonmetallic minerals, and energy minerals. China has rich mineral resources with complete varieties, which could be divided into four categories. The first is energy minerals such as coal, petroleum, and geothermal. The second is metal minerals like iron, manganese, and copper. The third is nonmetallic minerals such as adamas, limestone, and clay. The fourth is water and gas minerals like underground water, mineral water, and carbon dioxide. According to the *2008 China Mining Yearbook*, a total of 171 types of minerals had been discovered nationwide as of the beginning of 2007. Minerals with demonstrated reserves had reached 159 types, among which 10 were energy minerals, 54 metal minerals, 92 nonmetallic minerals, and 3 gas and water minerals. There were more than 20,000 locations rich in minerals. The total mineral resources with proven reserves accounted for 12% of the world's total, making China one of the few countries with a complete range of rich mineral resources. In addition, new minerals are being discovered in China. Taking energy minerals as example, proven reserves of various types were on the rise in 2012. The demonstrated reserves of coal were 1420.8 billion tons, and increased by 61.61 billion tons, a rise of 3.1% year on year. The proven reserves of crude oil were 3.33 billion tons, and increased by 1.52 billion tons, a rise of 2.8% year on year. The proven reserves of natural gas were 4379 billion $m^3$, and increased by 961 billion $m^3$, a rise of 8.9% year on year. However, the per capita figure is low, only 58% of the world's average, ranking 53rd worldwide. Moreover, the proportion of large and super large mineral deposits was small. Per capita reserves of 45 important mineral resources were less than 50% of the world average. Reserves per capita of petroleum, natural gas, copper, and aluminum were among the lowest, accounting for only one 25th of the world average.

Biological resources are an integral part of natural resources. They refer to animals, plants, and microorganism that are of certain economic values to human beings in the biosphere and the biomes they constitute. Biological resources include three hierarchies of gene, species, and ecosystem, which pose certain realistic and potential values to human beings. They represent biodiversity on the earth. There are assorted species of wildlife in nature. They have different shapes, appearance, and structures, and live in different regions. They are highly adaptable to natural environments as they grow and live on the plain, hill, mountain, plateaus, grassland, wasteland, and in freshwater and the ocean. China has some 480,000 categories of biological resources, among which there are some 30,000 higher plants, 200,000 spore-bearing plants, 150,000 insects, and more than 50,000 other species of animals. There are many types of plants in China, among which there are some 25,000 seed-bearing plants. Seed-bearing plants are further classified into angiosperm and gymnosperm. China has more than 200 types of gymnosperm, one-fourth of the world's total, and some 3000 categories of angiosperm. Among more than 7000 wooden plants, there are some 2800 arbor species. China preserves unique ancient paleontological species such as metasequoia, Ginkgo biloba, and Pseudolarix amabilis, and thus is dubbed a living fossil wild plant. In the monsoon region in east China, the vegetation types include tropical rain forests, tropical monsoon forest, central and southern subtropical evergreen broad-leaved forests, northern subtropical mixed forests of deciduous broad-leaved forests and evergreen broad-leaved forests, temperate deciduous broad-leaved forests, cold-temperate coniferous forests, subalpine coniferous forests, and temperate forest grassland. In northwest China and the Qinghai–Tibet Plateau, the vegetation types include steppe, semi-desert steppe scrub, desert steppe scrub, plateau fell-field, and plateau grassland and meadow thickets. China has the most species of wild animals in the world. It has some 4880 vertebrates, accounting for 11% of the world's total, among which there are 410 species of mammals, 1180 species of birds, 300 species of reptiles, 190 species of amphibians, and 2800 species of fish. Rare animals such as giant pandas, golden monkeys, white-flag dolphins, white-lipped deer, takins, brown pheasants, Chinese alligators, and crested ibis are unique to China. Red-crowned cranes in northeast China, golden pheasants in Sichuan, Shaanxi, and Gansu Provinces, and blue peacocks, paradise flycatchers, whooper swans, and green parrots are all species of rare birds. There are also many rare species of butterflies in Taiwan, Yunnan, and Sichuan Provinces. Meanwhile, mammals, birds, reptiles, and amphibians in China account for 10% of the world's total. However, China's species are becoming extinct at an accelerated speed. Among the 740 worldwide endangered species listed in the *Convention on International Trade in Endangered Species of Wild Fauna and Flora*, 189 species are in China, accounting for a quarter of the total.

In more than last three decades, with the extensive linear development focusing on output and growth rate, China has grown from a poor backward country into a prosperous nation. Meanwhile, consumption of natural resources has also being spiking, and the per capita quantity continues to shrink.

ii. Imbalanced distribution of natural resources

Due to influences of climate, topography, and historical reasons, the geographical distribution of China's natural resources is extremely imbalanced. Demands for resources and population distribution do not match the distribution of natural resources, which leads to imbalance in the supply of and demand for natural resources. In some cases, this seriously hinders economic development. This mismatching means the transporting of resources plays an important role in boosting regional economy. Land resources are unevenly distributed: arable land is mainly in the plains and basins in the eastern monsoon region; forest land is mostly in the northeast and southwest remote mountainous areas; and grassland is mainly in the inland plateau and mountainous regions.

Under the influences of climatic conditions, water resources are also unevenly distributed. As a result, despite vast water resource reserves, China faces difficulties in exploiting and utilizing the resources. Water resources in the drainage basin of the Yangtze River and south of the region take up 80% of the national total, but arable land in the regions only account for about 36%. Water resources in the drainage basins of the Yellow River, Huaihe River, and Haihe River are only 8% of the national total, but the arable land there takes up 40%. In terms of rainfall duration, most of the regions in China have few rainfalls in winter and spring, and abundant rainfalls in summer and autumn. The precipitation concentrates in May–September, accounting for more than 70% of the annual total. Intense rainfalls are frequent during this period. In the past 70 years, rivers like the Yellow River and Songhuajiang River underwent dry years for 11–13 consecutive years and wet years for 7 to 9 consecutive years. The amount of China's groundwater recharge is about 771.8 billion $m^3$ per year, of which the drainage basin of the Yangtze River has the most, standing at 213 billion $m^3$ per year. Water resources per capita in 16 provinces, autonomous regions, and municipalities are below the extreme water shortage threshold. And water resources per capita in another six provinces and autonomous regions, which are Ningxia Hui Autonomous Region, Hebei, Shandong, Henan, Shanxi, and Jiangsu, are below 500 $m^3$, meaning they are extremely water-deficient areas.

The uneven distribution of natural resources, together with imbalanced economic development and technological advances in different regions, leads to varying exploitation and utilization of minerals in different regions. The efficiency also varies greatly. Unreasonable exploitation and utilization result in relative shortage of mineral resources, and environmental pollution and destruction. It also breaks the ecological balance and further hinders the economic development and progress of human beings.

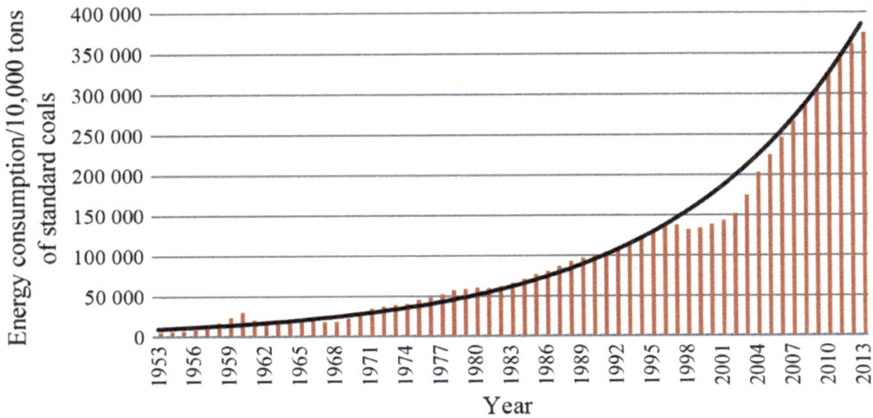


**Fig. 5.12** China's energy consumption (*Source* China Energy Statistical Yearbook)

iii. Huge consumption of resources

In 2013, China consumed 3.75 billion tons of standard coals, 6.5-fold of that in 1978 (Fig. 5.12). In 2012, China's primary energy consumption amounted to 2.735 billion tons of oil equivalent, accounting for 22% of the world's total and an increase of 7.28% year on year. China consumed most energy in the world for 4 consecutive years. Energy consumption per capita is also rising as shown in Fig. 5.13. The figure in 2013 climbed to 2.76 tons of standard coals from 0.5 ton of standard coals in 1978, an increase of 5.3-fold, slightly higher than the world average of 2.53 ton. During the same period, energy consumption per capita of the U.S. and Canada was both above 10 tons of standard coals. As people's living standard improves, energy consumption per capita is expected to continuously rise. In 50 years, China may face shortages of almost all minerals except for coal, and 50% of the resources are expected to be depleted.

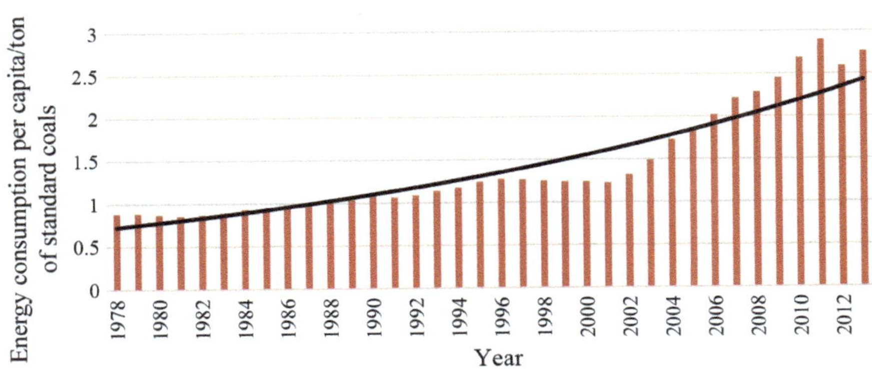


**Fig. 5.13** China's energy consumption per capita (*Source* Wind Analytics)

The consumption of nonferrous metals is climbing up. In 2006, the consumption of six nonferrous metals, refined copper, zinc ingots, refined lead, refined nickel, refined tin, and refined cadmium accounted for 27% of the world's total consumption. In 2012, the figure rose to 44%. The consumption of refined nickel and refined tin was close to half of the world's total as shown in Fig. 5.14.

In terms of water resources, industrial water consumption is growing at a faster rate, and the total water consumption continues to rise. According to *China Water Resources Bulletin*, with China's industrial development, the industrial water consumption has grown rapidly. After 2000, the annual growth rate has been close to 3%. The amount of industrial water consumption increased to 145.96 billion $m^3$ from 112.12 billion $m^3$ in 1997, accounting for 23.9% of the national total water consumption as shown in Fig. 5.15.

iv. Low utilization of resources

Low utilization of resources is the hallmark of China's extensive development path. Take the exploitation of minerals as an example. China has favorable ore-forming geological conditions, but minerals are not fully explored, with only less than one-third of the overall resources being exploited. Meanwhile, the utilization of minerals is extensive. In some places, it is an endemic problem that preference is given to rich and large mines in exploitation. In 2005, the overall coefficient of recovery in mining and the comprehensive utilization rate of associated mineral resources were 30% and 35%, respectively, 20% points lower than the world leading levels. Nearly 50% of large- and medium-sized mines didn't comprehensively utilize the minerals. As the exploitation of mines becomes more and more difficult, the challenge of environmental protection and restoration in mining areas is becoming ever increasingly daunting. If China failed to change the exploitation mode and the utilization

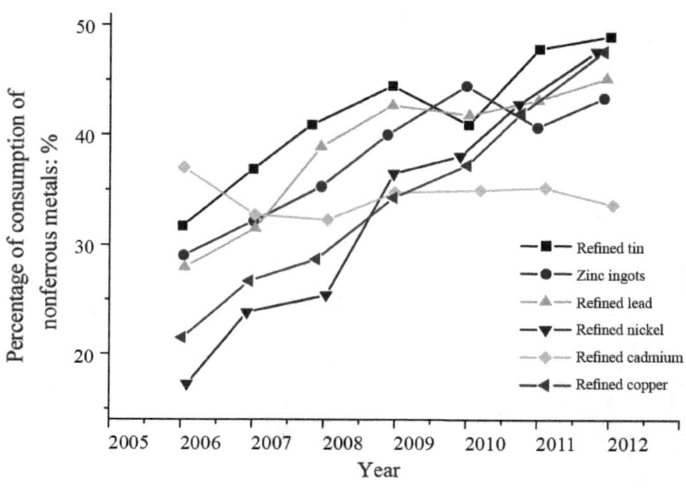


**Fig. 5.14** The proportion of China's consumption of main nonferrous metals to the world's total in 2006–2012

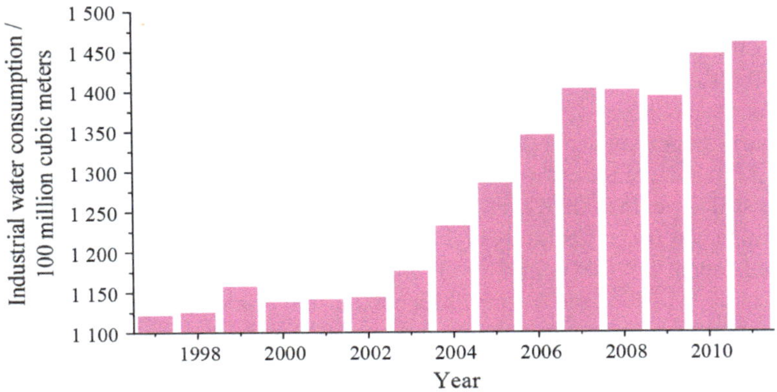


**Fig. 5.15** China's industrial water consumption since 1997 (*Source* China Water Resource Bulletin 1997–2011)

of minerals continued to be low, the destruction and pollution on the environment would exceed the environmental carrying capacity, and the economic growth would not sustain.

China's energy consumption per unit of GDP shows an evident downward trend. The figure fell from 13.51 tons of standard coals per RMB 10,000 in 1978 to 0.714 tons of standard coals per RMB 10,000 in 2012, a decline of one-third in real terms as shown in Fig. 5.16. But the energy consumption per unit of GDP is still high, showing China's utilization of energies remains low. In 2012, the figure was 2.5-fold of the world's average, 3.3-fold of that of the U.S., sevenfold of that of Japan, and higher than that of developing economies like Brazil and Mexico. By consuming

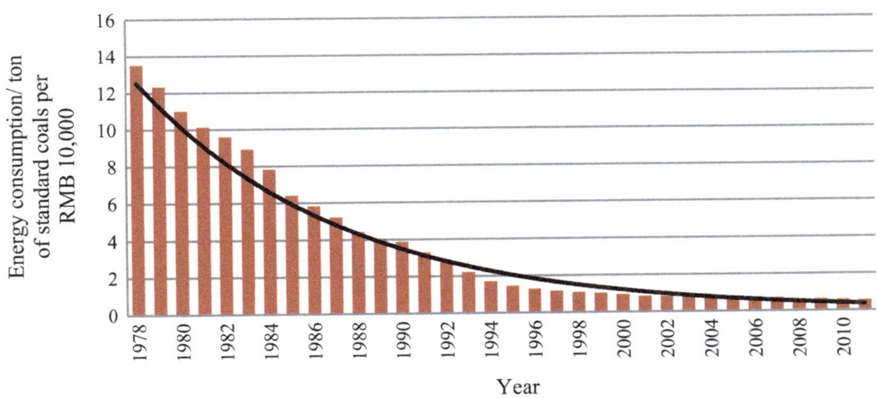


**Fig. 5.16** China's energy consumption per unit of GDP

every 1 ton of standard coals, China is only able to create RMB 14,000 of GDP, whereas the world's average is RMB 25,000, the U.S. is RMB 31,000, and Japan is RMB 50,000.

As for the efficiency of industrial water consumption, the amount per RMB 10,000 of industrial added value declined from 340.6 $m^3$ in 1997 to 77.4 $m^3$ in 2011 as shown in Fig. 5.17, a decrease of two-thirds in real terms. But the gap with world leading levels is still large. In 2009, China's water consumption per USD 10,000 of industrial added value was 603 $m^3$, slightly higher than the world's average of 569 $m^3$, 6.9-fold of that of Japan, 1.8-fold of that of Germany, and higher than that of countries like Brazil, India, Mexico, and Turkey. Meanwhile, the efficiencies of industrial water consumption vary greatly in different regions and provinces across China as shown in Fig. 5.18, which makes it difficult to introduce a unified water conservancy policy. In 2005, consuming every ton of water generated RMB 209.6 of industrial output in North China, more than fourfold of that in southwest regions where the efficiency of

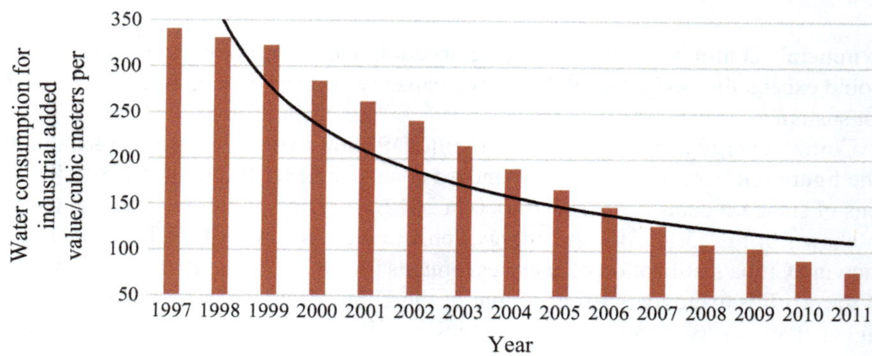


**Fig. 5.17** China's water consumption per RMB 10,000 of industrial added value

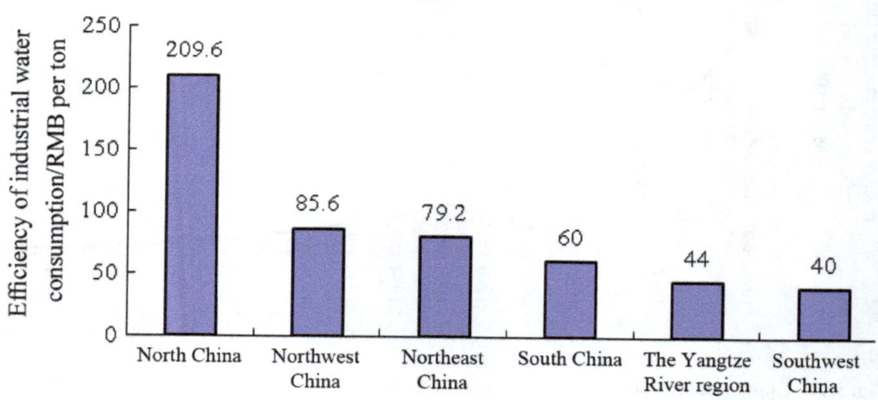


**Fig. 5.18** Efficiency of industrial water consumption in different regions in China in 2005

water consumption was lower. The efficiency varies in different provinces. In 2005, consuming every ton of water generated RMB 251 of industrial output in Beijing, 5.6-fold of that in Sichuan Province in the same period.

v. Unreasonable energy consumption structure

Compared with developed countries, China's consumption of fossil energies is higher and consumption of nonfossil energies rises at a lower rate as shown in Fig. 5.19. Specifically, the consumption of fossil energies with high-carbon emissions has been dominating China's energy consumption structure. The proportion slightly declined from 98.2% in 1978 to 90.6% in 2012 as shown in Fig. 5.20. Among it, the proportion of the most carbon-rich coal consumption decreased from 94.3% in 1978 to 69.5% in 2013. Despite a sharp decline, it remains a dominating position in the energy consumption structure and is far higher than the world historical record of 30.1% in 2013 since 1970 as shown in Fig. 5.21.

In terms of water resources, the use of nontraditional water resources remains low, and the future industrial water consumption faces great pressure. According to the 10th *Five-Year Plan for Industrial Water Conservancy* introduced by former China State Economic and Trade Commission, the amount of seawater and brackish water in industrial water consumption was 25.6 billion $m^3$ in 2001, 21.3% of that in Japan, and 12.8% of that in the U.S. The consumption of recycled water for industrial purposes was only 0.4% of all water withdrawals. As industrialization and urbanization advance, China's demands for water resources will be on the rise for a certain long period of time. Due to global warming, the contradiction between water supply and demand will be more striking.

vi. High degree of external dependence

The rapid consumption of energy for industrial purposes has gradually broken the balance of energy production and supply. As a result, China has imported more and more energies. Statistics by the World Bank show that from 1978 to the end of the twentieth century, China had produced enough energy to basically meet the demands. Since the beginning of the twenty-first century, energy consumption has outpaced the production. From 2005 to 2010, the average annual growth rate of energy consumption was 7.5%, 0.7% point higher than the growth rate of energy production during the same period. The external dependency of China's energy consumption is on the rise, and the trade surplus is shrinking as shown in Figs. 5.22 and 5.23. Despite the fast-growing economy, the proportion of energy import in GDP in 2013 was almost three times that in 2003. In 2013, 65% of petroleum consumptions were imported and 30% of natural gas consumptions were also imported, both hitting a record high. At the same period, the U.S. energy import still accounted for half of the country's trade deficit, but the deficit was shrinking rapidly as its imports of petroleum and natural gas reduced. The surplus of Russia's fossil fuels further expanded, making it the country with the largest energy surplus to date.

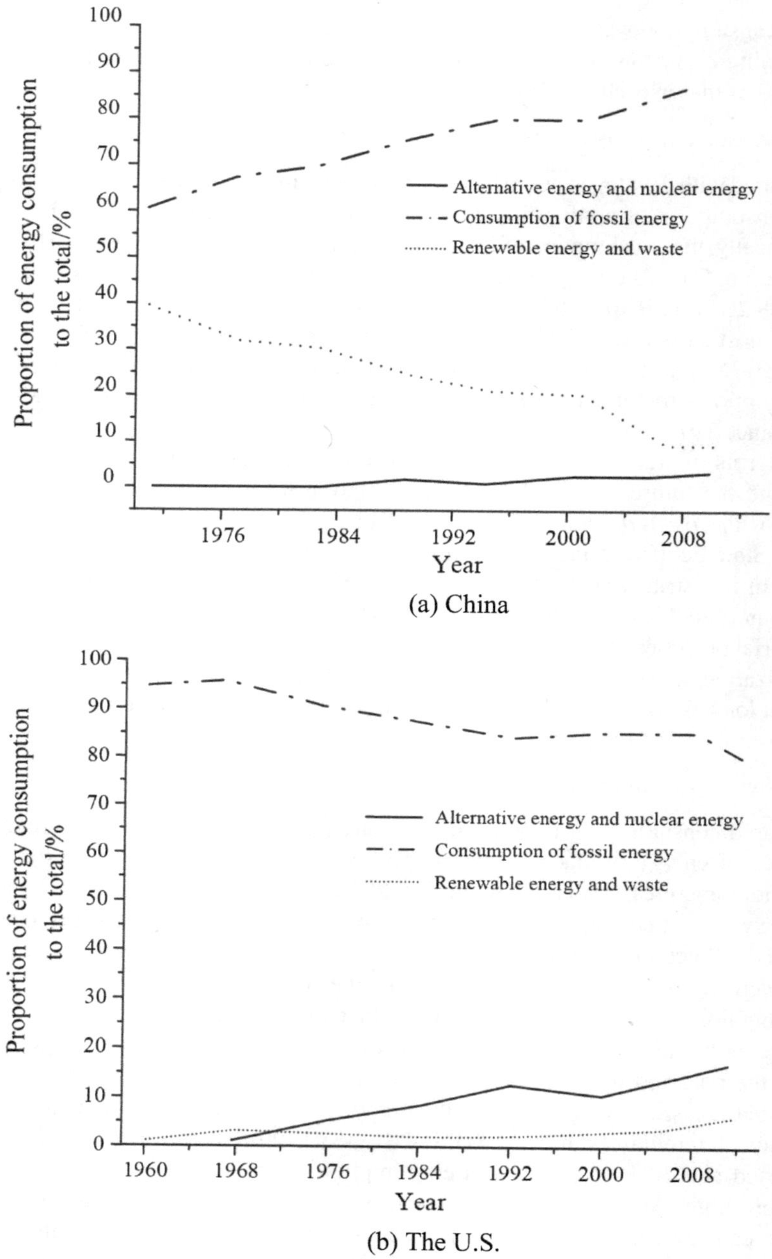


**Fig. 5.19** Evolution of the proportion of energy consumption to the total in China, the U.S., Japan, and South Korea

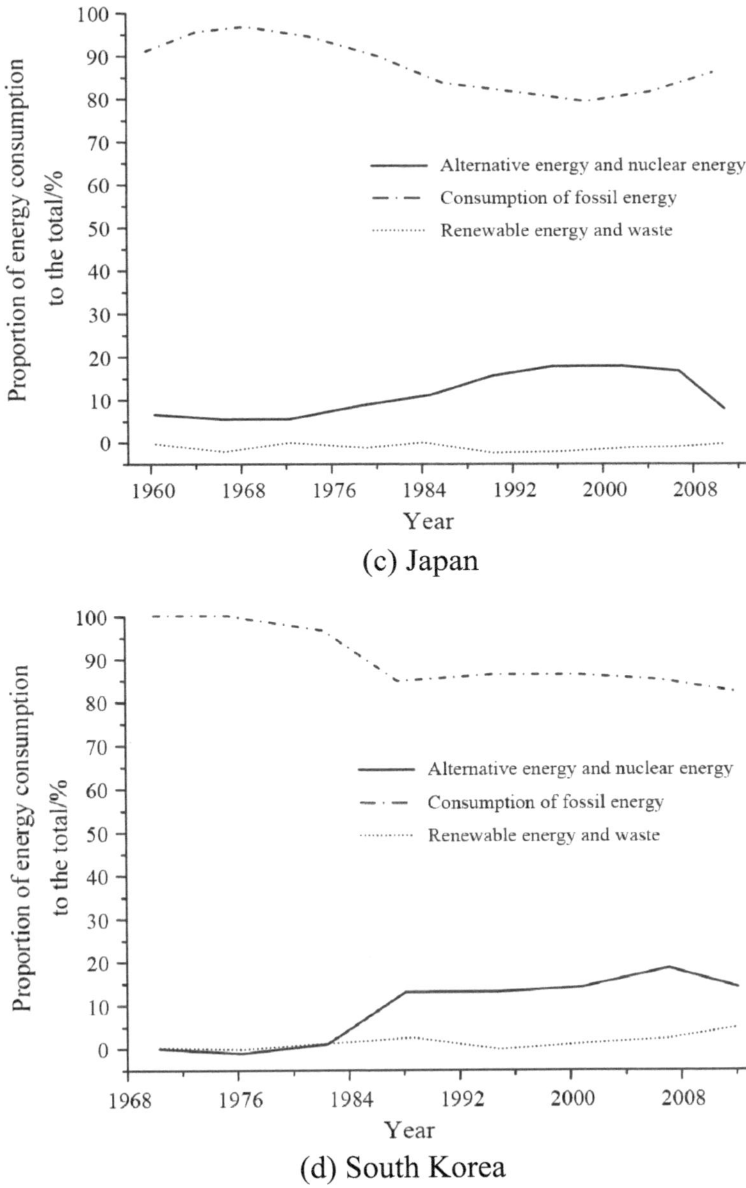


**Fig. 5.19** (continued)

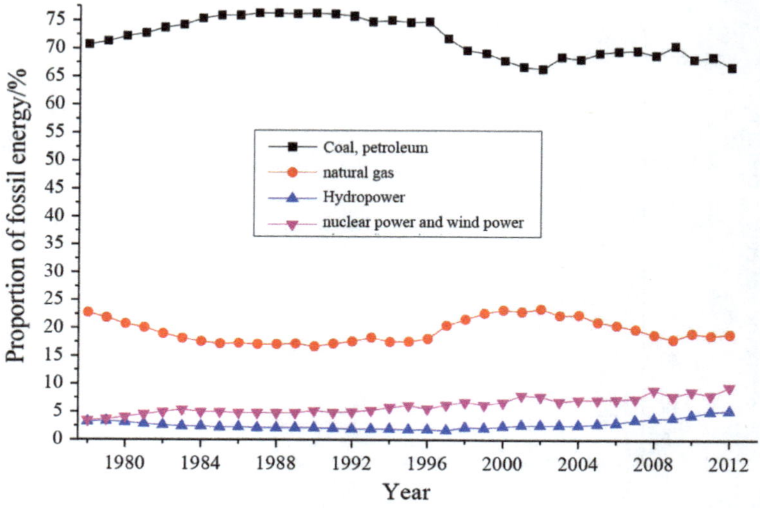


**Fig. 5.20** China's energy consumption structure (*Source* China Compendium of Statistics 1949–2008)

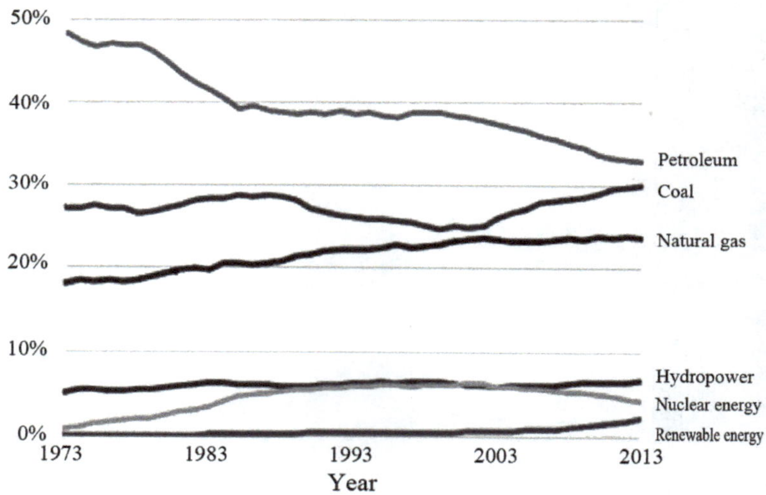


**Fig. 5.21** The proportion of primary energies worldwide. (*Source* BP Statistical Review of World Energy 2014)

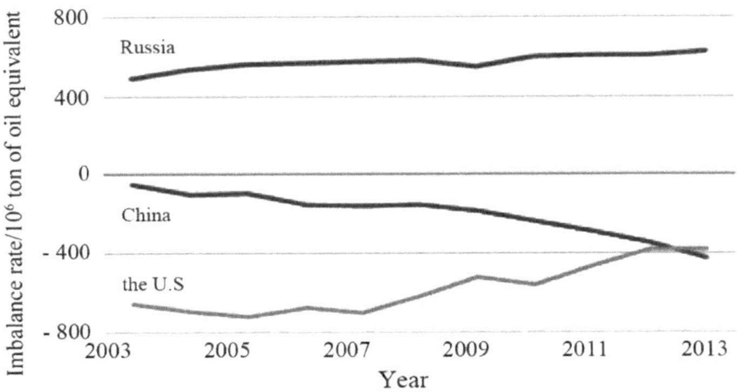


**Fig. 5.22** Supply–demand imbalance of fossil energies in China, the U.S., and Russia (*Source* BP Statistical Review of World Energy 2014)

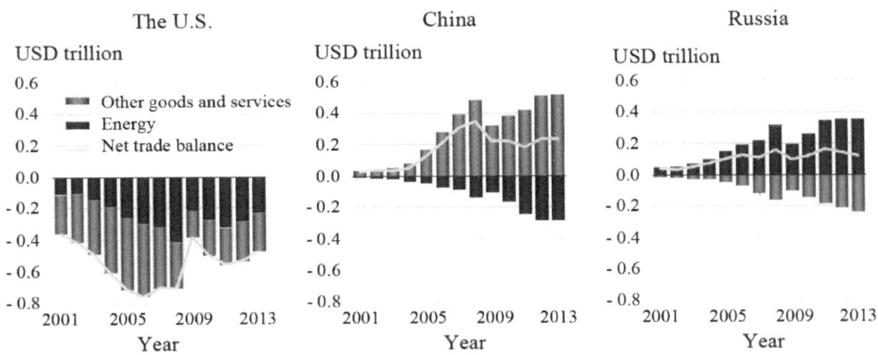


**Fig. 5.23** Trade balance and energy supply and demand in China, the U.S., and Russia

In terms of metal products, domestic self-sufficiency is gradually weakening. In 2012, the gap of zinc supply reached 110,000 tons as shown in Table 5.7. Other major metal products face the same shortage. By 2020, 25 minerals out of China's 45 main minerals will run short to varying degrees, 11 types of which are minerals of backbone significance to China's national economy.

**Table 5.7** Supply–demand balance of China's main metal products in 2006–2012

| Year | Lead: supply–demand balance/10,000 tons | Tin: supply–demand balance/10,000 tons | Copper: supply–demand balance/10,000 tons | Zinc: supply–demand balance/10,000 tons | Nickel: supply–demand balance/10,000 tons |
|---|---|---|---|---|---|
| 2006 | −8.00 | 376.00 | −21.70 | −3.00 | 0.60 |
| 2007 | 5.00 | 1111.00 | 30.30 | 0.00 | 4.20 |
| 2008 | 25.00 | 8600.00 | 3.80 | 25.00 | 5.00 |
| 2009 | 35.00 | 16,300.00 | 113.50 | 88.00 | 13.70 |
| 2010 | 28.00 | 1500.00 | 65.60 | 34.00 | 1.10 |
| 2011 | 25.00 | 7000.00 | 55.00 | 5.00 | 8.90 |
| 2012 | 20.00 | – | 80.00 | −11.00 | 5.30 |

### *5.1.3 Ecological Damages Further Reduce Environmental Capacity*

i. Resource loss and pollution exacerbate ecological pressure

1. Serious water and soil erosion

China's annual newly added soil erosion involving 11 provinces and autonomous regions measures more than 10,000 m$^2$, weighing more than 5 billion tons, among which about 2 billion tons flow into the sea. Regions with the most serious soil erosion include the Loess Plateau and the middle and upper reaches of the Yangtze River. Less-suffered regions include mountainous regions in north China (such as the Taihang Mount regions), the red soil hilly areas in south China, the black soil areas in northeast China, and the abutting mountainous areas where Sichuan, Yunnan, and Tibet meet. Soil erosion in arable land hit 600 million μ in area, 30% of the total. The annual lost soil weighs 1 billion tons, and the soil of 1.66 ton per μ in arable land is eroded, surpassing the ceiling as shown in Fig. 5.24. Soil erosion further causes the loss of topsoil, leading to land degradation and depletion of nutrients in the soil and no longer suitable for growing crops. The sands per cubic meter of the Yellow River weigh over 37 kg, the highest in rivers across the world. The figure in the Yangtze River is more than 1 kg, the fourth highest around the world.

2. Exacerbating water and land contamination

Under the influence of increasingly serious problems, such as atmospheric dust deposition, irrigating with sewage exceeding standard, applying sludge and garbage, dumping industrial residue, and excessive pesticides and fertilization, soil contamination is becoming an ever alarming issue and the soil quality is deteriorating continuously. The pollutions including heavy metal pollution, acid rain pollution, pesticide and organic pollution, radioactive pollution, pathogen contamination, and others are eroding the land resources every moment. Some pollutants have exceeded the standard by dozens of even hundreds of times. The area of polluted arable land has been

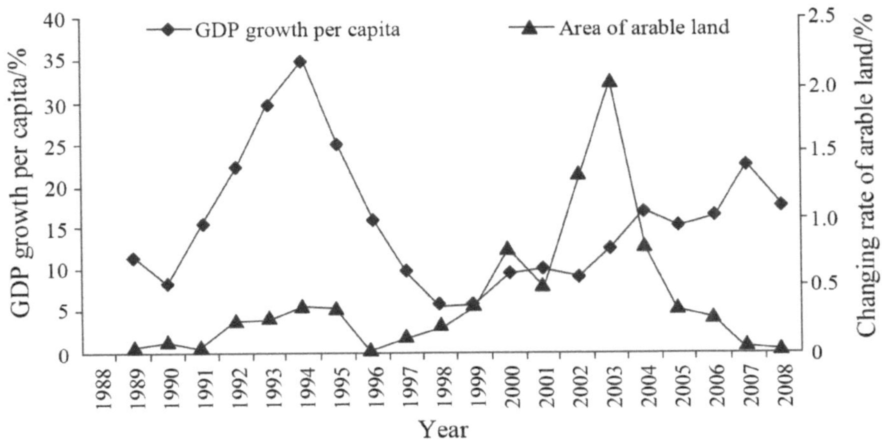


**Fig. 5.24** China's economic growth rate per capita and arable land changes in 1988–2008

on the rise since 1989 as shown in Fig. 5.25. In 2011, nearly 300 million mu out of 1.824 billion μ arable land nationwide were contaminated by heavy metals, 150 million μ of which were polluted by industrial wastewater, exhaust gases, and solid waste. Dumped solid wastes occupied and damaged some 2 million μ farmlands. More than 80 million μ of farmlands were contaminated by air pollution. The area of arable land irrigated by wastewater accounted for 7.3% of national irrigation area. Farmland contaminated by pesticides measures 140 million μ. At present, the use of fertilizers and pesticides per hectare in China outnumbers that in the developed

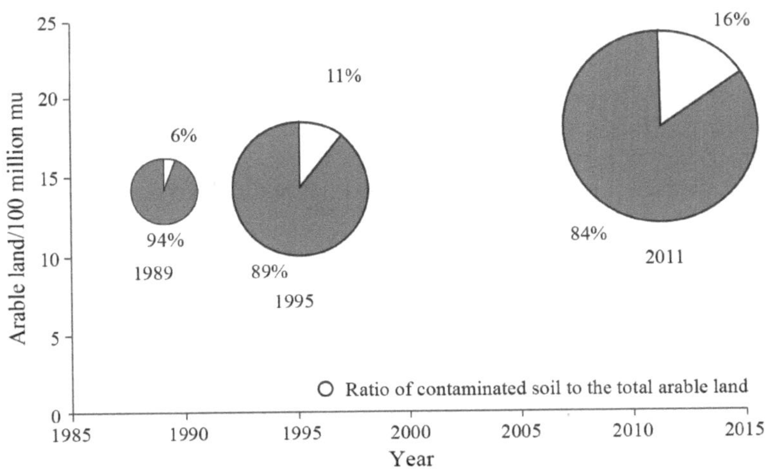


**Fig. 5.25** Ratios of contaminated soil to the total arable land area in 1989, 1995, and 2011 (*Source* Research report on the basic frame of thinking for the 12th Five-Year Plan for national environmental protection, Bulletin of China's Environment State)

countries by onefold, and the use is still on the rise. Many high-yield regions are also where fertilizers and pesticides are heavily used. The use of fertilizers and pesticides is the highest in east China, which is followed by central China and then west China. Southeast coastal areas with well-developed economy use especially more fertilizers and pesticides.

Water environment pollution has spread from the land to offshore waters, from surface water to groundwater, from single pollution to composite pollution, and from chemical pollution to water ecological degradation. Water environment faces dual threats of water quality deterioration and water ecosystem destruction, which is endangering the water environment safety and sustainable development of economy and society of the drainage basins.

China has established a monitoring system of water environment in 1984. Overall, the quality of surface water has been deteriorating over the last more than three decades. The discharge of major water pollutants has exceeded the environmental capacity. The structural, composite, and regional problems with pollution play out in concentrated form. In the 1980s, the water quality of China's major rivers was basically good, with only a few sections being heavily polluted. In 1978, there were only 5% of eutrophic lakes nationwide, and most of water had good water quality. Large-scale red tide happened only seven times in 1989. In 2012, sections with water quality of IV–V grade or inferior V grade accounted for 31.1%. The drainage basin of Haihe River suffered from heavy pollution, the basins of Yellow River, Songhuajiang River, Huaihe River, and Liaohe River were slightly polluted, the Yangtze River and rivers in Zhejiang Province and Fujian Province were generally in good conditions, and rivers in northwest and southwest and the drainage basin of Pearl River were in good conditions. Generally speaking, water quality in south China outperformed that in the north, and west China was better than the east. In 2012, lakes and reservoirs with water quality of IV–V grade and inferior V grade accounted for 38.7%, and eutrophic lakes took up 25%. Taihu Lake and Chaohu Lake were slightly polluted, whereas Dianchi Lake was heavily polluted. Seriously eutrophic lakes concentrated in urban areas, plains in east China, middle and lower reaches of the Yangtze River, and Yunnan–Guizhou Plateau. In the recent years, the marine environment is generally in good conditions, with some offshore water pollution and ecological damage remaining prominent and the polluted area of offshore waters being continuously expanded.

Besides conventional pollutants, drinking water is being contaminated by new types of toxic substances such as trace organics, algae, and algal toxins, and odorant components among others. About 360 million rural residents have no access to standard drinking water. Nearly 20% of concentrated groundwater sources have water quality inferior to III grade, meaning they are unsuitable for drinking. Except for conventional chemical indicators, carcinogenic, teratogenic, and mutagenic pollutants such as chloroform, toluene, tetrachloroethylene, benzopyrene, chlorobenzene, and benzene have been detected in drinking water sources of some urban areas. In recent years, a lot of "cancer villages" caused by water pollution have emerged. According to incomplete statistics, there have been more than 247 "cancer villages," whose number decreases from the east to the west in a gradient pattern, which is in line with the distribution of water resources and population and degrees of economic development.

3. Intensifying contradiction between population and land

The relations between population and land are reflected not only in the diminishing arable land per capita but also in land environment pollution and continuously low grain yield per capita. Under the pressure of increasing demands on agricultural products due to expanding population, farmers have to stress out the arable land, causing pollution and degradation. At present, the main solution to increase grain output is to apply more fertilizers and pesticides, which destructs the soil structure and emaciates and hardens the soil. The diminishing and degrading of arable land have become an unfavorable factor in China's agricultural production and economic development. In addition, with rapidly growing population, China's grassland experiences overgrazing and over-farming, directly resulting in desertification. Taking the typical survey area in Yulin, Shaanxi Province as an example, the total desertification area in the late 1980s was 11,042.98 km$^2$, 76.2% of the region. In the late 1990s, the area shrank to 9849.15 km$^2$, 67.9% of the region, a decrease of 1193.83 km$^2$. Among them, the area of severely and moderately desertification land reduced by 975.51 m$^2$, accounting for 81.72% of the total figure. Chinese experts on desertification warn that if no measures were taken, the desertification area by the end of the twenty-first century would have reached 80,000 km$^2$.

4. Declining utilization of mining and consequent pollution on the environment

The utilization of minerals is extensive. In some places, it is an endemic problem that preference is given to rich and large mines in exploitation. The overall coefficient of recovery in mining and the comprehensive utilization rate of associated mineral resources were 30% and 35%, respectively, 20% points lower than the world leading levels. Nearly 50% of large- and medium-sized mines didn't comprehensively utilize the minerals. As the exploitation of mines becomes more and more difficult, the challenge of environmental protection and restoration in mining areas is becoming ever increasingly daunting. If China failed to change the exploitation mode and the utilization of minerals continued to be low, the destruction and pollution on the environment would exceed the environmental carrying capacity, and the economic growth would not sustain.

5. Declining biodiversity under the influence of human activities

Although China has been making great efforts to protect forest resources by planting trees, due to historical reasons and natural conditions, its forest ecology is still fragile, and the imbalance of demand and supply is prominent. Expanding population and urbanization have generated new demands on forests: reclaiming construction land from forests, and more timber products. Statistics show that 61 out of 140 forest services nationwide witness forests being overcut under their jurisdiction, and 25 ones see the forest resources have been basically exhausted. At present, timber, fuelwood, pulp, and other forest products are all in acutely short supply. Meanwhile, increasing demands for food and arable land because of expanding population exacerbate deforestation.

Changes in urban environment as a result of human activities are reshaping urban vegetation in terms of input and output of matters, as well as conversion and utilization of energies. The changing microclimate is causing continuous degradation of its ecological functions. A study under China's state key basic research projects led by academician Zhang Xinshi showed that serious overgrazing on grassland resulted in a decline of 47% in the theoretical carrying capacity of grassland in the last 50 years, an annual decrease of 1%. The annual mortality of livestock is as high as 7%, and livestock loses one-third of their fat in winter and spring. The number of livestock increased from 9.686 million in 1949 to 51.769 million in 2002 by 430%. Those livestock exhale more carbon dioxide and cause desertification. 1.67 million $hm^2$ of grassland degraded every year. Academician Zhang noted that instead of using as the grazing land for thousands of years, the ecological functions of the natural grassland should be transformed into windbreak and holding sand, maintaining water and carbon and raising wild ungulate herbivores and reserving xerophyte.

Economic development leads to the evolution of urban landscapes, and further a sharp decline of biodiversity in the urban and surrounding areas as shown in Table 5.8. Urban plant diversity is supposed to increase from downtown to suburbs in a gradient pattern, and so is the wild and native plant diversity. Urbanization spurs the improvement of infrastructure. Vegetation coverage in the urban areas decreases as the green space has been occupied by buildings and roads. In downtown areas, one single species is planted for aesthetics, resulting in the unification of vegetation in downtown areas. Urban buildings destroy the natural landscape. As a result, there are less and less wild and native plants in central areas of the city, while there are more in suburb areas. According to a news report in 2005, there were less than 10 species of natural plans in the central areas of Beijing where buildings are densely distributed, the species in Zizhuyuan Park between the second and third ring road exceeded 50, those in the Old Summer Palace between the fourth and fifth ring roads were 287, and those in Beijing Cherry Valley beyond the fifth ring road were 433. The same trend was found in studies on Wisconsin in the U.S. animal diversity increases significantly from urban areas to suburb areas in a gradient pattern. Studies on birds in Beijing in 2005 showed that sparrows were the dominant species in downtown areas, where few birds of other species were spotted. However, there were 159 species of birds in the Old Summer Palace, accounting for more than 45% of birds in Beijing. Only sparrows were spotted in central Shanghai. Some seagulls can be seen on the Huangpu River in some seasons. The bird species in suburb Jinshan district reached as more as some 350. Due to contamination of water sources and food as a result of urban management, the number of carnivorous birds and beasts in urban areas has dropped sharply. Typical animals in urban Beijing include rats and bats, while there were 12 wild mammals in suburb Old Summer Palace and the Summer Palace, and as many as 18 species in Xiangshan Park outside the fifth ring road.

**Table 5.8** Some of the extinct animals

| | | |
|---|---|---|
|  |  |  |
| Yunnan box turtle (the species was listed as extremely endangered animal in 2009 and may have been extinct) | Small-toothed palm civet (extinct in the 1980s) | Spotted deer (extinct in the 1970s) |
|  |  |  |
| Clouded leopard (the last one was slaughtered by poachers in 1972) | Spruce grouse (extinct in 2000) | Crested shelduck (extinct in the middle of the twentieth century) |
|  |  |  |
| Yunnan lake newt (wild ones being extinct in 1996) | Painted stork (extinct in the 1950s) | White-flag dolphin (extinct in 2006) |

ii. More fragile ecology and more frequent meteorological disasters

1. Global warming is a global issue and seriously affects China's ecological environment
   See Chap. 2 on this topic.
2. Prominent urban heat island (UHI) effects accompanied by many ecological issues

UHI is an urban area or metropolitan area that is significantly warmer than its surrounding rural areas due to human activities. It is one of the most significant features of urban climate. UHI effect changes urban thermal environment, affecting regional microclimate, urban hydrology, air quality, the physical and chemical properties of urban soil, the urban biological distribution and behavior, and many other urban ecological processes such as metabolism and energy cycle. These further trigger a series of ecological environmental issues and threaten rural residents' health. Causes for intensifying UHI effect include the following:

(1) Changing underlying surface in urban areas. Larger impervious surface area and higher concentration of roads and buildings lead to larger heat capacity and thermal conductivity, low reflectivity, high absorptivity, and smaller aerial field angle, which result in an increase in temperature or surface temperature. Vegetation and water help significantly reduce urban temperature and surface temperature, but in a less effective way in suburban and rural areas. The increasing rate of pavement and the diminishing vegetation and water reduce the overall water retention and evapotranspiration in urban areas. As a result, there are more latent heats and the temperature and surface temperature climb up.
(2) High concentration of buildings and crisscrossing roads and bridges form the underlying surface of the city, which blocks wind, reduces wind speed, and prevents heats from dissipating.
(3) The larger the urban population and scale are, the more prominent UHI effect is as waste heat from automobiles, industry, air conditioning, and other sources also contributes to the UHI.
(4) High levels of pollution in urban areas worsen air quality and increase the concentration of soot, sulfur dioxide, nitrogen oxides, and carbon monoxide, which absorb infrared radiation and raise the temperature of cities.

As cities sprawl in an orderly manner, the UHI phenomenon is becoming more and more prominent. The maximum UHI value is 9 °C in Beijing, 7.2 °C in Guangzhou, and 6.9 °C in Shanghai. The UHI effects lead to the annual average temperature of 1 °C higher in urban areas than in suburban regions. In the context of global warming, it is much more intense in urban areas than in suburban areas when it gets warmer. According to statistics from Beijing Observatory, the average temperature in winter and summer over the past 60 years had increased by 0.4–0.3 °C every 10 years, whereas the figure in suburb Miyun was 0.2–0.1 °C as shown in Fig. 5.26. The UHI affects urban ecology in many ways. High intensity of UHI effects brings about hot weather and various abnormal meteorological phenomena, such as warm winter, hurricane, and storm. It also affects the urban climate, industry, and people's daily life (Fig. 5.27).

3. Uneven precipitation and frequent rainstorm disaster in urban areas

On the one hand, many cities face a severe shortage of water resources; on the other hand, they have suffered from rainstorm and flood disasters in recent years. Under the influence of climate changes in the last 60 years, the days of light rain in China decreased by 10%, whereas rainstorm days increased by 13%. Changes in

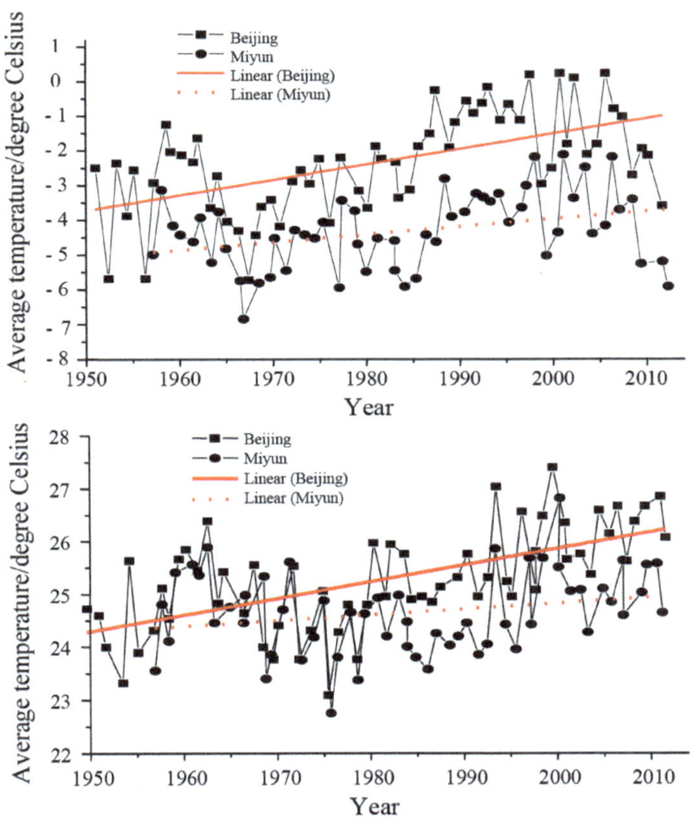


**Fig. 5.26** Temperature changes of Beijing Observatory and Miyun in winter (upper) and summer (lower) in 1951–2012

precipitation vary in different regions. For example, the annual rainstorm days in Wuhan increased by 0.4 days every 10 years, whereas the figure is 0.08 in Hubei as shown in Fig. 5.28. Pavement in the process of urbanization weakens the soil seepage capacity and makes it easier for water to be confluent and evolve into flood disaster. In recent years, Chinese cities have been frequently suffered from such disasters, such as the urban area being flooded by rainstorm in Wuhan, the rainstorms on June 23, 2011 and July 21, 2012 in Beijing, and the rainfall disaster on August 25, 2008 in Shanghai. They all have far-reaching effects.

In 2009, Guangzhou earmarked RMB 900 million to renovate streets that were vulnerable to floods. In 2010, the city was hit by a rare rainstorm in history. Although the conditions of some vulnerable sites were improved, some new ones were exposed. Particularly, many underground buildings were flooded, turning parking space to a "reservoir", and newly built residential communities to pools. Local subway was forced out of service, the urban transport was paralyzed, and some places were

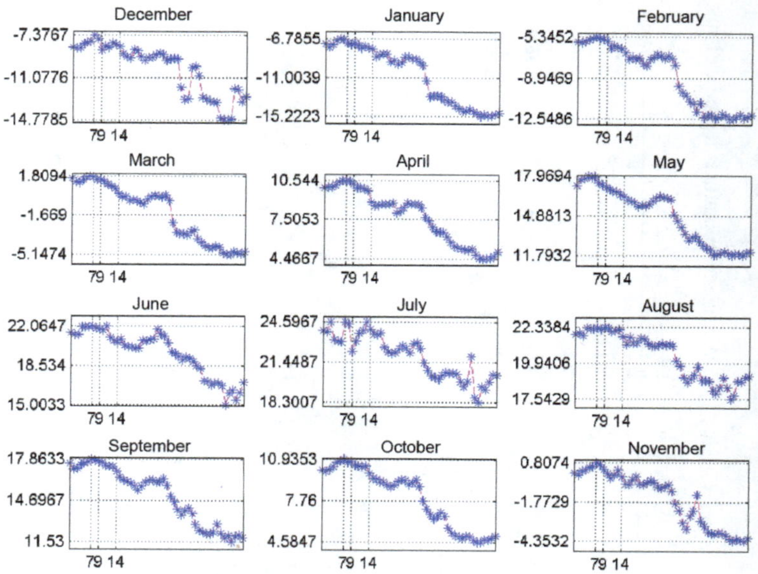


**Fig. 5.27** Hatching lines of the surface temperature in the east of Tiananmen Square in 2012 (7, 9, and 14 represent the third, fourth, and fifth ring roads of Beijing)

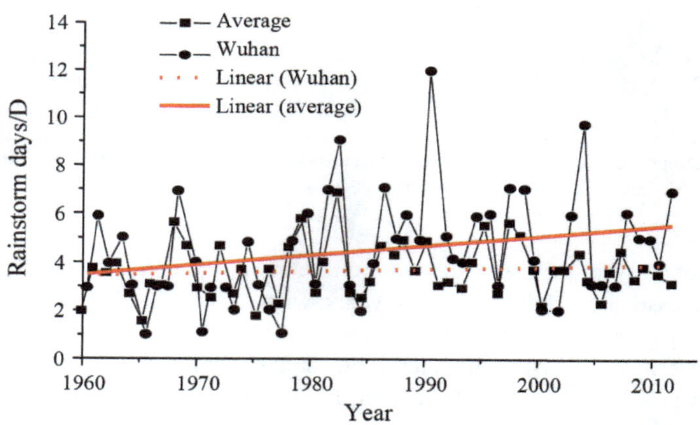


**Fig. 5.28** Comparison of rainstorm days in Wuhan and Hubei Provinces

blacked out, and water supplies were cut off. It can be said that the recent frequent rainstorms and floods highlight the poor urban planning in natural disaster response. In megacities, the UHI effects will trigger urban rain island effects. The average temperature in megacities is high, and the dust content in the air is large. When the warm air rises, surrounding airflow gathers in the urban center. Once the rising warm airflow encounters cold-air mass in high altitude, rainstorm will be formed, making the urban areas vulnerable to rainstorm disasters. This phenomenon is referred to as urban rain island effect. Therefore, the frequency and intensity of rainstorm in megacities are higher than surrounding suburban regions. This phenomenon will exist for a long period of time.

iii. Emerging complex environmental issues

Deepening economic reform and accelerating urbanization boost industrial development and agricultural modernization. The proportion of agriculture in GDP is becoming smaller and smaller, whereas the proportion of industries is becoming higher and higher. Big cities are facing more and more problems like population expansion, traffic congestion, housing shortage, high unemployment, and deteriorating air quality. Besides, township and village enterprises operate in small-scale and in an extensive manner, and cause heavy pollution to rural areas and the abutting areas of urban and rural regions. Air and water pollution and rampant discharge of garbage are endemic in many small towns, where the ecology is deteriorating rapidly. Agricultural pollution in the abutting areas of urban and rural regions is heavy. The chemical oxygen demand of industrial wastewater and the discharges of dust and solid waste from rural and township enterprises account for some 50% of the total discharge of industrial pollutants nationwide. As the agriculture develops, more pollutants are generated, like contamination to agricultural products by fertilizers and pesticides and the plastic pollution arising from agricultural production, the domestic sewage, garbage pollution, air pollution brought by burning straws, and contamination arising from large-scale farming and aquaculture. In particular, pollution resulted from replacing chemic fertilizers with organic fertilizers is becoming more and more serious.

1. Rampant garbage dumping

Solid waste has become one of the major sources of pollutants. The huge quantity and a wide variety of solid wastes made up of complicated components endanger the safety of cities (Fig. 5.29). At present, many large- and medium-sized cities are besieged by garbage. As a large number of urban industrial enterprises move to suburban areas, various solid pollutants have been left in the soil and endanger people's health. Hazardous wastes generated from industry and people's daily life haven't been treated appropriately. Medical wastes have been mingled with domestic garbage and even been recycled illegally. Illegal activities like dismantling and processing of waste materials, incineration, pickling, and soil smelting are rampant in many places, emaciating arable land, and heavily polluting the drinking water and air.

**Fig. 5.29** Solid waste pollution

2. Compound air pollution

Since 2012, China's central and eastern regions have been repeatedly blanketed in smog, and air pollution has become a serious issue, which affects industry, transport, and public health. The frequency and intensity in Beijing–Tianjin–Hebei region, the Yangtze River Delta region, and the Pearl River Delta region have been particularly high. The three regions occupy about 8% of China's land area, but consume 42% of the national total coals and 52% of gasoline, and produce 55% of China's total steel products and 40% of cement. The emissions of sulfur dioxide, nitrogen oxides, and soot, respectively, take up 30% of the national total. The emission per km is fivefold of that of other regions. The pollutions cause high PM 2.5 readings and the happening of smoggy weather. Therefore, the key regions to curb air pollution are those where economic activities and pollutants' emissions concentrate.

The regional air quality in China is deteriorating. While conventional pollution mainly caused by soot and coal combustion has not been controlled, regional compound air pollution mainly contributed by ozone, particulate matter 2.5, and acid rain becomes rampant. With the multiple-source coexisting pollutants combine and affect the environment with multiple agents. Nowadays, urban agglomeration has become the main spatial form of China's regional development. Dense population and rapid sprawling industries and other reasons lead to the diminishing distance for cushioning pollution. As a result, emissions of major pollutants rise sharply. Due to the dual influences of atmospheric circulation and atmospheric chemistry, pollutants cross borders of regions and spread to neighboring areas. Local pollution now turns into a regional issue (See Attachment VII).

Small climate capacity and fragile ecology is a basic national condition in China. Fast-growing economy driven by high carbon emission poses a daunting challenge to resources and environment. The environmental issues under such conditions are no longer purely about discharge of pollutants and destruction on ecology, but become compound under the interaction of various factors such as climate changes, resource exhaustion, and pollution migration. Without full knowledge of the mechanism of the occurrence, the environment and ecology cannot be restored overnight. With the emerging of new environmental issues and the increasing difficulty in restoration, it becomes more important to control pollution from the source. Therefore, economic growth model characterized by low carbon emission should gradually replace the high-carbon-emission model.

## 5.2 Significance and Meaning of Transforming Economic Growth Model

Development has been an eternal theme for human society. The history of social and economic development is the course to explore development path and to upgrade growth model. After the Second World War, the understanding of different countries on development has been evolving and upgrading. The focus of mainstream economic theories and practices in the West has evolved from economic growth to coordinated social and economic development and sustainable development. The purposes for economic growth gradually include noneconomic aspects, and the pursuit of growth rate has transited to efforts in attaining coordinated development of economy, society, environment, and the people.

### *5.2.1 Meaning of Transforming Economic Growth Model*

The core of new economic growth model remains as economic growth. But restrictions are added in terms of ecological carrying capacity. Attention was given to efficiency and sustainability, balance and coordination, and inclusiveness. The economic transformation is a process of reform and innovation in which following the underlying laws and principles, various economic driving forces will play out in a hurdle-free environment to unleash efficiency within the limits of ecological carrying capacity. Green and low-carbon industries and services will be the dominating driving forces for economy. Urbanization will be efficient, inclusive, and sustainable. Market will play a decisive role. The government will be law-based, service-oriented, and highly effective. Technological innovations will provide support. Cultural innovation will provide motivations for the public participation. The essential thing is developing the economy within the carrying capacity of the ecology, resources, and environment and ensuring ecological advancement.

i. The target and meaning of transforming economic growth model

Economic transformation has multiple targets ranging from pure economic growth to the overall progress of the society as shown in Table 5.9. Besides the model for economic growth, it includes constant improvements in industrial structure, income distribution, people's livelihood, urban–rural structure, regional development, resources utilization, and ecological improvement. The target is to realize overall, coordinated, efficient, inclusive, sustainable, and resilient economic development.

**Table 5.9** The comparison between the old and new development models

| Comparisons | Old model | New model |
|---|---|---|
| | Growth model | Development model |
| Contents | Growth in general sense (quantity) | Comprehensive, coordinated and sustainable development (quality) |
| Models | Extensive growth | Intensive growth |
| Features | Conventional | Modern |
| | Old | Innovative |
| | Unsustainable | Sustainable |

Compared with economic growth model, the economic development model realizes the evolution from conventional development to modern development, from rigid development to innovative development, from disharmony to harmony, and from unsustainable development to sustainable development. The previous conventional growth model is a set of approaches to economic growth: capital and labor-intensive, government-driven, foreign demand-driven, investment-driven, export-driven, and extensive growth. Modern and new development model is another set: technology-intensive, market-oriented, domestic demand-driven, consumption-driven, and intensive development.

1. The meaning of the new model

The new model of economic growth and social progress is the common goal. It is about improving people's life quality, fulfilling people's basic need, safeguarding the dignity of human beings, and enlarging their freedom to choose. To transform the economy, the extensive growth model should be upgraded into the intensive growth model, and the growth in general sense should be changed to comprehensive, coordinated, and sustainable development. Science and technology are the decisive forces to drive social and economic development.

2. Positioning of economic transformation: underlying guideline of the strategic choice

> In contemporary China, pursuing development in a scientific way best embodies the thinking that only development counts. Taking the pursuit of development in a scientific way as the underlying guideline and accelerating the change of the growth model as a major task is a strategic choice for promoting China's overall development.

3. Major direction of economic transformation: promoting strategic adjustment of economic structure

Carrying out strategic adjustment of the economic structure is the major goal of accelerating the change of the growth model. "We must strive to remove major structural barriers to sustained and sound economic development, with a focus on improving the demand mix and the industrial structure, promoting balanced development between regions and advancing urbanization."

To push forward strategic adjustment of the economic structure, we must focus on improving the quality of development on the basis of quantity growth and aim for coordinated economic growth and social progress. The key fields and directions for adjustment include the following: (1) Taking timely action to push forward reform on resource and factor process and ensuring the pricing mechanism to play its fundamental role in promoting the transformation of development model; (2) Strengthening social regulations on resource, environment, quality, and safety, and ensuring that the government plays its proper and effective role in promoting the transformation of development model; (3) Focusing on improving social security and the basic public service system so as to change the situation in which social development does not match the economic growth; (4) Deepening reform of the state-owned enterprises and monopoly industries and improving the surplus distribution mechanism in state-owned enterprises and monopoly industries; (5) Promoting technological development to upgrade the economic structure; and (6) Formulating and implementing reasonable policies for consumption to foster resource-saving and environmentally friendly consumption pattern.

4. Major targets of economic transformation: better structured, expanded in scope, and highly productive economy (comprehensive economic and social development)

"We should make China's open economy better structured, expand in scope and yield greater returns." "We should increase the vitality and competitiveness of China's economy." The target has been transferred from pure economic growth to the comprehensive economic and social development, creating a favorable environment for people's life and work.

5. The key to economic transformation: deepening reform and the rule of law

> Deepening reform is crucial for accelerating the change of the growth model. The underlying issue we face in economic structural reform is how to strike a balance between the role of the government and that of the market, and we should follow more closely the rules of the market and better play the role of the government.

Law-based governance is an important guarantee for deepening reform. Improving the legislative system and lawmaking in a scientific and democratic way, and strengthening lawmaking in key areas will provide institutional foundation for deepening reform. Impartial administration of justice and judicial credibility will be the guarantee for productive reform. By promoting law-based government administration and building a rule of law government, it will be conductive to ensuring the efficiency and legitimacy of deepening reform.

ii. Transforming economic growth model as the foundation of transforming development model

1. Changing the structure of factor input

The direction for changing the structure of factor input: instead of depending on increasing resource inputs, economic growth will be driven by technological advances and improving the quality of labor forces.

2. Changing demand mix

The key to changing demand mix lies in the internal structure of driving forces for economic growth, which are investment, consumption, and export. The driving forces should be changed from pure investment to investment and consumption, from productive investment to economic and social investment, and from foreign demands to foreign and domestic demands.

3. Changing industrial structure

The key and direction for changing industrial structure: the focus should be changed from the secondary industry to coordinated evolution of the primary and secondary industries; the focus of the secondary industry should be transferred from assembling products to manufacturing the key materials, core components, and other products with high added value and high-end equipment; the focus of the tertiary industry should be changed from life service industry to manufacturing service industry.

4. Changing the mechanism of economic driving forces

The central task in changing the mechanism of economic driving forces is shifting from input-driven growth to efficiency-driven and innovation-driven growth. Transforming economic structure requires following the underlying law for economic growth and development, making it clear that the development is government-led or promoted by government while giving full play to the role of market.

On the one hand, we should change the model of resource allocation, which is the primary content of economic development. The mix of factors of production and the supply and demand structure in production decided by the mix form the mechanism of driving forces for long-term economic development. The compensation mechanism for factors of production and demand mix closely relating to it form the mechanism of driving forces for short-term economic development.

On the other hand, we should change the driving forces for economic growth. The driving forces should be shifted from inputs to efficiency and innovation. To this end, the first is to improve administration and realize efficient growth. Measures include transforming the functions of the government to raise the efficiency of decision-making and administration and innovating business models to promote the efficient development of emerging industries. The second is to create a favorable business climate for innovation-driven development. Measures include theoretical innovation, as the basis and key for development and reform, institutional innovation as institutional basis and guarantee, technological innovation as intelligent support and driving force for enriching the material world, and cultural innovation, which helps to improve people's cognitive ability and provide them with motivation.

In addition, we should well understand the characteristics of the transition period for economic growth and respond to potential decline of growth rate with stable and quality economic development. Economic transformation will not be a plain sailing. There are inherent instability and uncertainty in transforming from old economic growth model to new development model. In face of the potential decline of growth rate, it is necessary to effectively maintain necessary and possible growth rate in the transition period. The expense of reform is relatively lower with moderate growth rate and conductive to tackling potential decline of economic growth rate while attaining the sustainable and coordinated economic, social, and ecological advancement.

5. Changing the emission structure

The first is to lower pollutant emission per unit of GDP. Measures include improving the utilization of energy and resources with technological advances, adjusting the energy structure by increasing the supply of clean and low-carbon energies, and reducing dependence on energy-intensive, polluting and high-emission industries, and intensifying governance and seeing businesses to bear the corresponding social cost of emission.

The second is to increase the carbon sinks. Efforts should be made to speed up afforestation and forest management, better protect the forest and wetland so as to enhance the capability of forests and wetland ecological system to mitigate and adapt to climate changes. In areas with suitable conditions, more trees should be planted for carbon sinks, and measures should be taken to enhance the carbon storage capacity of forest products. Forestry system for carbon sinks should be aligned in the national carbon emission trading system at a faster speed to boost the development of carbon sink forestry.

iii. Innovation-driven development and deepening reform are the driving forces for economic transformation

1. Innovation-driven development creates new sources of economic growth

    (1) Contents and positioning of innovation
    The contents of innovation include theoretical innovation as the foundation and key for development and reform, institutional innovation as institutional foundation and guarantee, technological innovation as intelligent support and technical guarantee, and cultural innovation as motivation.
    Technological and institutional innovation will push the boundaries of development and provide driving forces and clear barriers for economic transformation. Realizing innovation in low-carbon development and energy revolution are conductive to creating new growth areas.
    (2) Innovation-driven development strategy
    Innovation-driven development has been made a national strategy. The key points include "In response to changes in both domestic and international economic developments, we should speed up the creation of a new growth model and ensure that development is based on improved quality and performance. We should fire all types of market participants with new vigor for development, increase motivation for pursuing innovation-driven development, establish a new system for developing modern industries, and create new favorable conditions for developing the open economy. This will make economic development driven more by domestic demand, especially consumer demand, by a modern service industry and strategic emerging industries, by scientific and technological progress, by a workforce of higher

quality and innovation in management, by resource conservation and a circular economy, and by coordinated and mutually reinforcing urban–rural development and development between regions. Taking these steps will enable us to sustain long-term development." It can be seen that innovation will create new growth areas for a transformed economy and secure low expense and high yields of economic transformation.

(3) Technological innovation as the strategic support and guarantee
Technological innovation is the engine, or decisive forces, for social and economic evolution, and will provide sustained driving forces for economic transformation. Scientific and technological innovation provides strategic support for raising the productive forces and boosting the overall national strength, and we must give it top priority in overall national development.

(4) Cultural innovation as the motivation and public support for economic transformation
We should reach a consensus in the entire society on economic transformation and encourage microcosmic economic and social bodies such as public institutions, enterprises, and individuals, to voluntarily change their lifestyles and transfer motivations into practical actions. Cultural innovation is conductive to improving public awareness and can lay a solid foundation for low-carbon development, cultural advancement, and energy consumption revolution among the consumers.

2. Institutional innovation and reform is the institutional foundation and support for economic transformation

(1) Institutional innovation comes from deepening reform
As an important content of reform of China's system, institutional innovation helps to improve the structure and see social and economic mechanisms play their due roles to the fullest extent. Institutional issues are not only about the external conditions for economic transformation but also important contents of it.

(2) System reform is of diversity and inclusiveness
System reform includes overhauls on political system, economic system, cultural system, social system, technological system, and administrative system. It is the fundamental conditions to attain low-carbon development. From the perspective of economic growth and resource consumption, system reform helps economic restructuring better adapt to new developments and effectively curb the expansion of manufacturing businesses featuring high energy consumption, high emissions, and low efficiency.

Reforms of economic system, technological system, and administrative system are more closely related to low-carbon development especially in the primary stage of economic transformation or the transition period. We should "accelerate the improvement of the socialist market economy and the change of the growth model." "Deepening reform is crucial for accelerating the change of the growth model." "The underlying issue we face in economic structural reform is how to strike a balance between the role of the government and that of the market, and we should follow more closely to the rules of the market and better play the role of the government." "We will see the market plays the decisive role in resource allocation, and deepen economic reforms. We must uphold and improve China's basic socialist economic system, and step up efforts in improving modern market system, macro-regulatory system, and open economic system. We will accelerate economic transformation, build China into a country of innovators and promote economic development in a more efficient, fairer and more sustainable way." Deepening cultural reform will help release and develop productive forces and boost vitality in cultural creation among the entire nation.

Social structural reform aims for accelerating the establishment of social management system, basic public service system, a system of modern social organizations, and a social management mechanism. Such reform is conductive to the realization of operation of low-carbon society.

To deepen reform of the system for managing science and technology, we should "promote close integration of science and technology with economic development, and speed up the development of the national innovation system. We should establish a system of technological innovation in which enterprises play the leading role, the market points the way and enterprises, universities and research institutes work together."

The reform of the administrative system aims for "separating government administration from the management of enterprises, state assets, public institutions and social organizations, and to build a well-structured, clean, and efficient service-oriented government that has scientifically defined functions and satisfies the people." To push forward reform of the administrative system, we should "advance market-oriented reform with wider coverage and depth in a proactive and sound way, significantly reduce government's direct role in resource allocation, and see to maximize benefits and optimize the efficiency of resource allocation based on market rules, at market prices and through full competition."

In building ecological civilization, "resource consumption, environmental damage and ecological benefits should be covered by the system of standards for evaluating economic and social development, and related goals, evaluation methods and reward and punishment mechanisms should be adopted in keeping with the needs of promoting ecological progress." "We should establish a system for developing and protecting China's geographical space, conserving the energy and protecting the environment to promote a new pattern in modernization drive featuring harmony of nature and human beings."

### 5.2.2 *Significance of Economic Transformation*

i. Fundamental guarantee for achieving the "Chinese Dream" and the "Two Centenary Goals"

The report to the 18th National Congress of the Communist Party of China (CPC), titled *Firmly March on the Path of Socialism with Chinese Characteristics and Strive to Complete the Building of A Moderately Prosperous Society in All Respects*, officially put forth Chinese Dream and Two Centenary Goals. One of the milestone targets of Chinese Dream includes the Two Centenary Goals, which are to finish building a moderately prosperous society in all respects by the time the CPC marks its centenary and to build China into a modern socialist country that is prosperous, strong, democratic, culturally advanced, and harmonious by the time the People's Republic of China celebrates its centenary. The key to realize the dream of the Chinese nation lies in transforming economic structure, raising the national core competitiveness, strategic potentials and soft power, and overall improving the environment and quality for people to work and live. First, to raise the national core competitiveness and soft power, the economic structure has to be upgraded. China is a big but not powerful country. To address the impasse, we have to make breakthroughs in upgrading its economic structure, address the imbalances in development, strike a balance between efficiency and fairness, and pay equal emphasis on the quantity and quality of development. Greater leaps should be made in improving technological innovation so as to comprehensively enhance the core competitiveness and soft power of China. Second, to enhance China's strategic potential, the economic structure has to be upgraded. To settle the issue of excessively expensive expense in development, China has to optimize the allocation of natural resources, improve the quality of development, and secure sustained development through economic transformation so as to provide solid resource and environmental guarantee for improving its strategic potential. Third, to improve the environment and quality for people to live and work, the economic structure has to be upgraded. The Chinese Dream is about every Chinese people. To realize it, endemic environment issues affecting people's life and work should be effectively settled via economic transformation. The quality of environment should be continuously improved so as to see every Chinese benefit from the achievements of economic growth and reform and opening-up, including those in environmental protection and ecological civilization. The quality of life and work throughout society will be improved.

1. Accelerating economic transformation is an important way of exploring the path for modernization with Chinese characteristics

Building China into a modern country with Chinese characteristics is to achieve the goals of new type of industrialization, IT application, urbanization, and agricultural modernization. In this process, we should promote intensive integration of IT application with industrialization, sound reinforcing of industrialization and urbanization, and coordinated development of urbanization and agricultural modernization. Generally, we should synch the development of industrialization, IT application, urbanization, and agricultural modernization. Accelerating economic transformation is an important way of exploring the path for modernization with Chinese characteristics. It is a conclusion reached by drawing on the experience from long-term efforts in building China into a modern country. It is a move to act upon the guiding principle of scientific outlook on development, a major strategic thinking put forth on the basis of China's reality, and the direction China should uphold in fostering economic and social progress.

China is at the crucial stage in reform and also an important period in industrialization and modernization. New situations and problems arising in China's economy reflect, to a large extent, its national conditions and the characteristics of development in this specific stage. With the path to industrialization with Chinese characteristics, economic growth will be driven by consumption, investment, and export instead of only investment and export, by coordinated growth of the primary, secondary, and tertiary industries instead of purely the secondary industry, by scientific and technological advances, workforce of higher quality, and administrative innovation instead of merely material and resources' inputs.

To accelerate economic transformation, we must uphold innovation-driven development so as to provide powerful and sustained technological support for the change of growth model and upgrading of industrial structure, and accelerate the historical course of shifting from a big industrial country to an industrial power. We must strike a balance between urban development and rural development, and forge integrated development of the urban–rural economies for common prosperity. We must integrate energy conservation and environmental protection with building modernization and ecological civilization, and prioritize the building of a resource-saving, environmentally friendly society in the strategy of industrialization and modernization. We must keep in mind both the domestic and international markets and resources they have in fostering sound interaction between China's development and the common prosperity of countries across the world. We must place people first in development, pay more attention to improving people's livelihood, and work hard to achieve, safeguard, and promote the fundamental interests of the overwhelming majority of the people.

2. Accelerating economic transformation is practicing the guiding principle of improving socialism with Chinese characteristics

The report to the 18th CPC National Congress put forward the goals for reform of the administrative system in the following period of time. This is an important content of improving socialism with Chinese characteristics, which reads, "To reach the goal of establishing a socialist administrative system with Chinese characteristics, we should separate government administration from the management of enterprises, state assets, public institutions, and social organizations, and build a well-structured, clean, and efficient service-oriented government that has scientifically defined functions and satisfies the people." Economic transformation is conductive to continuing efforts in streamlining administration and delegating more power to lower levels, and accelerating the transformation of government functions. It also helps advance the reform to establish larger government developments and improve division of functions among them, and facilitates to exercise government administration in an innovative way, increase public trust in the government, and improve its competence. In the realm of reform of administrative system, economic transformation is about making the government better perform its functions and building a law-based, clean, and efficient government which serves for the people and be responsible to the people. Accelerating economic transformation is practicing the guiding principle of improving socialism with Chinese characteristics.

3. Accelerating economic transformation is the underlying requirement of socialist market economy with Chinese characteristics

In the realm of economy, economic transformation is about achieving socialist market economy with Chinese characteristics through economic upgrading. Therefore, accelerating economic transformation is the underlying requirement of socialist market economy with Chinese characteristics. Economic upgrading is reflected in the following aspects: first, the transformation of economic structure, which is to shift from planned economy to market economy through economic reform; second, the transformation of economic and social forms, which is to transform the society from backwardness to modern status through economic development; third, the transformation in the openness of economy, which is to turn a closed economy into an open one by integrating into globalization; and fourth, the transformation of economic growth model, which is to change from material-oriented model to people-oriented model, and from extensive development to intensive development. All of these, in fact, are achieved from the successful exploration and great practices of developing socialist market economy with Chinese characteristics.

There must be supportive mechanisms for economic transformation, including market mechanism, macro-regulatory mechanisms, and corporate mechanism, and adjustment concerning the rules on economic operation and the relationships between the government and microcosmic economic bodies. (1) Market mechanism. In socialist market economy with Chinese characteristics, the market plays a decisive role in resource allocation. It is necessary to accelerate economic transformation, adjust the relationships between the government and the market, increase the role of the market, and promote equality, rule of law, competitiveness, and openness in economic activities. First, we should improve the market mechanism so as to see market factors like the market, competition, pricing, supply, and demand run in accordance with the law of value. Second, we should establish market norms and regulations to foster market economy to run in an orderly and effective way. Third, we should build modern market, facilitate productive factors to enter the market, and form a complete market system for productive factors. (2) Macro-regulatory mechanism. Under the conditions of market economy, the macro-regulatory mechanism should specify the boundary between the government and the market, and change direct regulation to indirect regulation. It involves the following three aspects. The first is that the object of macro-regulation changes from previous enterprises to the market. The second is that the content of macro-regulation changes from direct pricing (setting interest rate) to overseeing the market price (interest rate) to maintain the order for competition. The third is the means of macro-regulation changes from issuing administrative orders of planned quotation to introducing overseeing policies, including market-oriented government regulation, and fiscal and monetary policies for the balance of macro-economy. (3) Corporate mechanism. Reform of corporate mechanism involves building the microcosmic foundation for socialism with Chinese characteristics. It is fundamental for socialist market economy to spur the vitality and efficiency of various microcosmic economic bodies by accelerating economic transformation. The reforms cover property rights system, corporate governance structure, and the mechanism for incentive to and restrictions on entrepreneurs.

ii. Important efforts in exploring new path for human progress

1. Opening a new chapter in human progress

    (1) First, it aims for settling Chinese issues. From a fresh perspective with economic transformation, we examine the laws for China's development, the building of socialism, and the progress of human society and offer a systematic answer to underlying questions of how to achieve prosperity, and harmonious development between human, social, economic, and ecological development in both theories and practices.
    (2) From a global aspect, economic transformation aims for common interests of mankind. While helping China achieves development amid the world situation, its economic transformation promotes global cooperation and increases the overall strength of emerging markets and developing countries so that the international balance of power will favor safeguarding world peace. Economic transformation also promotes win-win cooperation in international relations, the building of a shared community for mankind, and the common prosperity of all countries. It helps to build a more equal partnership for global development, in which countries sail on the same track and share rights and responsibilities for the common interests of mankind.
    (3) One of the goals of economic transformation is to achieve ecological civilization, a hallmark of the new stage for human progress. A society with ecological civilization features high efficiency, technology, low resource consumption, low pollution, coordinated and sustainable development. In essence, it distinguishes itself from other civilization in the history of human society as it makes up the drawbacks of high energy consumption and pollution in industrialized civilization we are undergoing. Ecological civilization is likely to become a new stage of human civilization.

2. Attempting to answer or validate core issues on sustainable development that are globally applicable
   To change the model of the economic growth, China has to address the following issues: (1) how the macroscopic trend of the environmental and socioeconomic development will affect the relationship between society and nature? (2) Which kind of incentive system can effectively promote the sustainable development of the nature–society complex system? (3) How to integrate and develop the existing operation mechanism so that it transits toward sustainability? These

are core issues on sustainable development that are globally applicable. China's practices of development will provide answers and validate them. The experience of China's economic transformation will offer inspiring exploration in settling common issues on global sustainable development.

3. Persistent driving force for world people to realize their dreams
Realizing the Chinese Dream by accelerating economic transformation is not about attaining modernization of Chinese characteristics and the great rejuvenation of the Chinese nation. It will be the persistent driving forces for the people in the world, including the Chinese people, to realize their dreams. The realization of the Chinese Dream will further expand the path for the people in the world to achieve the ideal life they long yearn for.[1]

iii. Voluntarily getting involved in global governance as a strategic choice

1. Adapting to new situations and proactively tackling new changes

In its future development, China has to adapt to new situations amid economic globalization and proactively cope with new changes of multi-polarization, cultural diversity, and continuous advancement of information technologies. Economic transformation will help China better cope with the new changes and adapt to new situations. Technological innovation and intensive growth are compatible with the continuous advancement of information technologies, and cultural innovation will enable China stand firm in the world where diverse cultures flourish.

2. Seeking development opportunities while proactively tackling global changes

Global changes affect the world in an unprecedented scope and scale. Economic globalization, climate change, the third industrial revolution, and energy revolution all presented China development opportunities. Via economic transformation, China should proactively seize the commanding heights for future development, integrate itself into the global unified market as an approach to economic globalization, promote green growth as an approach to global climate change and shortage of resources, encourage scientific and technological innovation as an approach to the third industrial revolution, and transfer the way of energy production and consumption as an approach to energy revolution.

3. Shifting from passive response to active participation to expand global influence

Under the old development model, China had to passively respond to world development. Economic transformation provides China an opportunity to voluntarily integrate into world development and participate in global governance as it helps China strengthen exchanges and cooperation with other countries, bring forth changes in global governance mechanisms, and promote world peace and development. In this way, China's voice will be further intensified in international affairs, and China will have a favorable international climate for further reform and opening-up.

---

[1] Xi Jinping's speech delivered during his state visit to Tanzania, March 25, 2013.

### 5.2.3 *Significance of Economic Transformation to Low-Carbon Development*

i. Relationship between economic transformation and low-carbon development

Economic transformation is an approach, and low-carbon development is one of the ultimate goals. Specifically, changes in factors input, demand mix, industrial structure, and driving forces give occasions to improvement in efficiency and reduction in carbon emission. Low-carbon development will be realized. But low-carbon development is not just about reducing carbon emission brought by technological advances but also about intensive development which includes continuous improvement in industrial structure, income distribution, people's living standard, urban–rural structure, regional development, resource utilization, and ecology and environment.

ii. Significance of economic transformation to low-carbon development

Economic transformation is the prerequisite and main way to achieve low-carbon development. Meanwhile, it is conductive to the achievement of low-carbon development, a new economic growth pole[2] in future global development. China can achieve low-carbon development only by changing its growth model. Economic transformation will help settle development impasses and break the ecological bottleneck. Extensive practices of low-carbon development have shown that low-carbon economy is a realistic choice to get rid of the development impasses and break the bottlenecks of climate change, environmental pollution, energy shortage, and ecological imbalances.

## 5.3 Economy and Environment Must and Can Thrive Simultaneously

The traditional development path which stresses quantitative growth and expansion in scale is exhibiting more and more drawbacks. It is imperative to transform such a growth model. From the guiding principle for economic evolution, it can be seen that the focus of economic development is shifting from quantitative increase to the coordinated development of economy, environment of human beings in themselves. Objectively, economy and environment are both players in the game, but also partners. Only by observing the guiding notion of placing people first and promoting the positive interaction between economy and environment can the two soundly reinforce each other.

[2]The idea of economic growth poles was put forth by French economist Francois Perroux in 1955.

### 5.3.1 *The Role of Economy in Environmental Preservation*

How to handle environmental pollution in economic development of human beings? Ecological conservation, such as sewage treatment, air pollution control, and urban waste treatment, requires immensely huge amount of investment by the government. In terms of energy consumption, improving the utilization of energies needs the support of science and technology as well as inputs in environmental technologies. Without economic growth, there will be shortage of technologies and funds for environmental protection. The emerging and solving of environmental issues are closely related to the stage of economic development and advancement of technologies. Only when both economic growth and investment in environmental protection are equally emphasized, the issues can be properly solved.

i. Economic growth dominates changes in environment

As society evolves, technology advances, and the population expand, human beings have gradually improved abilities to intervene with nature and remake nature as they want. When people utilize and remake nature in accordance with the laws of nature, they could continuously improve the quality of the environment, and vice versa. When they act arbitrarily upon their willingness and ignore the laws of nature, they would give occasion to vicious circle in the ecological system and worsening environment quality. This shows that economic development dominates changes in the environment.

Improper economic growth or development model will lead to damages to the environment. Excessive carbon dioxide emissions and other pollutions in industrialization forced people in impoverished regions to get the land deforested for crops, overgraze the pasture, and grow on steep gradients, contributing to soil erosion and land desertification. Extensive development model results in external environmental costs, regardless of the renewal of resources and the value of ecology. Low-cost industrial expansion is the root cause of serious environmental pollution, and waste and shortage of resources.

ii. Environmental issues should be solved by economic means

The key to attain the targets of environmental protection lies in investment in the sector. China's investment in environmental protection has been significantly improved over the past 20 years. In 2012, China pumped RMB 825.3 billion into environmental conservation, a spike of some 48-fold from RMB 17 billion in 1991 as shown in Fig. 5.30. The figure was 24 times the GDP increase during the same period. The proportion of funding for environmental protection to GDP increased from 0.78% to 1.6%, with the highest record of 1.9% in 2010. The investment in environmental protection accounted for 1.77 to 3.14% of the total fixed asset investment, reflecting the real situation of material investment in the sector. Although the total amount was on the rise, it was far enough to meet the actual demands to curb the pollution issues in China and far less than the amount invested by western countries when environmental pollution was a primary policy focus. This is also the root cause of why China's pollution issues had not been effectively addressed. Experiences of the

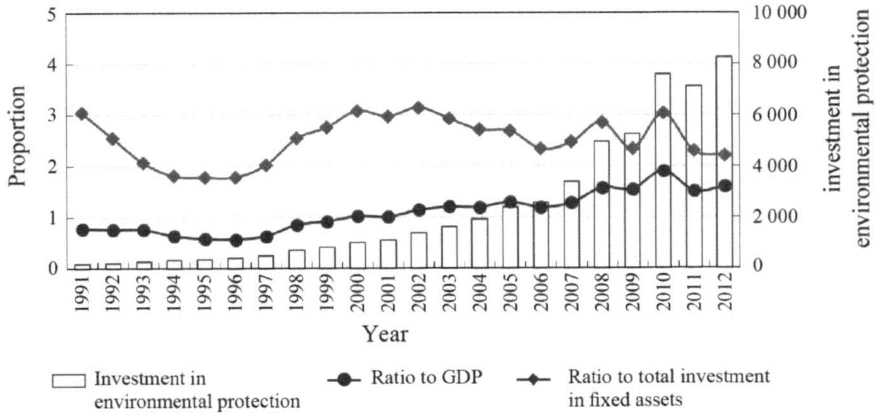


**Fig. 5.30** Changes in China's investment in environment protection

U.S. and Japan told us that the proportion of investment on environmental protection to total investment in fixed assets should stand around 5–7%. Therefore, there is room to improve in this regard.

The Chinese Government has been committed to environmental protection and achieved primary results as it has been increasing investment in the sector year by year. However, we should not ignore some underlying issues behind superficial phenomena. Actions for environmental conservation have been enforced by the government with administrative means, whereas economic means fall short. Even though the government pays a high regulatory cost, it stuck in the mire in face of increasingly worsening environmental issues. It is tough to address them. The root causes of pollution are socioeconomic activities. Therefore, solely depending on the government to solve them is far from effective, but the key lies in economic means. By changing the cost–benefit balance of economic entities, such means will indirectly contribute to an environmentally friendly outcome. Common measures include environmental taxes and fees and tradable discharge permits.

iii. The relationship between environmental protection and low-carbon economy

Low-carbon economy features low energy consumption, low emission, and few pollution as a response for mitigating the influences of carbon-based energy on global warming. It aims for sustainable development of society and economy. The essence lies in improving energy efficiency, energy-saving technologies, renewable energy technologies, and technologies for emission reduction to develop low-carbon products and maintain global ecological balance. This is an economic development model as the economy shifts from high energy consumption to low energy consumption.

While boosting low-carbon economy, environmental protection is also one of the top priorities. The Chinese government encourages developing the economy rapidly while conserving the ecology. If the economy is likened a blood system, ecology is the internal environment. Ecological conservation is a task with equal importance to economic development. Buoyant low-carbon economy is conductive to environmental protection as it drives industries to be more environmentally friendly and beneficial to the public interests. Low-carbon economy brings a cleaner society and economy, and avoids unintentional pollution in consequence of growing the economy. As it evolves, the public will have better understanding of environmental protection.

To realize harmonious development of the economy and society, we must adjust China's energy structure and reduce emissions of carbon dioxide. Energy structure adjustment is one of top priorities in developing China's energy and an integral part of ensuring the state energy security. Stepping up efforts in strategically adjusting the energy structure is to reduce demands and consumption on fossil fuels and dependence on oil imports, lower the proportion of coal consumption, and develop renewable and clean energy. The second measure is to improve energy utilization. Contributing factors for poor energy utilization include structural reasons, consumers' behavior, and efficiency management among others. The key lies in restructuring the energy industries to optimize it and accelerating technological advances. Measures should be taken targeting both the supply and demand sides to promote the efficiency in energy utilization. Inputs in energy technologies should be increased, and research and development on energy-saving technologies should be encouraged. An energy aid program should be implemented to help low- and middle-income families gain access to efficient energy services. Industry associations should play their due roles to foster businesses to promote energy utilization. Efforts should be made to strengthen international cooperation and promote technological exchanges.

### 5.3.2 Coexistence of Economy and Environment

After long-term development of human society, economic growth and environmental pollution take on varying relations. Setting economic growth speed and environmental pollution extent at two levels of high and low, four outcomes are likely to appear as shown in Fig. 5.31.

Quadrant A1: Slow economic growth and low level of environmental pollution and damage. Slow economic growth poses light pressure on the environment, and the influence is slight. Relationships between economy and environment circulate at a low level. World economy followed such a pattern before the industrial revolution. As economy grew slowly, it could not fulfill people's ever increasing demands for materials. The development model could not satisfy the demands of modern economy and society.

Quadrant A2: Slow economic growth and severe environmental pollution and damages. This model shows that the relationships between economy and environment are in a vicious circle. The worsening environment can not shore up economic development but only stands as a restraint. Some people call such development model as "impoverished pollution." The root causes include sluggish economy, low-level industrialization, but expanding population, which exacerbates environmental issues.

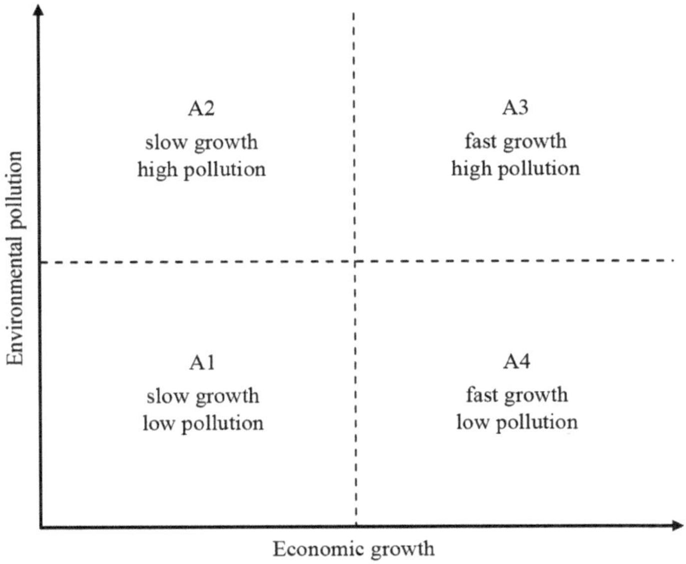


**Fig. 5.31** Patterns of economic growth and environmental pollution

Quadrant A3: Fast economic growth with severe environmental pollution. This model features extensive and predatory economic growth driven by high energy and resources inputs, high economic yields, but result in worsening environmental quality. Deteriorating environment is bound to jeopardize the economy. Therefore, such development is not sustainable. It is not the target we pursue. Some call such development model industrialized pollution or affluent pollution. A typical case is the development model of Western developed countries from the 1940s to the 1960s.

Quadrant A4: Fast economic growth and slight environmental damage. In such a model, economy is sustained while the environment maintains in sound condition. A virtuous cycle between economy and environment is attained. Therefore, such development model is sustainable as it takes both the economic development and ecological conservation into account, and realizes coordinated development of the economy and environment. It is an ideal development model we pursue.

It is evident that we pursue high-quality economic growth and zero emission of pollutants. Can we give considerations to both the tasks? Objectively speaking, environment and economy compete but coordinate with each other. As long as their relationships are well handled and the synergy of economy and environment is fully used, it is possible to achieve coordinated development of economy and environment.

i. Analysis of economy and environment from the perspective of game theory

On the premise of limited resources, protecting people's living environment conflicts economic development at least in the short term. How to choose between economic development and environmental protection in allocating resources is an issue human beings have to settle. Increasing inputs in one side will result in losses on the other side in the short term. This contradiction is extremely striking for cash-strapped developing nations.

Environment constrains economy in two aspects. First, the exploiting of resources and energies from nature for economic development is restricted by the supply capacity of ecology system. For renewable energies, the amount of exploitation should be within the renewal limit. For example, the number of trees cut down should be less than that of the planted trees; otherwise, it will cause damage to forest resources. For nonrenewable resources, the amount of exploitation should be within the natural reserve. Second, pollutants discharged because of economic activities should be subject to the capacity of ecological system. The environment is unable to handle wastes emitted from economic activities. The total amount is subject to the capacity of the environment. Exhausts exceeding the capacity will accrue in the environment and lead to environmental pollution ultimately.

Many countries undergo different stages in industrialization where economic development and environmental protection constantly conflict with each other. China develops its economy to satisfy people's ever growing material and cultural demands. As a country started from a semi-colonial and semi-feudal agricultural society, China's economic and industrial foundations used to be weak, and suffered from conflicting economic development and environmental protection. Both people's living environment and the natural ecology should be protected. In urban areas, economic activities especially industrial development adversely affect residents' living environment. In rural areas, damages to nature and ecology are even more endemic. The ultimate goal of economic development is to improve people's living standard. Damaging the ecology in the process and adversely affecting people's life go against the original intention of growing economy. Contradictions and problems are inevitable, but we should not sit passively for our end, but proactively think about how to settle them. We should not develop at the expense of environment, but should be more active in seeking a path for sustainable development.

ii. Synergy of economy and environment

The relations between economic development and environmental protection are fundamentally about the relationships between human and nature. Addressing environmental issues, in essence, is to balance the relationships between human and nature, between people and people, and between economic development and environmental protection. In the course of human society, the relations between human and nature evolved from harmony in ancient times to conquest and confrontation in modern industrial revolution, and voluntary adjustment in contemporary era for building modern civilization featuring harmony of human and nature. This is the reflection of the law of unity of the opposite. Some maintained that ecological conservation and economic growth are antagonistic, noting environmental protection has to be at the expense of economic growth. Practices over the years have shown that it is possible to strike a balance between economic growth and environmental protection, which could reinforce each other and coordinate with each other.

Objectively, economic development and environmental protection support and reinforce each other. On the one hand, sustainable development, which has been promoted in the twenty-first century, features including environment into the economic costs. In consequent, environmental protection has become an effective way to reduce the costs and raise the efficiency of the economy. The sustainability and stability of economic development depend on the abundance and sustainable production capacity of natural resources. From this perspective, protecting and improving the environment provides material foundation and conditions for the steady and sustainable economic development. On the other hand, environmental protection here is not about passive restoring damage but about reasonable development and utilization of the environment on the premise of not damaging it. It is ridiculous to require human beings to preserve the environment at the expense of degrading the civilization. Rather than stagnating the economy, environmental protection today asks for sustainable financial and technological supports from economic growth. From this point of view, economic development reinforces and depends on each other. They could advance simultaneously. Environmental protection is the fundamental requirement of economic growth and aims for improving people's living standards.

High-quality economic development improves people's livelihood and raises requirement on environmental conditions. People began to voluntarily preserve and improve the environment for a better life. Meanwhile, sound environment provides better conditions for economic activities and more resources for the economic systems, and allows more waste discharged from the economic system. In this way, the environment fosters economic development. Improved economic strengths enable more surplus values to be invested in environmental protection and control, such as building nature reserve or treating wastes.

The ecology and natural resources provide fundamental conditions for economic growth. This is shown in the following aspects. First, the ecology provides necessary resources and energies to the economic system, and with economic activities, various resources are turned into products to satisfy the needs of human beings. The variety, quantity, and quality of resources and energies provided by the ecology determine, to a certain extent, the nature and direction of economic development. Without the support of resources and energies from the environment, any economy would collapse. Second, the ecological system could accommodate wastes from the economic system. Economic activities inevitably generate certain amount of wastes, which could not be all preserved in the economic system, but will be eventually discharged into the environment. The environment is capable of dispersing, storing, and discomposing these wastes. Such functions will reduce costs of treating wastes with manual labors. Third, the economic system is the outcome of ecological system which exists before the emergence of mankind. After human beings were born, they began to utilize and transform natural environment for survival. When such activities evolved to a certain level, economic system came into being. So economy is the outcome of human beings utilizing and transforming the environment. Fourth, besides providing resources to economic activities, the environment satisfies people's demand for comfortableness.

Clean water and air are essential factors for industrial and agricultural production, and the basic necessities for a decent life. Wonderful natural environment refreshes people and relaxes their minds, and helps improve their health.

iii. Comprehensive evaluation on coordination of economy and ecology

As one of management functions, coordination means organizing the activities of different groups in an organization centering on its mission so that they work smoothly together for the same goal. There are complicated interactions between the economic system and the ecological system as shown in Fig. 5.32. Only when the interactions generate coordinating effects, the economic and environmental systems can advance in an orderly and coordinated manner. The essence of coordinated development of the compound systems of the economy and ecology is to fully use and foster the positive dynamics between economy and ecology for a virtuous cycle of the two. In this way, the economy progresses steadily, resources are efficiently utilized, and the environment stays in a good condition. Coordination is the approach to sustainable development, and sustainable development is the ultimate goal.

The complex system of economy and ecology (2E system) is an open dynamic system with unified functions and structure formed by subsystems of different nature through interaction and interdependence. The coordination of the 2E system for win-win results means a harmonious coexistence of various economic and ecological subsystems with reasonable economic development as the foundation and preservation of environment to the utmost extent. The original intention of people-oriented development is to achieve win-win results with the 2E system.

2E system is an open, nonlinear dynamic complex system and involves multiple goals. It is vulnerable to external influences. Different samples used will directly affect the results of empirical analysis. Based on the above analysis on economy and ecology from the perspective of game theory and their coordination, a mathematic model or quantitative method can be established to reflect the status and extent of

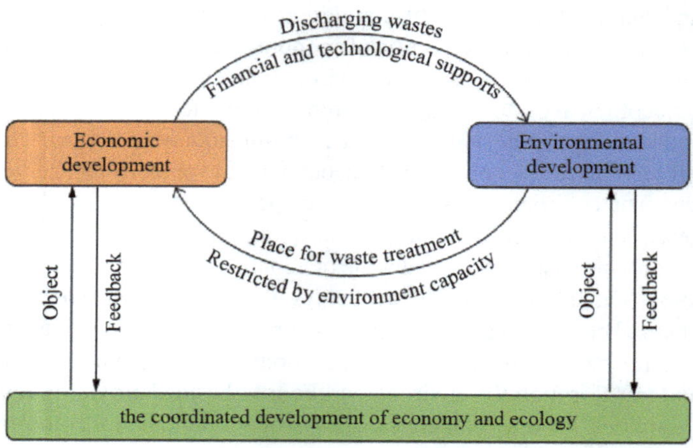


**Fig. 5.32** Chart flow for coordination of economy and ecology

coordination in the system. The key to the coordination model lies in how to determine the development levels of economic and ecological systems and the overall development of the 2E system. Here, a comprehensive evaluation system for 2E system is made to realize the process. In line with principles of being comprehensive, representative, quantitative, operable, and scientific, appropriate indicators from the economic and ecological subsystems are chosen, as shown in Fig. 5.33, to construct the comprehensive evaluation system for 2E system. After collecting and standardizing related data on China's economic and ecological subsystems in 2001–2012, the researcher analyzes the standardized data, determines the weights or characteristic quantity of each indicator chosen from the subsystems, and calculates the development value of each subsystem. Finally, to illustrate the inherent coordination of each system, coordination coefficients are employed to calculate the extent of coordination among each system so as to evaluate the status of coordinated development.

Figure 5.34 exhibits the comprehensive coordination coefficients of the 2E system. The icon -■- represents the coordination coefficient of the overall development of economy and environment, and the icon -▼- represents the coordination coefficient of pollutant emissions. The results show that the overall coordination coefficient of China's 2E system maintained at the developing stage in 2000–2012 and reached the primary level of coordination in 2012. Besides, from the coordination of pollutant emissions and economic development, the coordination coefficient has become even after 2010, showing that it reached the primary level of coordination. This demonstrates that the emission of various pollutants broke the limits of planning, seriously undermining the coordination between economy and ecology. If pollutant emissions were not effectively curbed, it would have increasingly negative effects on the overall coordinated development of economy and ecology.

All in all, the evaluation system for 2E system enables effective calculation and comprehensive evaluation of the coupling and coordinating state of China's economy

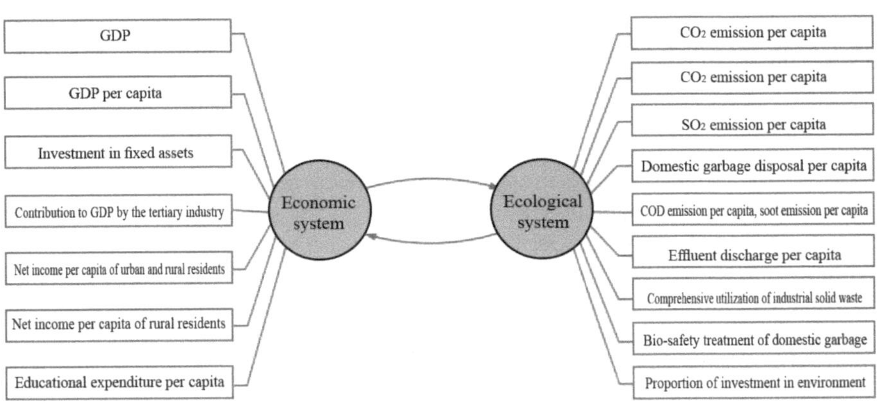


**Fig. 5.33** The evaluation system for 2E system

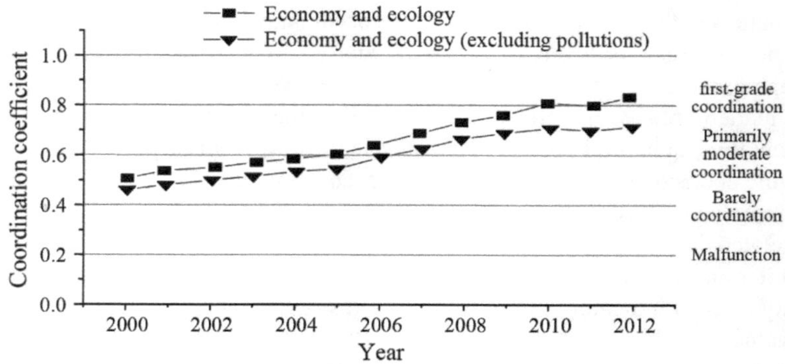


**Fig. 5.34** Comprehensive coordination coefficient of economy and ecology

and ecology. The evaluation results are referential for China to make more scientific and effective policies for economic development and ecological conservation in the future.

### 5.3.3 *International Experiences on Win-Win Results of Economy and Ecology*

i. Social and ecological market economy in Germany

Drawing on experience from its development path in the 1970s when the country suffered from a series of environmental emergences in the consequence of relentlessly developing the economy at the expense of environment after the Second World War, Germany began to place environmental issues on the top of its national political agenda, restructured its economy, and established a unique development model of social and ecological market economy. Social and ecological market economy changes the conventional development model where gains in one side result in losses of the other between economy and ecology. It aims for ecological balance and virtuous cycle of ecological conservation, economic growth, and social justice via technological innovations. Going for win-win results of social justice, ecological balance, and economic prosperity, the German Government carried out a raft of policies for a complete supporting system, among which there were over 2000 regulations and laws on environmental protection. In 1998, the country formulated and released the twenty-first-century environmental protection guideline, in which ecology was made an important factor for economic growth and job creation, and also a priority in economic development. Changed mindset, established legislation, improving policies, and booming environmental protection businesses together laid a solid foundation for Germany to establish the accountability competitiveness.

Germany's policies to foster social and ecological market economy were very successful. The first achievement is the building of ecologically friendly economy. Germany develops the economy in an ecologically friendly manner throughout the source to the whole operation and results of economic activities. It gradually obsoleted heavily polluting energies and raw materials but made full use of renewable ones, and adopted advanced technologies to improve operation, save energies, and lower pollutants. Great efforts were also made to control environmental pollution. Such a development model provides potent support for the country to realize the goal of sustainable development. Second, a new buoyant sector, environmental protection industry, was generated, thanks to the support to ecologically friendly economy, which enables Germany to maintain leading position in global market of environmental protection. At present, German has one of the most dynamic markets for environmental protection businesses, and its trade volume of environmental protection technologies accounts for one-sixth of the world's total. Germany's approach to social and ecological market economy shows us a path to win-win results of economy and environment. By transforming conventional mindsets, and fostering technological innovation and ecologically friendly economy, environmental protection gradually become a deeply rooted notion shared by the public, an integral part of business operation, and the core of national development strategy.

ii. Recycling-based social management for economic sustainable development in Japan

Japan used to face catastrophic environmental issues and then achieved huge success in environmental protection and building recycling-based society. Its strategy for environment management evolves as its economy grows. Japan's unique approach to environmental management is instrumental to China.

From end-of-pipe control to environmental protection for improving life quality and prevention management on the whole process of operation, Japan established recycling-based social management for economic sustainable development as shown in Fig. 5.35. The focus of the social management model is the coordinated development of economy and environment. In 1993, Japan enacted the *Basic Law of Environment of Japan* and upgraded environmental protection to a national strategy. The importance of environment management was highlighted. The manage-

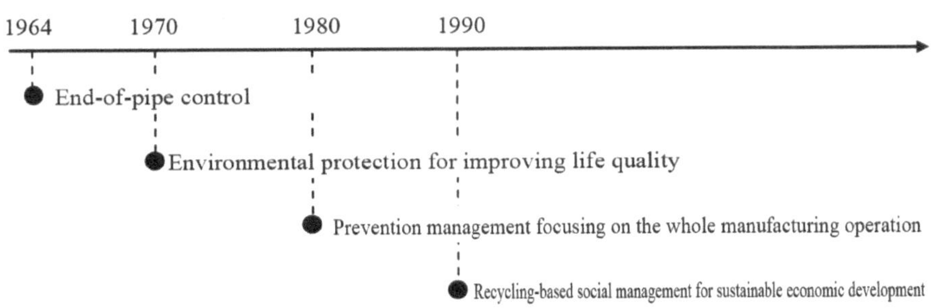


**Fig. 5.35** Environment management evolution in Japan

ment guiding principle shifted from purely protecting natural resources to pursuing sustainable development, and the focus changed from passive control to active prevention. Environmental impacts of economic activities were highlighted, and the notion of sustainable development was established. With higher awareness on environmental protection, and penetrating green consumption and purchase throughout society, the environmental performances of businesses are brought to the spotlight of public attention, which prompts businesses to actively implement measures for environmental management. Environmental laws and regulations are no longer just about reining in polluting activities but evolve for fostering multiple market-oriented policies.

iii. Extended Producer Responsibility (EPR) model in Sweden

Sweden is one of the first countries promoting EPR through legislation. The idea of EPR dates back to a bill regarding waste recycling and management enacted in 1975 in Sweden. The concept was first formally introduced in a 1988 report to the Swedish Ministry of the Environment by Thomas Lindhqvist, an environmental economist. EPR is an environmental protection strategy to reach an environmental objective of a decreased total environmental impact of a product by making the manufacturer of the product responsible for the entire lifecycle of the product that especially for the take-back, recycling, and final disposal. EPR makes up the void of product responsibility after sales and defines the responsible entity for the take-back, recycling, and final disposal of products. After the theory was put forth, the U.S., European Union and OECD and other countries and regional organizations revised and improved it to adapt to their own conditions. Despite different focuses, the EPR policies aim to specify the responsible entities for the take-back, recycling, and final disposal of products.

In conclusion, the models adopted by Germany, Japan, and Sweden generate coordination effects in the interactions of economy and environment, and realizes a virtuous cycle between economy and environment. In consequence, economy evolves steadily, while resources are fully and efficiently utilized and the environment stays in good conditions. These models offer us instrumental experiences and approaches for the coordinated and orderly development of China's economic and environmental systems.

# Chapter 6
# Strategic Goals of Low-Carbon Development in China

Proactive tackling climate changes and promoting low-carbon development are inherent requirements of attaining sustainable development and advancing the building of ecological civilization. It presents great opportunities for China to transform the model of economic growth, adjust the economic structure, and push forward the new industrial revolution. It is also an international obligation China should take as a responsible power.

China has been placing combating climate change high on its government agenda and incorporating it as an integral part of efforts in building ecological civilization and a Beautiful China into the national development plan. A host of measures to voluntarily mitigate and adapt to climate changes have been rolled out. Following the emission reduction targets agreed on in the 2009 Copenhagen Summit, which is to cut the greenhouse gas emissions per unit of GDP by 40–45% on 2005 levels by 2020, China enacted *National Strategy for Climate Adaption* at the end of 2013 and *National Plan on Climate Change 2014–2020* in September 2014, which formulated the guiding principles, targets, requirements, policy orientations, key tasks, and supporting measures for climate change mitigation. In *China–U.S. Joint Statement on Climate Change* in 2014, China made public its carbon dioxide emission targets in 2030, the peak period for its greenhouse emission. On June 30, 2015, China submitted the Secretariat of the United Nations Framework Convention on Climate Change a document: the *Enhanced Actions on Climate Change: China's Intended Nationally Determined Contributions*, pledging to achieve the peaking of carbon dioxide emissions around 2030 and making best efforts to peak early, lower carbon dioxide emissions per unit of GDP by 60–65% from the 2005 level, increase the share of nonfossil fuels in primary energy consumption to around 20%, and increase the forest stock volume by around 4.5 million cubic meters on the 2005 level. These measures represent China's strengthened efforts in mitigating climate change after 2020. The country also requires that provinces, autonomous regions, and cities nationwide pursuing low-carbon development should explore an effective path to achieve the peaking of carbon dioxide emissions early, and specifies targets for both the peak and total amounts.



X. Du et al., *Overview of Low-Carbon Development*,
https://doi.org/10.1007/978-981-13-9250-4_6

Meanwhile, we should be aware that China still has no clear mid- and long-term plans for climate change mitigation. Despite the efforts for and achievements in saving energy, reducing consumption, developing renewable energies, and increasing carbon sink, no breakthroughs have been made in shifting the economy from the track of high-carbon expansion. There are potential risks that the energy system and infrastructure system China is making great efforts in would be highly dependent on high-carbon technologies and consumption pattern.

Given the tremendous importance of China's response to climate change and low-carbon development on global efforts in this regard and safeguarding ecological safety, Chinese President Xi Jinping made a solemn promise that China will make new contributions to global climate change in the 2013 APEC Summit. He stressed that mitigating climate change was not required by others, but China's voluntary response. In 2014, Xi made remarks on the new normal of Chinese economy on many occasions, reflecting the resolution of Chinese Governments and competent departments in transforming the economy from extensive expansion to highly efficient, low-cost, and sustainable growth.

Against this backdrop, setting a long-term strategic target for low-carbon development and voluntarily making carbon dioxide emissions restrictive conditions for socioeconomic development are conducive to transforming China's development model and consumption pattern, adjusting industrial structure, shifting the economy from extensive growth to intensive development, and bringing out the full potential of the economy. It is also necessary for China to integrate into global low-carbon development, untie economic growth from any connections it may have with carbon emissions, and attain the goal of promoting low-carbon green development and building a Beautiful China proposed in the report to the 18th National Congress of the CPC.

## 6.1 Overall Situation of and Scenario Analysis on China's Low-Carbon Development

Generally speaking, energy consumption and carbon emissions are combined results of population growth, economic development level and path, urbanization model, and changes in residents' lifestyle among other factors. For a comprehensive and systematic analysis on affecting factors on China's low-carbon development, this book employs the method of scenario analysis. Based on the differences on inherent requirements and targets of China's shifting to low-carbon development (as shown in Table 6.1), the book builds standard scenario, low-carbon scenario, and intensified low-carbon scenario to conduct quantitative studies on the possible trends of energy, industrial, construction, and transportation sectors.

**Table 6.1** Overall description on standard, low-carbon, and intensified low-carbon scenarios

| Scenario | Overall description |
|---|---|
| Standard scenario | The original intention is to fulfill demands for domestic development and to improve people's living standards. On the basis of reducing greenhouse emission by 40–45% on the 2005 level by 2020, relating policies and mainstream technologies will be continued, while no additional measures will be taken to mitigate climate change |
| Low-carbon scenario | Voluntary measures are to be taken to transform the economy and reduce energy consumption and greenhouse emissions by taking factors related to energy, environment, and ecological safety into account. Low-carbon development is remarkably speeded up. Efforts are made to fulfill greenhouse reduction commitment made in the 2009 Copenhagen Summit, and the promise proposed in *China–U.S. Joint Statement on Climate Change* on peaking carbon dioxide emissions by 2030 and target of developing nonfossil fuels. The technology of CCUS will develop gradually after 2030, and large-scale application in power generation and industrial activities will be realized around 2040 |
| Intensified low-carbon scenario | In achieving the goal of containing global temperature increase within 2 °C through global cooperation, the potentials for energy saving and low-carbon energy development have been further tapped in the consequence of intensifying transfer of capital and technologies worldwide. Since the 13th Five-Year Plan, China has been incorporating climate change mitigation into plans and strategies for national development. With improving lifestyle and consumption pattern, and accelerating CCUS development as well as the widespread application in power generation and industrial activities, related plans and commitments of greenhouse reduction are expected to be fulfilled ahead of schedule |

### *6.1.1 Methodology of Scenario Analysis*

Scenario analysis is the major methodology employed in this book. The basic research framework is shown in Fig. 6.1.

The predictive analysis of future socioeconomic trends is based on scenario analysis. On the basis of analyzing the historical trend of China's economic and social development, and referencing predicating data on sectors by 2050 released by major research institutes at home and abroad, the book builds parameters for social and economic development including population, GDP, and urbanization rate.

Technological advances, the speed of economic transformation and shifts in the pattern of energy consumption, and changes in residents' lifestyle and consumption all profoundly affect economic activities and even national low-carbon development. This book takes the historical development trends of major energy-consuming and

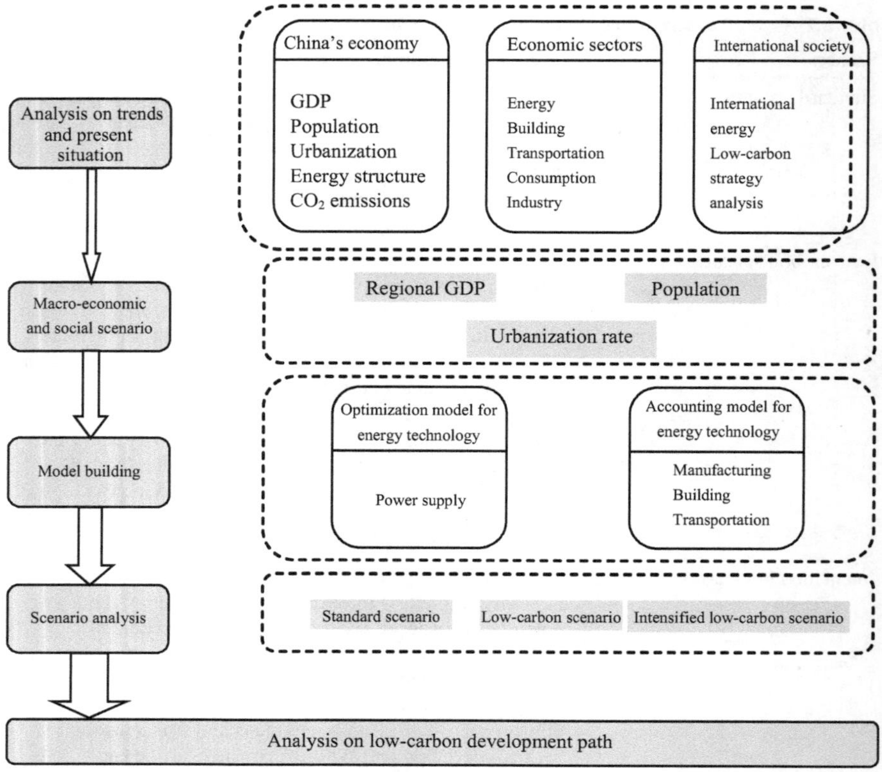


**Fig. 6.1** Framework for scenario analysis on China's low-carbon development

greenhouse gas-emitting sectors such as energy, construction, transportation, and industrial activities as the important basis to set up parameters for future industries. It employs optimization model and accounting model for energy technology as approaches to studying changing trends of energy production and conversion sectors, as well as energy consumption level, structure, and efficiency in energy-consuming sectors of manufacturing, construction, and transportation against different scenarios. The focus of analysis lies in the changing trends of carbon dioxide emissions in energy activities (note: the overall scenario excludes carbon dioxide emissions in industrial activities, and the carbon dioxide emissions mentioned below specifically mean those in energy activities). Possible outcomes of carbon dioxide emissions from present to 2050 in China will be reached.

**Table 6.2** Major parameters and features of relating scenarios for China's low-carbon development

| | 2010 | 2020 | 2030 | 2040 | 2050 |
|---|---|---|---|---|---|
| Population/100 million | 13.41 | 14.00 | 14.20 | 14.02 | 13.53 |
| GDP growth rate/average rate in the last 10 years | – | 7.2% | 5.2% | 3.8% | 2.8% |
| Urbanization rate | 49.3% | 60% | 68% | 73% | 75% |

### 6.1.2 Overall Situation of Social and Economic Development

Economic development is the core of China's Two Centenary Goals, and also the economic foundation to enhance its overall national strengths, the quality of people's life, and its say in the international community. When the second centenary goal is attained and China becomes a modern country, Chinese people's average income per capita will be close to or equivalent to the level of moderately developed countries after adjusting for exchange rates. According to Xi's remarks on new normal in economic development, China's economy will continue to rise in the overall trend, but it is also gradually shifting gears in terms of growth rate. Studies on the future trend of China's economy at home and abroad have basically reached an agreement that its economy will continue to expand as the growth rate slows down. The most conservative estimate says that China's economy is expected to double between 2010 and 2020, and China is to emerge as the world's largest economy around 2030. From now to 2050, the proportion of China's population to the world's total will gradually decline, down to 17.2% around 2030. Given a medium level of fertility rate, the proportion of senior citizens over 60 years old will continue to rise, up to 23.8% by 2030. In terms of urbanization, the future global average rate is expected to be 0.4%, and urbanization in developing countries is to accelerate. China's urbanization rate will outnumber the world average level. By 2030, urban population in China is expected to account for 68% of the total population, and the figure will continue to rise and reach 75% by 2050. Forecasts on China's economy are detailed in Table 6.2.

### 6.1.3 Comparison on Different Scenarios of Carbon Emissions

Generally speaking, the standard scenario, low-carbon scenario, and intensified low-carbon scenario represent three typical development paths for China's future development.

i. Standard scenario

In this scenario, although measures have been taken to save energy and reduce emissions, demands for improving people's living conditions continue to drive rapid increase in energy end use. In 2050, public building area per capita and residential building area per capita in China will increase 1.7-folds and 0.5-fold, respectively, on the 2010 level, reaching 16 and 43 $m^2$, which are close to the current higher levels of EU members; car ownership per capita is expected to rise by 7.5-folds on the 2010 level, with 400 among 1000 people owning cars. Growth in end-user demands leads to high level of energy demands in major sectors and manufacturing high energy-consuming products. Energy end use in industrial sector is expected to peak around 2030, and that in construction and transportation sectors will be on the rise by 2050. In this context, China's primary energy consumption will continue to rise, climbing from 5.2 billion tons of standard coal in 2020 to 6.5 billion tons in 2030 and then to 7.2 billion tons in 2050. Coal will remain as China's main source of energy on a long-term basis. The consumption will gradually rise from 4.3 billion tons of coals produced in 2020 and peak around 2030 at the amount of 4.6 billion tons. And the figure is expected to stay high above 4 billion tons for a long period of time. Correspondingly, the share of coal consumption in the primary energy consumption will fall from 59% in 2020 to 50% in 2030, and further down to 35% by 2050. Nonfossil energy has been developed rapidly, but the production is still unable to fulfill the demands for low-carbon development. The share of nonfossil energy in the primary energy consumption will climb to 18% by 2030 from 14% in 2020, and further up to 29% by 2050. No major breakthroughs have been made in CCUS (carbon capture, utilization, and storage) technologies. Total stock of carbon dioxide captured and stored by 2050 will remain at only 600–700 million tons. By 2050, energy consumption per capita in China will reach 5.4 tons of standard coal, carbon emission intensity will be down to 1.6 tons of carbon dioxide per ton of standard coal, and the energy consumption per unit of GDP will be 0.28 ton of standard coal per RMB 10,000, which is close to the average level of OECD countries in 2000.

Under such conditions, carbon emissions from China's industrial activities will peak around 2030; that from the construction sector will do around 2040, and emissions from the transportation sector will keep fast growing around 2050. By 2020, China's carbon discharges will reach about 10.9 billion tons, and intensity of carbon emission will be cut off by about 46% on the 2005 level, signifying the fulfillment of commitment to Copenhagen conference for climate change in 2009. By 2035, the total carbon dioxide emissions will reach the peaking of 13 billion tons, exceeding the total discharges of all OECD countries in 2010; emission per capita will stand at 9.2 tons, close to Japan's peaking figure of 9.6 tons. By 2050, the total emissions will fall down to 11.5 billion tons, equivalent to the aggregate emissions of the U.S., EU, and India given all of them continue current commitments to emission reduction; emissions per capita will be down to 8.5 tons, close to the level of the EU in 1990. To achieve these targets, all walks of life should do their respective contributions. The government, businesses, and the public should transform their ways of consuming on the current basis. This scenario presents relatively challenging targets.

ii. Low-carbon scenario

In low-carbon scenario, energy consumption in various sectors will be contained within a reasonable limit by means of regulation, increasing popularity of low-carbon lifestyle, and drastic improvement in energy efficiency. In consequence, energy consumption demands in the industrial activities will reach the peaking around 2025, and the demands in construction and transportation sectors will keep on the rise at a gradually slowing down growth rate by 2050. In this context, the total primary energy consumption will climb from 5.1 billion tons of standard coal in 2020 to 6.1 billion tons of standard coal in 2030, and peak at 6.4 billion tons by 2040, after which the figure is expected to slightly fall down. Policies containing coal consumption will achieve remarkable results, with the figure gradually falling back to 2.7 billion tons by 2050 after the peaking of 4.1 billion tons around 2020 as a consequence of more consumption of clean energy. The dominating share of coal in China's total consumption will continuously decline. The share in the primary energy consumption will reduce to 47% by 2030 from 57% around 2020, and further down to 30% by 2050. The consumption of nonfossil energy will meet the set target. Its share in the primary energy consumption will rise to 21% by 2030 from 15% around 2020, and gradually climb to 36% by 2050. Around 2030, the CCUS technologies applied in power generation and the industrial sector will deliver results, and the amount of carbon dioxide stored will rise to 1.8 billion tons by 2050 from 10 million tons around 2030. By 2050, energy consumption per capita will be 4.7 tons of standard coal, with the intensity down to 1.19 tons of carbon dioxide emissions per ton of standard coal, and the energy consumption per unit of GDP standing at 0.24 ton of standard coal per RMB 10,000, which is close to the U.S. level around 2020 given the country continue its commitment of INDC (intended nationally determined contributions).

Under such conditions, carbon dioxide emissions from sectors of manufacturing, construction, and transportation will peak around 2025, 2030, and 2040, respectively. By 2020, the total emissions will reach 10.5 billion tons, and intensity of carbon emission will be cut off by about 48% on the 2005 level, signifying the fulfillment of commitment to Copenhagen conference for climate change in 2009 ahead of schedule. By 2030, the total carbon dioxide emission will reach the peaking of 11.4 billion tons, with emissions per capita standing at 8.1 tons, close to EU's 2005 average level. By 2050, the total emissions will fall down to 7.5 billion tons, with the figure per capita down to 5.6 tons, close to the level of China before 2010, but still falling behind the EU's 2050 level.

iii. Intensified low-carbon scenario

In the intensified low-carbon scenario, energy demands are further contained, the research and development and application of low-carbon technologies accelerate, and people live a more environmentally friendly life. Against this backdrop, energy demands of all sectors further slow down, and will reach the peaking earlier around 2020, 2040, and 2045 in the sectors of manufacturing, construction, and transportation, respectively, as shown in Fig. 6.2. In this context, China's primary energy consumption will climb to 5.9 billion tons of standard coal by 2030 from 4.4 billion

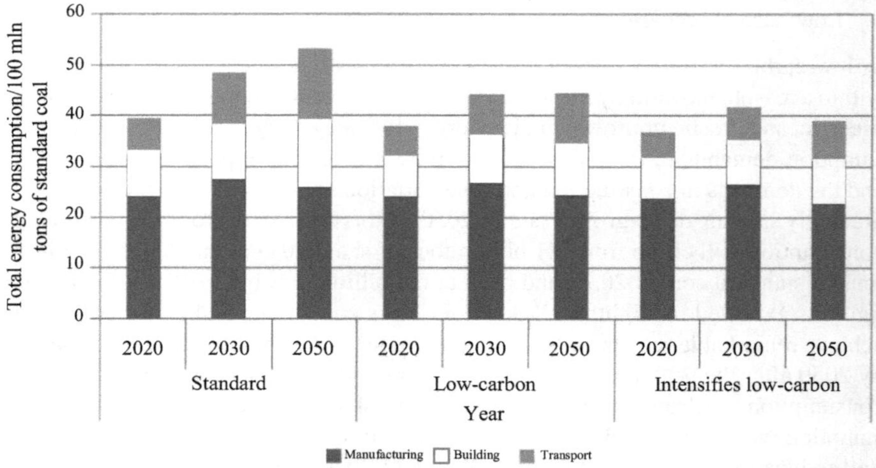


**Fig. 6.2** Total energy end uses in different sectors of three scenarios

tons of standard coal by 2020, peak at 6.1 billion tons by 2040 and then slightly fall as shown in Fig. 6.3. Policies containing coal consumption will achieve remarkable results, with the consumption sharply falling back to 2.0 billion tons by 2050 after the peaking of 4.0 billion tons before 2020 as a consequence of even more consumption of clean energy. The share of coal consumption in the primary energy consumption will reduce to 45% by 2030 from 57% around 2020, and further down to 24% by 2050. The consumption of nonfossil energy will outperform the set target. Its share in the primary energy consumption will rise to 25% by 2030 from 16% around 2020, and eventually climb to 46% by 2050. After 2020, the application of CCUS technologies will start in pilot projects and then spread rapidly to the power generation and

**Fig. 6.3** Primary energy consumption in different scenarios

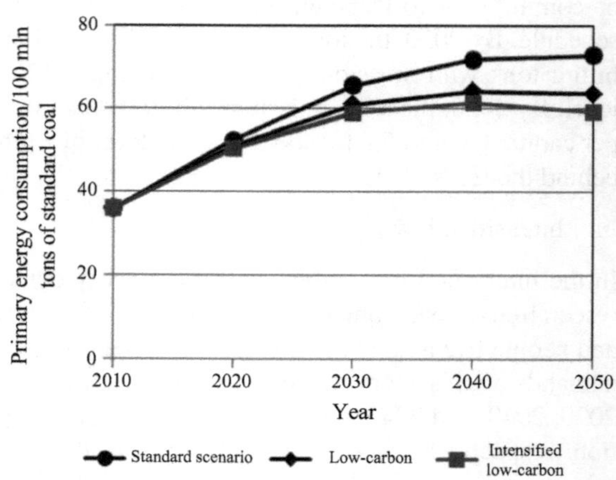

the industrial sector. As a result, the amount of carbon dioxide stored will rise to 2.7 billion tons by 2050 from 10 million tons around 2020. By 2050, energy consumption per capita will be 4.4 tons of standard coal, with the intensity down to 0.75 tons of carbon dioxide emissions per ton of standard coal, and the energy consumption per unit of GDP standing at 0.23 ton of standard coal per RMB 10,000, signifying the gap with the level of developed countries is further narrowed down.

Carbon emissions from sectors of manufacturing and building will peak around 2020 and 2030, respectively. Emissions from the transportation sector will reach the peaking around 2040. By 2020, the total emissions will reach 9.5 billion tons, and intensity of carbon emissions will be cut off by about 50% on the 2005 level, signifying the fulfillment of commitment to Copenhagen conference for climate change in 2009 ahead of schedule. By 2025, the total carbon will reach the peaking of 10.4 billion tons, with emissions per capita standing at 7.5 tons. This means China will reach the peaking on the condition that its emissions per capita are lower than the level of major developed countries when they peak in terms of carbon dioxide emissions. By 2050, the total emissions will fall down to 4.4 billion tons, with the figure per capita down to 3.3 tons, close to the level of China before 2000 as shown in Fig. 6.4.

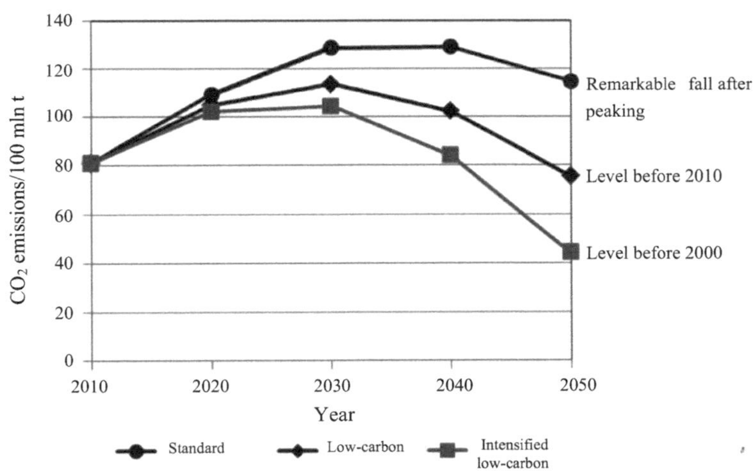


**Fig. 6.4** $CO_2$ emissions in different scenarios

### 6.1.4 Low-Carbon Development Path for Different Sectors

i. Reducing carbon emissions in energy sector is significant to reach the peaking sooner at a lower level

In the three scenarios, China's total energy consumption remains an upward trend, with the figures by 2050 up 1.6-folds, 0.8-fold, and 0.6-fold on the 2010 level, respectively. The goal is to reach the peaking of carbon dioxide emissions sooner at a lower level. Besides controlling service amounts and upgrading industrial structure, we need to depend on building a low-carbon system for energy production and consumption so that the emission intensity per unit of GDP decreases at a rate far lower than the annual growth rate of energy consumption. Lowering carbon emissions in the energy sector is of great significance to China's low-carbon development, but it faces daunting challenges. This can be seen from comparing the targets of China's low-carbon system for energy production and consumption with the targets set in the *U.S. Clean Power Plan*. Provided the U.S. realized the targets of reducing carbon dioxide emissions from power plants by 30% by 2030 on the 2005 level, which is set in the *Clean Power Plan* released in June 2014, it would mean that 120–150 million kW of powers supposed to be produced by coals would be generated by clean energy, and the newly added installed nonfossil fuel generating capacity would be 100–200 million kW. To reach the peaking of carbon dioxide emissions around 2030, China has to install more than 1 billion kW of new nonfossil fuel generating capacity by 2030 on the 2010 level and supply over additional 900 million tons of standard coals. The task of optimizing energy mix for China is unprecedentedly tough seeing from the globe. As a result, China has to proactively integrate itself to global low-carbon development of the energy sector, strengthen international cooperation across the board, and promote energy revolution in terms of energy production, consumption, and technologies so as to ensure the low-carbon sustainable development of its energy, economy, and society.

ii. Containing increment, optimizing stocks, improving efficiencies, and replacing power with clean energy remain key tasks for low-carbon industrial development

Industrial activities consume the most energy. Since 2000, energy consumption in industrial activities has accounted for around 70% of the total. In 2010, as high as 73% of total carbon emissions were from industrial activities. Heavy industry and manufacturing remain the dominant industrial sectors in China, and the penetration rate of advanced technologies to improve energy efficiency is still below 50%, with some sectors remaining as low as 10% and only a few of sectors reaching 70%. Given such conditions, there is room to further upgrade the industrial structure and improve energy efficiency. Moreover, the current energy consumption per capita in China is one-third of the U.S. level and 60% of the EU and Japan level. As urbanization advances, improving living conditions will boost demands for services and transportation. In consequence, energy consumption and carbon emissions in these sectors will climb up rapidly. In this context, the sooner emissions in industrial activities peak and go downward, more room can be left for other sectors to develop. Then, it is possible to reach the peaking of total carbon dioxide emissions sooner. For low-carbon industrial development, major industrial production should be contained and backward production should be eliminated. High energy-consuming sectors like steel and cement making should reach the yield peaking by 2020. China is speeding up efforts in integrating IT application into industrialization and fostering high-tech and strategically emerging industries. By 2020, strategically emerging industries will become the leading driving forces for national economy, with the value accounting for up to 15% of the GDP. By 2030, strategically emerging industries will have been well developed corresponding to the world's leading level, and backboned the sustainable development of economy and society (Fig. 6.5). China will devote greater efforts in reducing emissions in industrial activities. Giving full play to the late-mover advantages, China will upgrade technologies to contain increment of energy consumption in industrial activities within 30%, and reduce energy consumption per unit of industrial added value by more than 75% on the 2010 level by 2050, while maintaining rapid growth of pillar industries. China will promote technological advances for emission reduction in industrial activities and replace power with clean energy. The shares of natural gas and electricity in energy end use will rise over 60% by 2050 from 27% in 2010. On this basis, total energy consumption in industrial activities will peak at 2.6 billion tons of standard coal around 2030, and carbon

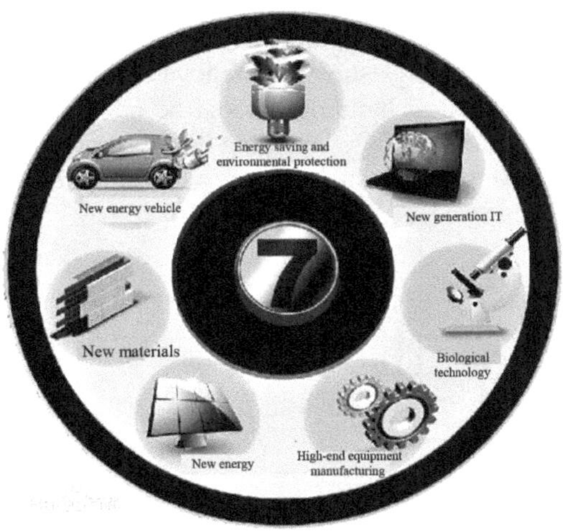


During the 12th Five-Year Plan period, seven sectors of energy saving & environmental protection, new-generation information technology, renewable energy, new materials and new-energy vehicle were made strategically emerging industries in an effort to boost China's indigenous innovation.

**Fig. 6.5** Strategically emerging industries during the 12th Five-Year Plan (*Photo source* http://js.people.com.cn/html/2012/05/31/113425.html)

dioxide emissions will peak at 7 billion tons. With further decline in the output of high energy-consuming products and wide application of CCUS technologies in industrial activities, total carbon dioxide emissions will fall back around 70% of the 2010 level by 2050.

iii. Containing total amount, improving efficiency, and using more renewable energy are the focus of efforts in boosting low-carbon development of the construction sector

The energy consumption maintaining per unit of area of buildings is only one-fourth to one-third of the developed countries' level as shown in Fig. 6.6, but it is based on the current poor services. Meanwhile, the efficiency of energy utilization in China's construction sector falls far behind that in developed countries (Fig. 6.7). The biggest reason is that the level of energy utilization in China's construction sector still lags behind the international leading level. Given urbanization rate is far lower than the average level of developed countries, and that people in many regions in China still have no access to heating system in winter, the improving housing conditions and services in future urbanization will further drive rapid increase of energy consumption in the construction sector, which will emerge as the main source of increment in energy demands.

To promote low-carbon development of the construction sector, we need to first of all effectively curb destructive demolition and reconstruction. An inclusive and green lifestyle should be advocated so that by 2050, the area of public and residential buildings per capita can remain below and even lower than the average area per capita of the 2005 level in developed countries in the low-carbon scenario. Meanwhile, the existing buildings should be renovated for energy saving, and the energy-saving standard for new buildings should be improved. These efforts aim at aligning energy-saving efficiency of building envelope and other structures with international leading levels, and reducing energy consumption by 30–40% on the

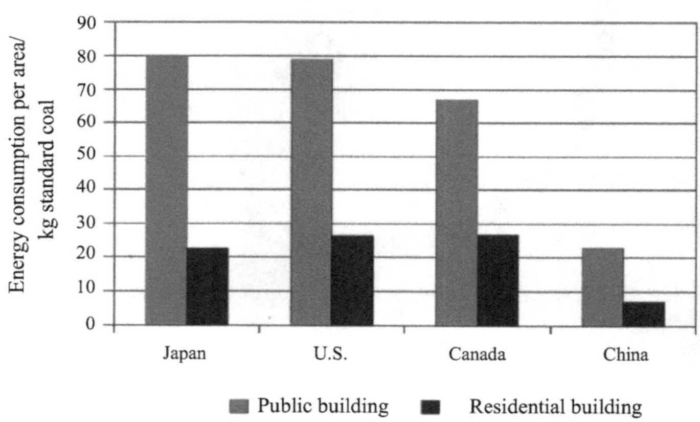


**Fig. 6.6** Energy consumption per unit area of buildings in China and foreign countries (*Source* China's Ministry of Housing and Urban-Rural Development)

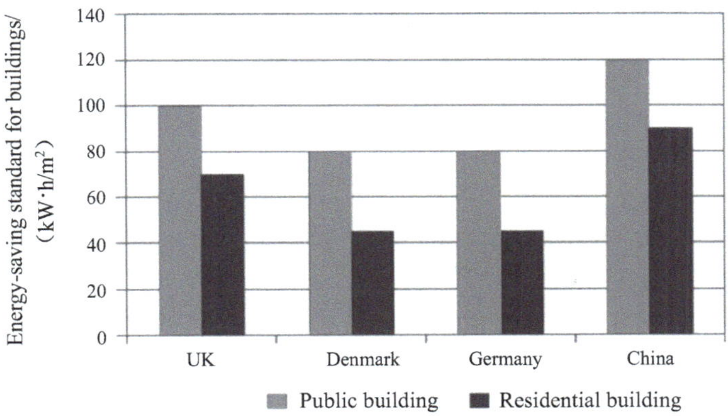


**Fig. 6.7** Energy-saving standard of buildings in China and foreign countries (*Source* China's Ministry of Housing and Urban-Rural Development)

2010 level across northern urban areas in winter. Residents are encouraged to use energy-saving home appliances, and foster a green lifestyle so that electricity consumption per capita and energy consumption per capita can be contained below 1300 KW/h and 300 kg of standard coal, respectively, when the heating system is not on (Fig. 6.8). In addition, distributed energy resources including photovoltaic power,

Renovating existing buildings for energy conservation is the main approach to reducing energy consumption on buildings. The focus is to renovate the building envelope. Measures include adding insulating layer on the exterior wall and roof, remodeling windows, installing heat insulating doors in stairwell, transforming heat metering and temperature control for heating system, and rebalancing heating system for heat converting system and pipelines.

The progress of and reduced cost of renewable energy technology make it affordable to be applied in buildings. Household PV power station generates electricity for domestic use and the surplus can be sold to power grids for profit.

**Fig. 6.8** Renovating existing buildings for energy conservation and applying distributed renewable energy in buildings (*Photo credit* http://cn.hc360.com/zj/company-44376400/product-288395.html)

solar power, and geothermal energy will be widely applied in public and residential buildings through technological innovation. Carbon emissions in the construction sector are expected to reach peaking around 2030 and begin to fall down after that. By 2050, the share of coal consumption in energy end use will be reduced to 15% or even lower from 56% in 2010.

iv. Optimizing transportation structure, increasing efficiency of automobile fuels, and improving energy supply are the focus of efforts in promoting low-carbon development in the transportation sector

Energy consumption per capita of the transportation sector was 0.25 ton of standard coal in 2010, far lower than those of the U.S. and Japan in the same period, which were 3.04 tons and 0.91 tons, respectively. In China, every 1000 people only have 58 vehicles, less than half of the world's average (150 vehicles per 1000 people) as shown in Fig. 6.9. Given increasing demands for transportation as a consequence of future population expansion, rising urbanization rate, and people's living conditions, as well as rapidly growing demands for passenger and freight transportation and more ownership of private cars, energy consumption, and carbon emissions in the transportation sector will maintain rapid growth for a long period of time (Fig. 6.10).

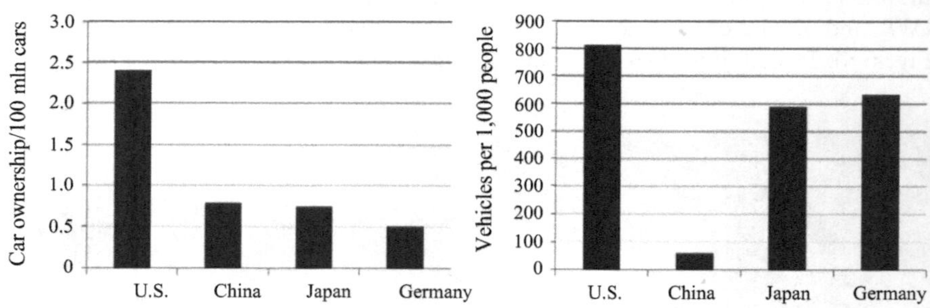


**Fig. 6.9** Car ownership and vehicles per 1000 persons in major countries

**Fig. 6.10** Emission level of new passenger vehicles in major countries

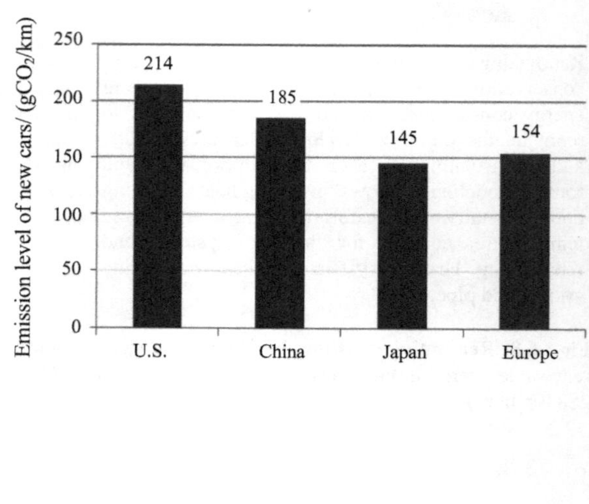

To promote low-carbon development of the transportation sector, we should advocate green lifestyle and consumption among the public, and encourage people to take bicycles and buses for transportation in an effort to contain the increasing car ownership and popularize transit trips. By 2050, 50% of residents will take public transit, with the rate in super big and megacities reaching the level of international metropolis; car ownership per 1000 people will be contained within one-third or half or even lower than the level of Japan and Europe. On the other hand, while ensuring the reasonable growth in people's transport demands, China will curb the excessively rapid increase in rotation rate of goods and passenger transport. By 2050, the figures will be contained within tenfold and fourfold on the current level. Besides, energy-saving technologies like highly efficient gasoline-fueled vehicles will be promoted so as to raise the efficiency of passenger and freight transport. By 2050, energy consumption per unit of passenger and goods transport will be reduced by about 25% on the 2010 level; the ownership of electric vehicles will account for 60% of the total automobile, reaching a leading level worldwide (Fig. 6.11).

v. Encouraging low carbon consumption is the key to attain national low-carbon development

Increasing demands in end-use sectors like construction and transportation in consequence of advancing urbanization are the major source of energy consumption and carbon dioxide emissions in the future, which should be well regulated and curbed. At present, 80% of China's population lives east of the Aihui–Tengchong line, the area of which is less than one 30th of the world's total. The population density is therefore five times higher than the world's average. This special national condition determines that

China used to be known as a bicycle kingdom thanks to its once record high bike ownership hitting 500 million. People riding bicycle used to be ubiquitous in cities across China. Cycling was the most common daily exercise for Chinese. As the living conditions improve, traffic flows made of automobiles replaced the ubiquitous scene of riding crowds, also bringing about surge in energy consumption and pollutant emissions. Calls for returning to green transit from the era of automobiles have been on the rise.

**Fig. 6.11** The used-to-be bicycle kingdom calls for the return of green transit (*Photo credit* left photo: Beijing Traffic Management Bureau; right photo: //0755.s1979.com/auto/20150407/07142420107.shtml)

energy consumption per area in east China has far outstripped the world's average, and the loading capacity of natural resources and environment. Calculations show that coal consumption per area in east China, referring as coal consumption density, is 12-fold of the world's average. As high as half of the consumption is generated by direct combustion, which pollutes the environment the most. Meanwhile, thanks to the rapid increase in travels by car fueled by growing demands for transportation in urban areas, fuel consumption density has reached threefold of the world's average. In this context, the density of carbon dioxide emissions in east China has reached six-fold of the world's average, and the environment load exceeds more than five times of the world's average. Considering the condition that the population and natural resources are unevenly distributed, China cannot continue the current development path any more in terms of the loading capacity of natural resources. If it continued the prevalent high carbon consumption, encouraged the common pursuit of bigger cars and houses, and set the consumption per capita in developed countries like the U.S., Europe, and Japan as the targets for China's modernization, its energy consumption would have magnified by four fivefold or two threefold on the current level, resulting in energy consumption, as well as pollutants and carbon dioxide emissions hitting the loading ceilings of environment, further restraining the sustainable development of China's society and economy. If we adopt a low-carbon intensive lifestyle and consumption pattern, transform the development path, lifestyle, and consumption on the premise of satisfying the reasonable demands for rising living conditions, we can attain modernization with lower energy consumption per capita while maintaining relatively high living conditions. As energy saved in end uses will have magnifying benefits in energy production side, China's pressure of energy shortage is expected to be relaxed greatly. It is more likely that China could finish the building of ecological civilization and sustainable economic and social development.

### 6.1.5 Conditions for Attaining Low-Carbon Scenario

Shifting to low-carbon development will greatly promote the sustainable economic and social sustainable development, but the inertia forces of high-carbon development remain strong. Besides, challenges in industrial development, policies, technologies, mindsets, fundamental capabilities, and other areas remain daunting, being the main restraints and uncertain factors affecting the attainment of low-carbon scenario, and even the intensified low-carbon scenario. In terms of industrial structure, through unremitting efforts for upgrading the industrial structure, the share of the tertiary sector in the total GDP outnumbered the share of the secondary sector for the first time in 2013, but the latter is expected to remain a dominating share in a short run. High energy-consuming sectors of mining and steel, building materials, and cements dominate the secondary industry, which aggravates high carbon emissions in China's economy. Excessive dependence of economic growth on resources and factors input has not fundamentally changed. In terms of policies, existing energy-conserving policies are not fully enforced; special policies like carbon emission trading have not been instituted; the long-term funding system by governments at all levels, financial institutes, and social entities for low-carbon technologies, projects, and works has not taken shape; and there is no sound pricing mechanism to properly reflect the scarcity of resources, costs of environmental damages, and market demand mix, nor market systems for carbon trading. Core technologies for energy saving in traditional industries like power supply, transportation, construction, metallurgy industry, chemical industry, and petrochemical sector are still under the grip of developed countries. International technology transfer and exchanges for low-carbon technologies are not fully open. Poor research and development and innovation capabilities and backward technologies in low-carbon development have constrained the development of low-carbon industries and economy in China. In terms of mindset, amid sluggish global economy and slowing down national economy, local governments prefer export-oriented and investment-driven development for short-term economic growth. As a result, they tend to resist transforming to low-carbon development, which exacerbates the difficulty of economic transformation. In terms of fundamental capabilities, the systems for supervising carbon emissions and evaluating the performances of component departments are incomplete, resulting in incomplete statistics on each industry. Management for energy saving and emission reduction at the primary level is not well established. Law enforcement in this regard is poor, and the problems of high cost in abiding by the regulations in striking contrast with low cost in breaking them have not been effectively solved.

In order to promote low-carbon development and reduce emission to the maximum extent, the following basic conditions should be achieved:

(1) Maintaining the growth rate at a reasonable level. From the perspective of low-carbon sustainable development, it is unnecessary and impossible to maintain high-speed growth for the purpose of attaining the targets of completing the building of a moderately prosperous society by 2020 and achieving modernization by 2050. The realities require us to voluntarily raise the awareness of low-carbon development, change the mindset of placing GDP growth at the top position, and get rid of obsession with growth rate or worries about gearing down economic growth. Instead, we should gradually lower annual GDP growth rate down to the decrease rate of carbon emissions per unit of GDP, and maintain the energy consumption elasticity coefficient to and at a lower level of 0.4.

(2) Making significant progress in promoting low-carbon development for end-use sectors. The development of end-use sectors determines China's future energy demands. The peaking of carbon emissions in these sectors should be based on primary completion of China's industrialization and its urbanization reaching more than 65%, when energy consumption in industrial sectors and the construction sector becomes stable and that in transportation sector slows down. China's low-carbon development should satisfy the demands of building a modern industrial system and sustainable construction and transportation systems. With technological innovation as the focus of efforts, we should improve energy efficiency, development of renewable energies, and large-scale

low-carbon businesses. We should build a statistical and assessing system for energy consumption and carbon emissions covering all areas, sectors, and major energy consumers so as to realize effective monitoring, accounting, and evaluation. Guided by the government, enterprises and the public should actively practice low-carbon manufacturing and a green lifestyle.

(3) Attaining substantial advances in optimizing energy mix. The energy sector is the core for low-carbon development, which requires the share of natural gas, nuclear energy, and renewable energy, which accounts for and exceeds 50% of the energy mix and keeps on an upward trend. Newly added energy demands should be mainly satisfied by renewable energy, and fossil fuels consumption should be maintained at a stable level. It is no easy task to maintain high-speed development of nonfossil energy. To achieve it, there must be effective supporting fiscal, financial, industrial and investment policies, and a market-based pricing system truly reflecting the scarcity of resources and environmental costs should be established. The reform of energy pricing should be properly linked with the innovation of fiscal and taxation policies.

(4) Furthering enhancing carbon sequestration of forest ecology. Carbon sequestration of forest features advantage of low cost that energy-saving efforts carried out in each sector have not. While attaining minus emissions, it will improve the entire ecology, making carbon sequestration an important means for China to tackle climate changes. In the future, it is necessary to further expand forest areas, improve the forest quality, and enhance carbon sequestration. By 2020, China's forest coverage rate will reach 23%, and forest stock will reach 14 billion $m^2$. Both the area and stock of forest will be significantly increased. On this basis, the forest coverage rate will total 28% by 2050, and sustainable forest management will be attained.

## 6.2 Stage Goals for China's Low-Carbon Development

In the process of pursuing low-carbon development, we should establish a philosophy of promoting ecological civilization, work actively to explore a new path for low-carbon industrialization and urbanization, accelerate efforts to transit to low-carbon development, improve or upgrade the structure of industry, realize efficient, low-carbon and clean energy production, supply and consumption, blaze a new path for low-carbon development that is better than that of major developed countries, reach the peaking of carbon dioxide emissions per unit of GDP and per capita at as low figure as possible, and take innovative measures to address the regional pollution and global warming at the same time. To achieve these goals, we must coordinate efforts of related industries in transiting to low-carbon growth, curb excessively rapid growth of total carbon emissions as soon as possible, and promote the peaking of emissions by 2030 and gradually falling of it to 2010 level by 2050 on the basis of fulfilling the emission commitment made at the Copenhagen meeting. In the process, we should improve the ecology in an all-around way, raise the global competitiveness of related industries significantly, ensure that China occupies a leading position in the world in sustainable development, and strive to achieve the Chinese Dream featuring economic prosperity, national rejuvenation, people's happiness, and harmonious society ahead of the schedule.

Specifically, we may make the following considerations and arrangements for targets of each stage for low-carbon development:

Between 2010 and 2020, the target is to transit the growth path from extensive to intensive development. With gradually declining share of highly energy-consuming industries, the industrial energy will be constantly improved, and GPD will steadily grow at around 7.2%. By 2020, China's GDP will reach about RMB 81 trillion, with GPD per capita reaching RMB 58,000, signifying that a moderately prosperous society in all respects has been built. With growing economy at a reasonable speed and improving living conditions, energy consumption and carbon emissions slow down and will reach the peaking by 2020. The output and carbon emissions of highly energy-consuming industrial sectors will peak before 2020. By the year, total primary energy consumption will be contained within 5.1 billion tons of standard coal, total carbon dioxide emissions will remain below 10.5 billion tons, carbon dioxide emissions per unit of GDP will reduce by 48% on the 2005 level, the share of nonfossil energy in the total energy consumption will account for 15%, carbon dioxide emissions in some economically developed regions will peak, and the national environmental quality will remarkably improve.

Between 2020 and 2030, with the basic completion of industrialization, urbanization will be the main driving force for economic growth. At this stage, annual GDP growth rate will remain below 6%; the shares of the secondary and tertiary sectors in the economy will be improved to around 40% and 55%, respectively; urbanization rate will be improved to 68% and China will become a high-income country. The annual growth rate of energy consumption will show a downward trend and energy consumption elasticity will gradually decline to the level of developed countries when they finished industrialization. In this context, by 2030, the total consumption of primary energy will be contained under 6.1 billion tons of standard coal; consumption of fossil fuels will peak; nonfossil energy will replace fossil fuels as the main sources to satisfy energy supply, and account for more than 20% of total primary energy consumption. On the premise of securing sustainable social and economic development, carbon dioxide emission intensity per unit will fall by 20% on the 2010 level; carbon dioxide emissions per unit of GDP will reduce by some 65–70% on the 2005 level, and total carbon dioxide emissions will be contained below 11.5 billion tons.

Between 2030 and 2050, China will become an innovation-driven and energy-saving country, and its main driving forces will shift to technology innovation and robust domestic demands. By 2050, China's urbanization rate will reach around 75%, signifying the completion of the process; the share of the tertiary sector in total GDP will be around 65%; GDP per capita will correspond to the current level of developed countries; and the second centenary goal of building China into a prosperous, democratic, culturally advanced, harmonious, and modern country will be attained.

By that time, China will truly realize the transformation toward low-carbon economy and society on the basis of high-degree development; low-carbon energy will be developed on a large scale; the share of low-carbon energy including natural gas and nonfossil energy will account for 50% of total primary energy consumption; and energy-saving and low-carbon production and lifestyle will be fully established. The ecology will be improved in an all-around way and carbon emissions will fall back below the 2010 level. By that time, any connections that China's economic growth has with energy consumption will be untied. Low-carbon energy development will be further accelerated, and a green, low-carbon, and productive energy mix will be established by the end of the twenty-first century.

## 6.3 Basic Thoughts on China's Low-Carbon Development

According to the general laws of low-carbon development in developed countries, the timing and road maps for the peaking of carbon emissions in China will vary by region and sector. From a geographical perspective, carbon dioxide emissions in a few economically developed regions will peak earlier than the east region, which will outpace the national peaking. By sector, the industrial sector will outpace the construction sector and the transportation sector in peaking carbon dioxide emissions. On the premise of fulfilling the promise of peaking carbon emissions by 2030 made in the *China–U.S. Joint Statement on Climate Change* in 2014, industrial sectors can reach the peaking around 2025 or even earlier. Before the peaking period, the national energy consumption and carbon dioxide emissions will slow down, and then will be maintained at a stable level slightly lower than the peaking value for a period of time, and then gradually decline.

The timing and plans for low-carbon development are shown below:

(1) Between 2010 and 2020, most of the economically developed regions in east China have the carbon dioxide emissions approaching to the peaking value. In 2015, the emissions of industrial sectors in economically developed regions in east China's Beijing–Tianjin–Hebei region, the Yangtze River region, and the Pearl River region first reached the peaking values. By 2020, the majority regions in east China will have the emissions peak or approach to the peaking; total energy consumption and greenhouse emissions will continue to increase, but should be contained within 10.5 billion tons.

To achieve those targets, the key lies in maintaining a reasonable growth rate and improving the quality of economic growth. While attaining the goal of doubling GDP by 2020 on the 2010 level put forth in the report to the 18th National Congress of the CPC, China should curb the impulses of local governments in pursuing GDP growth and investment, contain the expansion of highly energy-consuming industries, and solve overcapacity. During the period, the focus on reduction of carbon emissions will lie in structural carbon reduction and improving the energy mix. Years from 2010 to 2020 will be a period of drastic changes in China's industrial structure. With accelerating adjustment of industrial structure, and reducing share of highly energy-consuming sectors, more emissions will be avoided as a result of improving energy mix. At the same time, it is also a period during which nonfossil energy grows rapidly. By 2020, the share of nonfossil energy in the primary energy consumption will rise to around 15%; the share of natural gas in fossil energy will be further improved. With abovementioned measures putting in place, energy consumption and greenhouse emissions per unit of GDP will significantly decline. It is expected that energy consumption per unit of GDP will reduce by more than 25% on the 2010 level, and the carbon dioxide emissions per unit of GDP will decrease by nearly 45–50% on the 2005 level by 2020, signifying exceeding the established target made in the Copenhagen climate change conference.

(2) Between 2020 and 2030, China's total carbon dioxide emissions from energy consumption will be close to the peaking value. Between 2021 and 2025, carbon dioxide emissions in industrial sectors will peak, and the emission increase in construction and transportation sector will be effectively contained. Around 2025, total carbon dioxide emission from energy consumption in economically developed regions will reach the peaking. Efforts will be made to peak the emissions from national consumption before 2030, with the figure being contained within 11.5 billion tons, and emissions per capita within 8 tons. Before 2030, carbon dioxide emissions from energy consumption in more economically developed regions will peak. In this way, the total carbon dioxide emissions from energy consumption will basically reach the peaking value around 2030.

To achieve the abovementioned goals, the key lies in transiting China's economy to intensive growth after completing industrialization. Around 2020, China will basically realize industrialization. Total carbon dioxide emissions in the industrial sectors will be stable, while the construction sector, transportation sector, and domestic consumption will become major sources for increase in energy consumption and carbon dioxide emissions.

During the period, structural carbon reduction and optimizing energy mix will continue to play an important role. As China basically achieves industrialization, it will boost low-carbon businesses and technologies, and foster emerging industries and new momentum for economic growth as an effort for restructuring the economy and intensive development. Economic restructuring will create immense high-quality jobs and improve the international competitiveness of Chinese economy. During the period, energy consumption per unit of GDP will continue to decline. Between 2020 and 2030, rapid development in renewable energy and nuclear energy will increase the share of nonfossil energy in the primary energy consumption to 20% before 2030. Moreover, introducing policies to control the total amount of fossil energy as soon as possible will ensure the carbon dioxide emissions from energy consumption to reach the peaking around 2030 or as early as possible.

(3) After 2030, carbon dioxide emissions in construction and transportation sectors will be close to the peaking. After reaching the peaking, the total energy consumption and carbon dioxide emissions will remain at a level slightly lower

than the peaking value and then gradually decline. During the period, while propelling a steady decline in emissions in relatively economically developed regions, China should make efforts to ensure the peaking in less developed regions and stabilize emissions in the construction and transportation sectors.
To achieve the abovementioned goals, the key lies in basically completing the transition to low-carbon economic development. This includes two aspects. First, the economic growth will be driven by industrial innovation and low carbon consumption. Second, newly added energy demands will be satisfied by nonfossil energy, and energy consumption demands in construction and transportation sectors will be stable.
During the period, innovation and nonfossil energy will play a decisive role. High-tech and high-intelligence industries with low- or zero-carbon emissions will rise to be dominating industries. The effects of industrial innovation will be more prominent in reducing carbon dioxide emissions. With the progress of technologies, there is a possibility that minus-carbon business will flourish based on carbon absorption and storage technologies. At the same time, newly added energy demands will be fulfilled by nonfossil energy. Clean and renewable energy will break the restrictions of difficulty in storing and become safer and easier in use. Besides, China will emerge as a consumption-saving and environmentally friendly country, and enter the stage of innovation-driven development. Emissions in consumption sectors like construction and transportation will become the main growth area of carbon emissions. As the public has stronger awareness of low-carbon development and in practicing green lifestyle, low carbon consumption will become a major approach to reducing carbon emissions.

# Chapter 7
# Low-Carbon Energy: Foundation of Low-Carbon Development

## 7.1 Concept and Meaning of Low-Carbon Energy

Energy is the essential material basis for economic growth and social development. Energy production and consumption generate about two-thirds of global greenhouse gases (GHGs) emissions. The key for transiting to a low-carbon society lies in the transformation of the energy sector toward low-carbon development. It involves all sectors from energy production to processing and consumption in multiple areas of industry, construction, and transport. Closely linked with industrialization, urbanization, globalization, and IT application, low-carbon development of the energy sector is a protracted and complicated systematic project.

In a narrow sense, low-carbon energy development mainly refers to the shift of dominating energy supply from fossil energies to nonfossil and low-carbon energy, or the shift of energy supply system from those with high carbon emissions to relatively low emissions and the reduction of carbon emissions intensity of energy production and consumption per unit (Fig. 7.1). Among various primary energies widely used worldwide, renewable energies including hydropower, wind power, solar power, geothermal power, and ocean energy generate zero greenhouse gas emission, so does nuclear energy. Biomass energy, as a type of renewable energy, combusts and generates carbon dioxide, but seeing from the life cycle of biomass, the emission is basically equivalent to the amount discharged in photosynthesis, and the emission can be regarded as zero. As an exception, using some kind of biomass energy will produce methane, which is polluting. In general, biomass energy can only be regarded as a type of energy that produces few greenhouse gas emissions.

Low-carbon energy development is a relative concept with no absolute standards. According to the *Guidelines for National Greenhouse Gas Inventories* released by the Intergovernmental Panel on Climate Change released in 2006, the emission factor of coal combustion was about 94.6–101 tons, petroleum was 73.3 tons, and natural gas was 56.1 tons in producing every trillion joules of power. Therefore, fossil energies fall into the category of high-carbon energy compared to renewable energy and nuclear energy, but natural gas can be regarded as low-carbon energy compared



X. Du et al., *Overview of Low-Carbon Development*,
https://doi.org/10.1007/978-981-13-9250-4_7

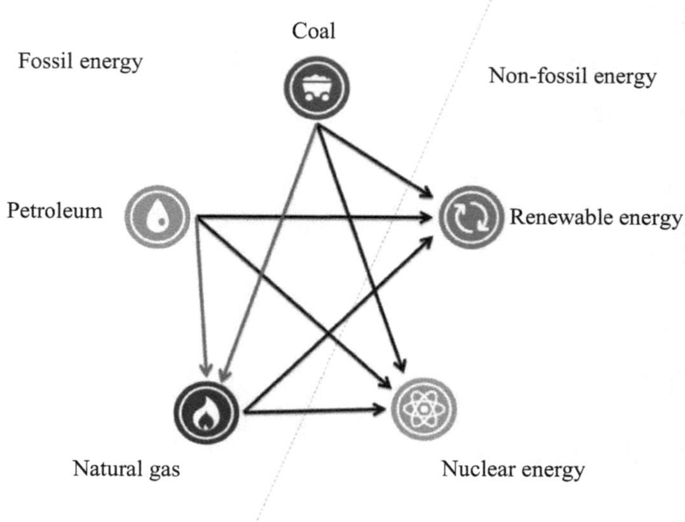


**Fig. 7.1** The path to low-carbon energy development

to coal and petroleum. Besides, low-carbon energy development is not only about reducing GHG emission in a specific section but also about GHG emission in the entire life cycle. For example, such nonfossil energy as nuclear energy, wind power, and solar energy generates no carbon emission, but its production, construction, and operation including materials manufacturing and transporting produce a certain amount of GHGs. Power consumption in end-use sectors of construction and transport does not directly produce GHG emissions, but emissions are brought indirectly in power generation. In sectors like coal liquefaction and coal gasification, the processed petroleum and natural gases are relatively low carbon compared to coals, but the production process still brings large amount of GHG emissions. For a specific country or region, low-carbon development should be planned and advanced from a systematic point of view to ensure the comprehensive optimization of the energy system.

In a broad sense, low-carbon energy development refers to low-carbon production and consumption of energy. On the one hand, energy production and supply is closely connected and interacted with the consumption pattern. Satisfying unreasonable high-carbon energy consumption with relatively low-carbon energy supply is still likely to increase the total GHG emissions, and it is not low-carbon energy development in real sense. On the other hand, the energy production and consumption system is closely linked to development path, economic structure, technological level, industrialization, and urbanization model; therefore, it is characterized as public goods and generates externality effects, and has potent lock-in effects. For a specific country and region, while promoting low-carbon energy mix, more importantly, it should lower the GHG emissions in the entire economy and society to improve the efficiency of energy use. This puts forward higher requirements for efforts for advocating low-carbon production and consumption patterns, promoting low-carbon urbanization and industrialization, and advancing the integrated development of low-carbon energy development and the application of IT and intelligent technologies. Particularly, against the backdrop of active global response to climate changes, new patterns of economic growth, accelerating social transformation, ever-changing technologies, and integration between globalization and IT application, all countries will make more efforts in exploring low-carbon paths for energy, social, and economic development.

## 7.2 Current Conditions and Trends in Global Low-Carbon Development

### *7.2.1 Continuous Progress of Low-Carbon Energy Development in Major Developed Countries*

With rapid growing world economy since the 1960s, the world's total energy consumption and carbon dioxide emissions have also soared. Meanwhile, carbon dioxide emissions in primary energy consumption per capita showed overall downward trend as shown in Fig. 7.2(a). The figure of 2013 declined to a varying extent in major regions on the 1965 level. Among them, carbon dioxide emissions in primary energy consumption per capita in Europe and Middle East continued to decrease; in North America and Africa, after a temporary rebounding in the 1990s and 1980s, respectively, the figure returned to the downward trend. By the extent of development, OECD member states witnessed overall declining, whereas non-OECD states experienced certain degree of rebounding after the entry into the twenty-first century.

Among major countries, the carbon dioxide emissions in primary energy consumption per capita continued to decline in general as shown in Fig. 7.2(b). France had the most drastic decline with the figure in 2013 halved on the 1965 level; Japan has witnessed rebounding emissions in the recent 10 years, with the figure in 2013 rising by nearly 10% on the 2003 level and by 16% on the lowest 1998 level. Carbon dioxide emissions in primary energy consumption per capita in Germany and the UK also significantly declined. China's figure continued to decrease, with 2013 level declining by 4.3% on the 2003 level, similar to that of the U.S. in terms of trending and decrease rate.

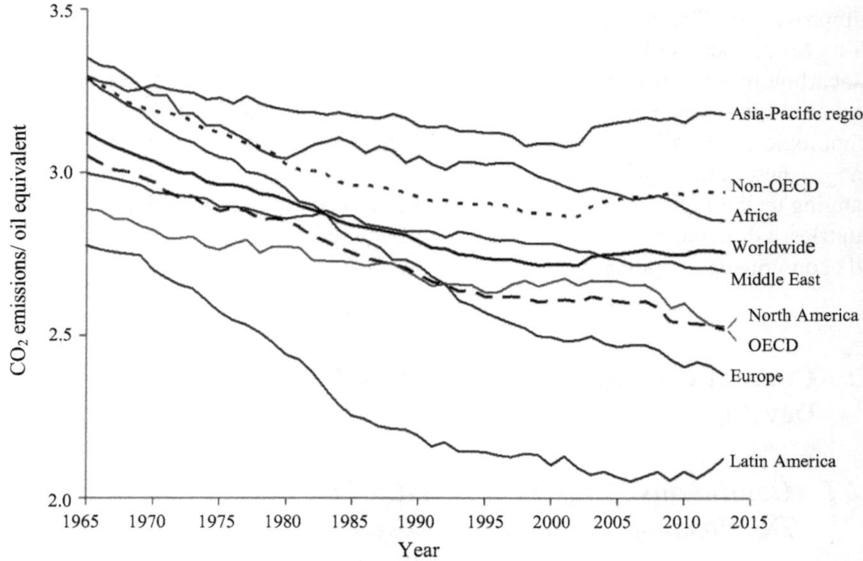


(a) $CO_2$ emissions in primary energy consumption per capita in major regions

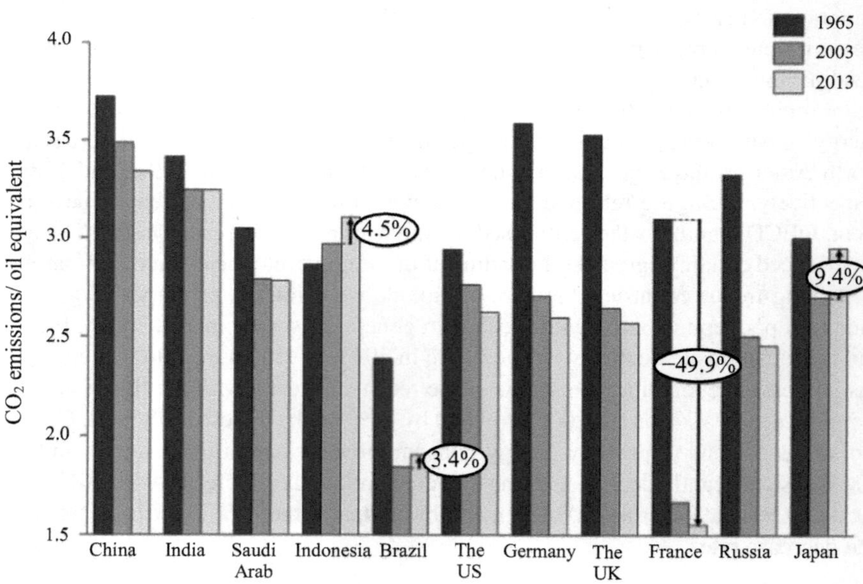


(b) $CO_2$ emissions in primary energy consumption per capita in major countries

**Fig. 7.2** $CO_2$ emissions in primary energy consumption per capita in major countries and regions

### 7.2.2 *Rising Proportion of Renewable Energy in Primary Energy Mix*

Against the backdrop of global response to climate change, the increase in energy demands for human development should not rely on fossil energies, but instead on nonfossil energies, of which renewable energy will play an essential role. Major EU countries, the U.S., Canada, and Brazil have achieved remarkable progress in developing renewable energy.

i. Increasingly developing hydropower

Hydropower is the main way of using water energy. It is the clean and low-carbon energy with the most mature technologies and highest competitiveness and can be exploited on a large scale. There are now 159 countries with hydropower projects worldwide. In 2013, hydropower consumption worldwide totaled around 860 million tons of oil equivalent, accounting for 6.7% of the world's primary energy consumption, and 16.4% of world total generated power. Hydropower generation in developed countries was 13.0% of the world's total, and developing nations 19.3% as shown in Fig. 7.3(a). In some developed countries with abundant water resources, hydropower dominates power generation nationwide. Specifically, the proportions of power generated by hydropower stations in Norway in national total power generation and primary energy consumption are the highest, accounting for 96.1% and 64.9%, respectively. More than 50% of power in Canada, New Zealand, and Switzerland are generated by hydropower stations as shown in Fig. 7.3(b).

Seeing from the development trend, against the backdrop of overall saturated and even declining energy demands, the installed hydropower capacity in many developed countries remains stagnant. Countries like Norway, New Zealand, France, Italy, Spain, and Australia have installed no new hydropower projects since 2012. The increase rate of newly installed hydropower capacities in the U.S., Canada, and Japan grew by 0.2%, 0.9%, and 1.1%, respectively, as shown in Fig. 7.4(a). However, as the energy mix is being optimized, and the energy storage capacity of the power system is increasing, pumped storage power station projects have become the new focus of investment in the power supply sector in Europe. Just in Thuringia in Germany, 13 pumped storage power stations have been planned with a total installed capacity of 5.1 GW. The installed capacity of pumped storage power stations being planned and under construction reaches 4 GW. The technology of pumped storage accounts for 99% of large-scale energy storage technologies, making it the important support for large-scale renewable energy exploitation. Therefore, it is highly valued by European countries.

The International Energy Agency (IEA) estimates that global remaining hydropower has a great potential and concentrates in Africa, Asia, and Latin America. By 2050, global installed hydropower capacity is expected to double on the current level to reach 2000 GW with annually 7000 TW/H of power generated. Given the demand of storage peaking in large-scale development of renewable energy, in the

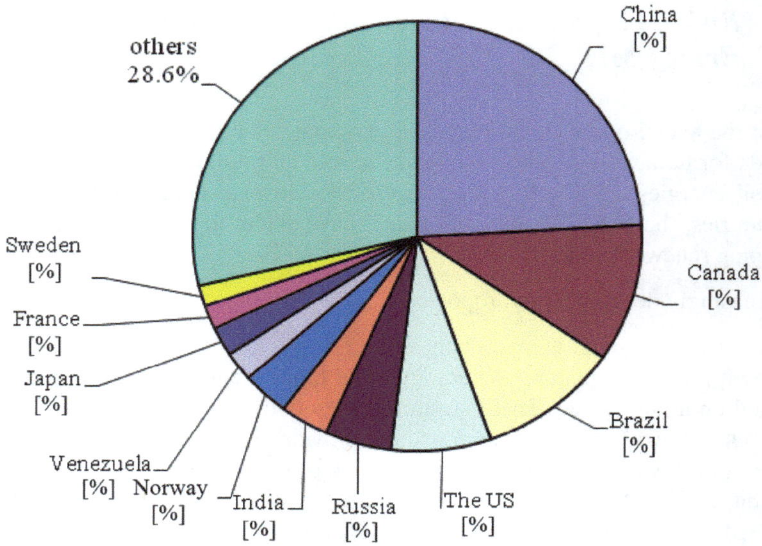


**(a) Proportions of hydropower generation in major countries in the world's total**

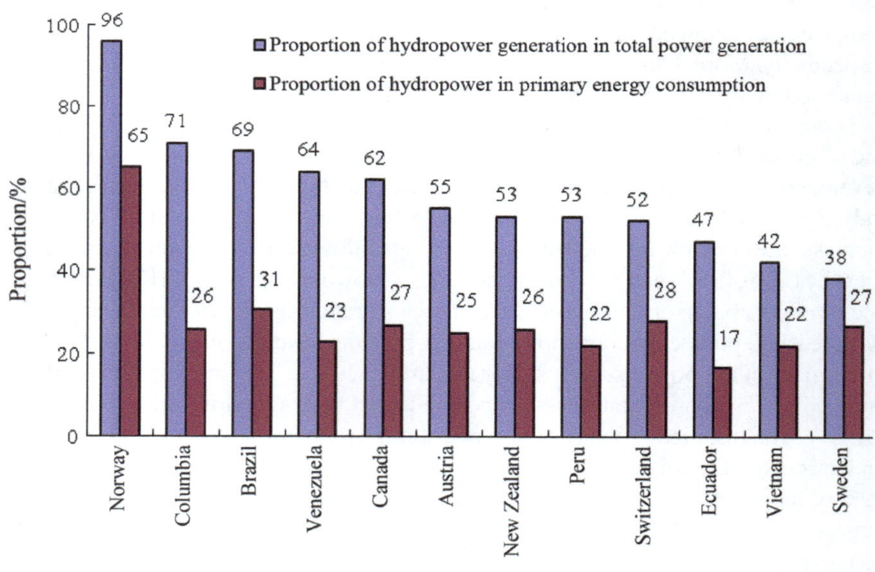


**(b) Proportions of hydropower generation in major countries in the world's total**

**Fig. 7.3** Hydropower generation in major countries

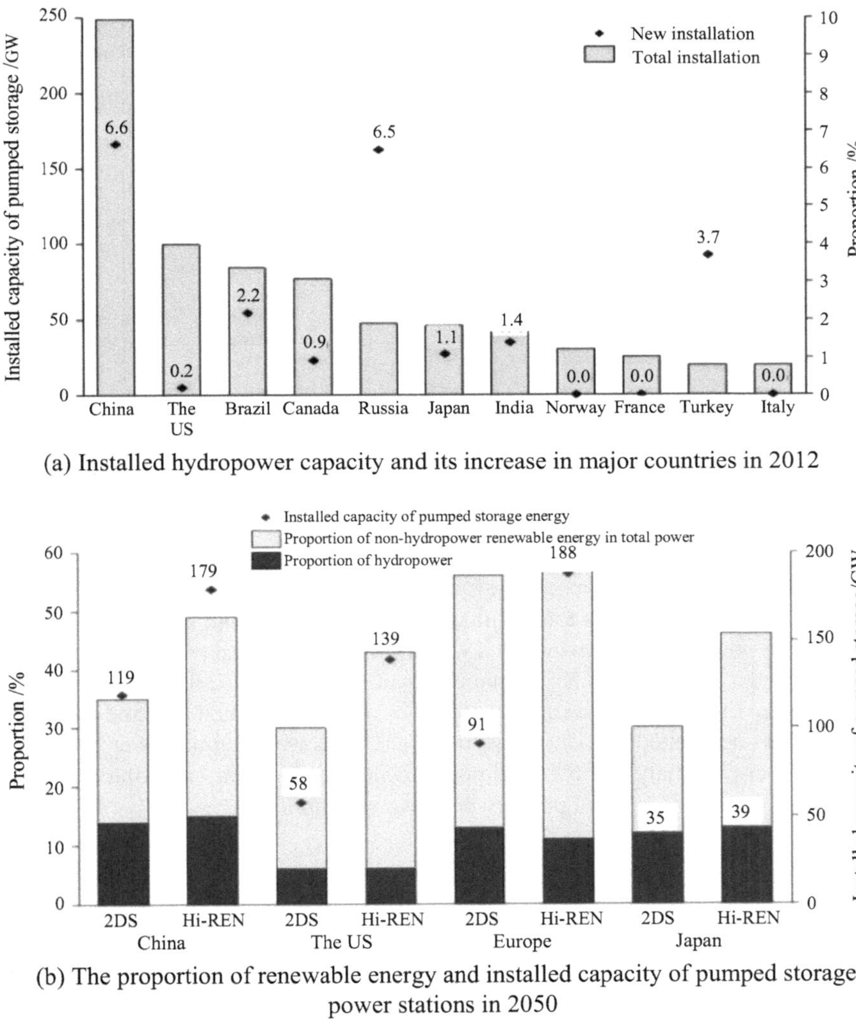


**Fig. 7.4** Installed hydropower capacity

IEA's 2DS and Hi-REN scenarios, the installed capacity of pumped storage power stations is expected to triple even quintuple to reach 412–700 GW as shown in Fig. 7.4(b).

ii Rapid growing exploitation of solar energy and wind power in major countries

Since the late 1980s, the solar energy and wind power have been rapidly growing in major countries over the past years. Between 2003 and 2013, the installed capacities of solar power and wind energy projects worldwide increased by 49 and 23% annually, and power consumption increased by 51 and 26% annually as shown

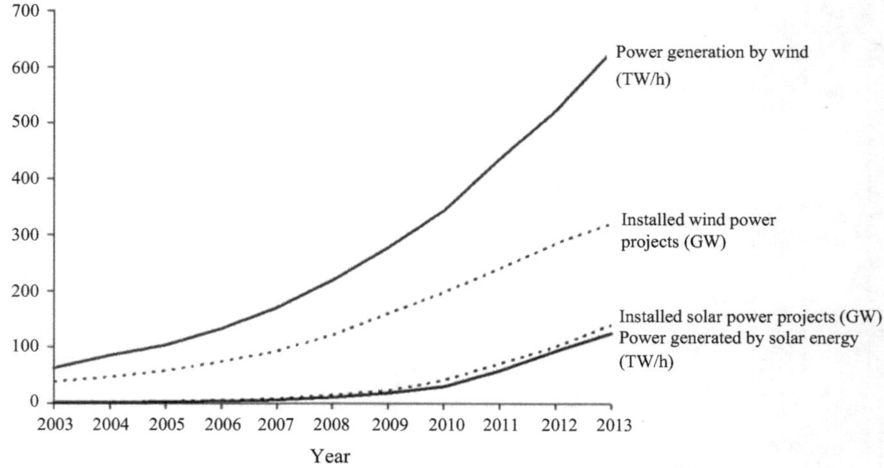


**Fig. 7.5** Power generated by solar and wind energy worldwide in 2003–2013

in Fig. 7.5. In 2013, the proportion of power generation by solar and wind energy reached 1.5% of the world's total primary energy consumption. By region, 80% of global solar power generation installed wind power projects and power consumption concentrated in China, the U.S., Germany, Japan, Spain, the UK, Italy, India, France, Australia, and Belgium. In developed countries, power generated by wind energy in Portugal has accounted for 10% of primary energy consumption; power generated by solar energy in Italy is 2.6%; and power generated by solar and wind energy in Spain is close to 10% of total primary energy consumption.

### *7.2.3 Nuclear Energy Plays an Important Role in Low-Carbon Energy Development*

Nuclear energy is a type of clean energy featuring stable operation, reliable supply, efficient cost, zero discharges of pollutants and GHG emissions, small occupation area, and little ecological impact. As long as safety measures are strict and nuclear wastes are properly disposed, nuclear energy can be the best solution for low-carbon energy development. At present, nuclear energy accounts for 4.4% of global primary energy consumption, a surge on the 1960s level, which is less than 0.5%. In global power generation composition, the proportion of nuclear energy outstripped 15% in the 1990s. However, with the rapid growing nonnuclear power in developing countries, this proportion has declined to current 10% as shown in Fig. 7.6(a).

Major developed countries have adopted nuclear energy in succession. Nuclear energy consumption accounts for the largest proportion of total energy consumption in the U.S., with the figure in 2013 reaching 188 million tons of oil equivalent.

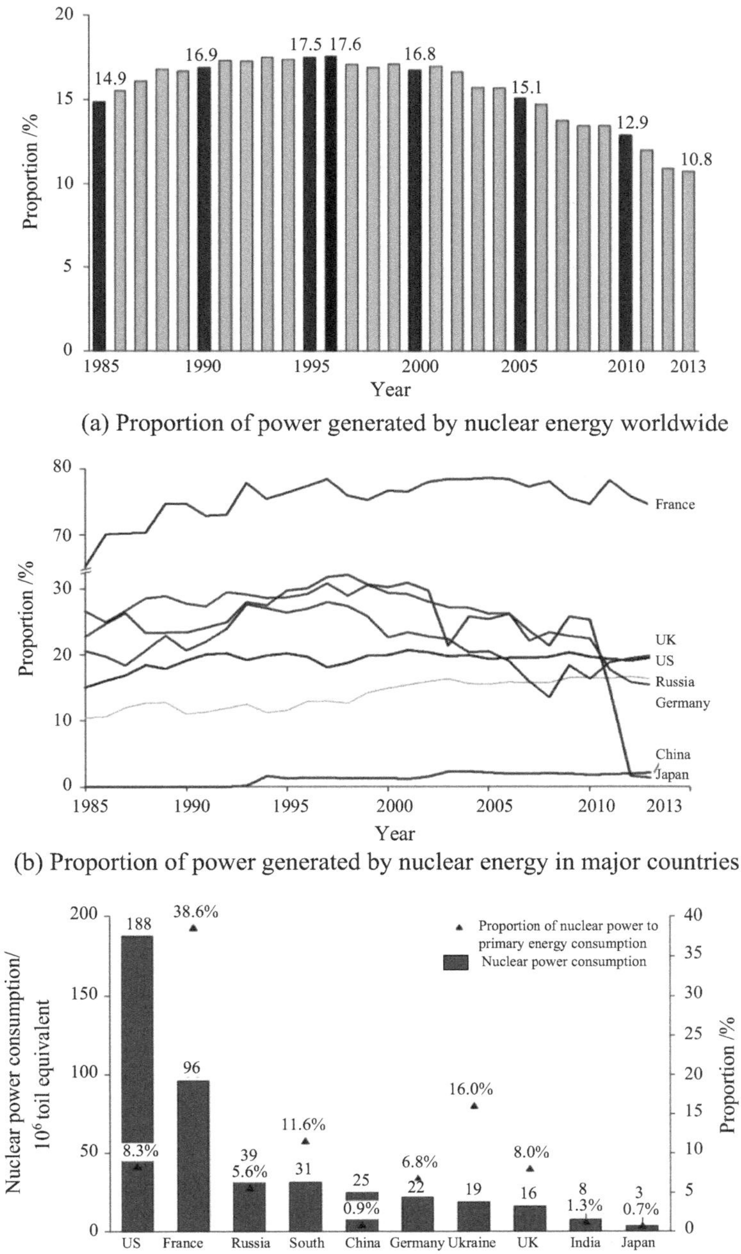


(a) Proportion of power generated by nuclear energy worldwide

(b) Proportion of power generated by nuclear energy in major countries

(c) Nuclear power consumption in major countries in 2013

**Fig. 7.6** Nuclear power generation and consumption worldwide and in major countries

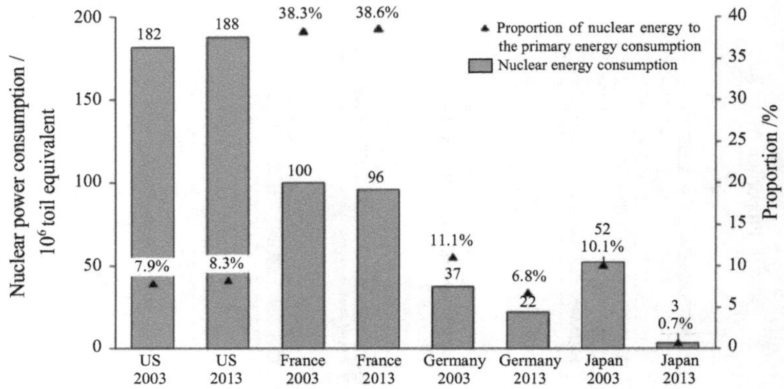


(d) Changes of nuclear consumption in typical nuclear-energy-consuming countries

**Fig. 7.6** (continued)

France follows, consuming nuclear energy equivalent to 96 million tons of oils. France's share of nuclear energy in primary energy consumption is the highest, accounting for 38.6, and 75% of total power generation, far higher than other countries as shown in Fig. 7.6(b) and (c).

After the 2011 Fukushima nuclear accident in Japan, major countries have been divided on developing nuclear energy. While ensuring the safe operation of existing nuclear power stations, the U.S. and France continue to support the development of nuclear energy. Germany immediately revised the *Nuclear Energy Law* to radically stop using nuclear energy. In 2013, Germany's nuclear energy consumption declined by 40% on the 2013 level, and the proportion to national primary energy consumption was also down by 4.3 percentage points as shown in Fig. 7.6(d). Japan, where the accident happened, shut down all nuclear power stations during Democratic Party's Noda Yoshihiko administration. However, after taking office in 2012, Shinzo Abe pledged to approve the building of more nuclear reactors.

## *7.2.4 Remarkably Accelerating Use of Natural Gases Including Unconventional Gases*

Although natural gas is not a type of energy with zero-carbon emission, it is cleaner than coal and petroleum. With its abundance and mature technologies in development, transportation, and use, natural gas has become an important alternative for many countries to shift toward low-carbon energy development.

i. The development and use of natural gases are an important basis for low-carbon energy development in the UK

With rapid growing economy and energy consumption, particularly coal consumption, since the first industrial revolution, the UK has been plagued by environmental pollution. Since the 1970s, the replacement of coal with natural gas has been the most remarkable change in the UK energy mix. In 1964, the country adopted the *Continental Shelf Act* as the legal foundation for exploring oil fields in the North Sea. During 1964 and 1973, Brent field and Amethyst natural gas field were discovered and exploited in succession. The two oil crises in 1973 and 1979 accelerated the process of exploiting oil resources in the UK. While gradually improving self-sufficiency rate of oil resources, the efforts increased the proportion of oil consumption to the primary energy consumption. The UK had emerged from an importer of petroleum and natural gas in 1973 to an exporter of petroleum in 1990, and an exporter of both petroleum and natural gas in 2000. However, with increasing use of natural gas, it had to import the resources to support domestic consumption. Its proportion to primary energy consumption increased from less than 5% of coal consumption in the 1960s to a doubling figure. As carbon dioxide emissions per unit heat provided by natural gas are only around 60% of that by coal, replacing coal with natural gas avoids 120 million tons of GHG emissions, 22% of the net GHG emissions in the UK in 2011.

Replacing coal with natural gas is the result of many influencing factors. The key contributing factor is the privatization of the power industry in the late 1980s and early 1990s in the UK. Changes in the market structure directly led to rapid growth in power supply. Huge fuel demands outstripped the supply of coal. In the face of limited coal resources and soaring price, a large number of power companies began to seek for alternative energy. At that time, natural gases were discovered in the North Sea oil field. Due to the huge supply, the price of natural gases went down by more than 50% in the 1990s. With mature technologies of combined cycle gas turbine plant, natural-gas-fueled power plants requiring less investment began to replace coal-fired power plants. At the same time, manufacturing industries and services sectors gradually increased the use of natural gas and electric energy under the circumstances of shortage of coal supply, declining price of natural gas and sufficient power supply.

ii. Shale gas revolution greatly improves low-carbon energy development in the U.S. but also sparks great controversy

Shale gas is a type of unconventional natural gas that is trapped within organic-rich shale and its formations. It is attached or found free within shale. Mainly composed of methane, shale gas is a type of clean and efficient energy. In the recent 10 years, the U.S. has become the world's only country exploiting shale gas for commercial uses on a large scale, thanks to its mature technologies and well-developed network of pipe facilities. It replaced Russia and became the largest producer of natural gas in 2009. This process is dubbed as shale gas revolution. Statistics by the U.S. Energy Information Administration shows that the domestic natural gas output in the U.S. exceeded coal production for the first time in 2009; as of 2013, the proportion of natural gas production to the total primary energy production had been 35% as shown in Fig. 7.7(a). Cross-referencing the conclusion from a 1998 EIA prediction that the U.S. carbon dioxide emissions will increase with an annual growth rate of 1.3% beginning in 2020, the actual carbon dioxide emissions did not maintain growth but began to drastically decline after 2009 as shown in Fig. 7.7(b). According to the latest *Annual Energy Outlook 2014* released by EIA, carbon dioxide emissions in the U.S. will decrease by 9% in 2020 on the 2005 level, and the country will attain energy independence by 2035. The shale gas revolution will play a significant role in this process.

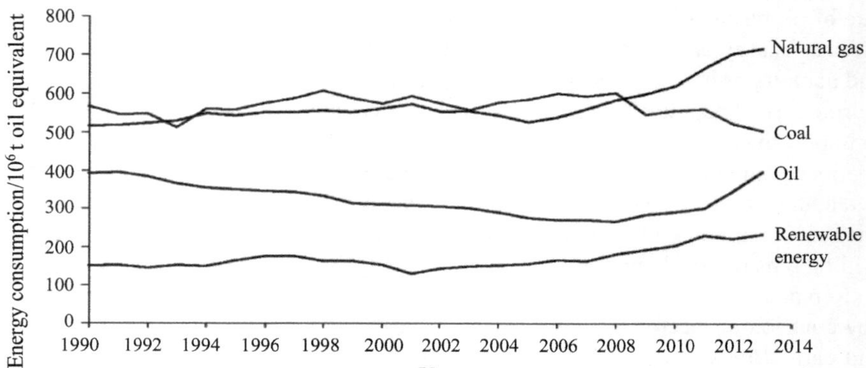


(a) Primary energy consumption in the U.S.

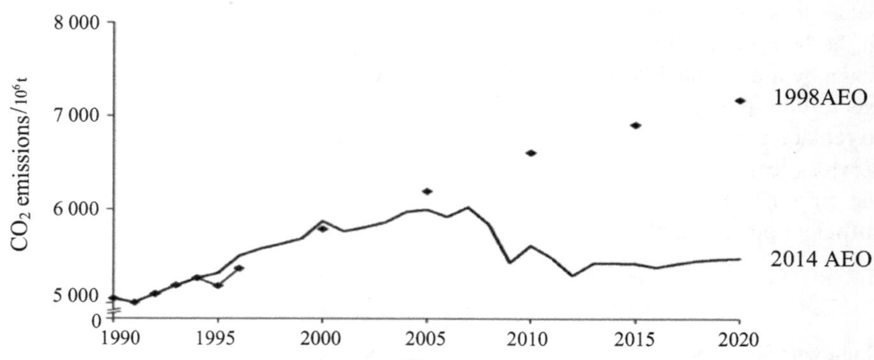


(b) Carbon dioxide emissions in the U.S.

**Fig. 7.7** U.S. energy consumption and related carbon dioxide emissions. *Note* Energy production data between 1990 and 2013 and carbon dioxide emissions data between 1990 and 2011 come from EIA Primary Energy Production by source, 1949–2012, and Carbon Dioxide Emissions From Energy Consumption by Source Selected Years, 1949–2011, and http://www.eia.gov/totalenergy/data/annual/index.cfm. Data in 2012–2020 come from AEO 1998 and 2014

Table 7.1 Proven reserves and actual production of shale gas in the U.S. Unit: 1 billion $ft^3$

| Year | Proven reserves | Actual production | Proportion (%) |
|---|---|---|---|
| 2007 | 23,304 | 1293 | 5.55 |
| 2008 | 34,428 | 2116 | 6.15 |
| 2009 | 60,644 | 3110 | 5.13 |
| 2010 | 97,449 | 5336 | 5.48 |
| 2011 | 131,616 | 7994 | 6.07 |
| 2012 | 129,396 | 10,371 | 8.01 |

Shale gas revolution greatly promoted low-carbon energy development in the U.S. First, shale gas is cleaner than coal and petroleum. Second, the U.S. boasts abundant shale gas reserves, mature technologies, and commercial development, which makes it possible to use the energy on a large scale. Its proven reserves in 2012 were sixfold of that in 2007. The actual production was as high as 10,371 $ft^3$, which was only 8% of proven reserves as shown in Table 7.1. According to AEO 2014, the natural gas production will expand rapidly at an annual growth rate of 1.6% between 2012 and 2040, making the country an exporter of natural gas by 2020. Third, the U.S. government supports replacing coal-fired power plants with natural-gas-fueled plants. In 2014, the U.S. Environmental Protection Agency issued a new clean energy plan, requiring all domestic power plants to reduce carbon emissions by 30%. Meanwhile, the AEO 2014 predicated that the U.S. will phase out coal-fired power plants with installed capacity of 60 GW by 2020. Tightening regulation on newly built coal-fired power plants and the elimination of previous ones, together with the lower price of natural gas (its price is only higher than that of coal), make natural-gas-fueled power plants the best alternative.

### 7.2.5 *Further Development of Low-Carbon Construction and Transport Sectors*

Energy consumption in the construction and transport sectors in developed countries and regions has accounted for some 60% of their total energy consumption. Carbon emission reduction in these sectors of developed nations is mainly achieved by applying low-carbon and energy-saving technologies. In this regard, many practices have been conducted and positive progress has been made, inspiring future development and application of low-carbon and energy-saving technologies.

Closely associated with working and living of the people, construction is a sector that consumes a large amount of energy. The heating system, air conditioning and lighting facilities, home appliances, office equipment, hot water supply and cooking facilities, elevators, and ventilation system among others need to consume energies such as power, coal, thermal energy, natural gas, oil, and renewable energy. Energy consumption in construction depends on the building area, performance, and service level and energy efficiency. Significant increase in the building area tends to bring about rapid growth of energy consumption. Better insulation performance could effectively lower energy demands for buildings with the same service level. People's increasing demands for service levels relating to buildings, such as higher heating temperature, cooler environment provided by air conditioners, longer comfortability in rooms, and more home appliances, will all increase energy consumption. Improving energy efficiency for facilities and systems in the building is conducive to reducing energy consumption. For example, the Australian government formulated *Building Energy Rating Act* on March 18, 2010, which provides for energy efficiency in major national commercial buildings. According to the act, large commercial buildings should explicitly provide energy ratings in delivery, sales, leasing, and advertising so as to encourage developers and operators to adopt effective technologies for efficient energy use.

GHG emissions from the transport sector account for one-fourth of the total GHG emissions in European Union (EU), the second largest source of emissions next to the energy sector. To promote low-carbon transport, EU enacted a test standard for motor carbon dioxide exhausts as early as in 1980. The standard has been revised several times since then. The *Regulations* on automobile carbon dioxide exhausts was incorporated in national policies for low-carbon transport in 1999, when the EU required automakers to disclose the fuel economy and carbon dioxide emissions for consumers to choose vehicles. Afterward, the EU gradually improved low-carbon transport policies, which mainly include incorporating the aviation sector into the administration of EU emission trading system, formulating a new road map for new-vehicle exhausts and standard for exhaust per unit, and reviewing energy consumption and carbon dioxide emissions throughout the whole life cycle of vehicles.

Low-carbon development in the construction and transport sectors can be promoted through the following measures: adopting high-performance and low-carbon energy-consuming equipment, like promoting energy-saving home appliances, electric vehicles and hybrid vehicles, and new energy-saving building materials; planning urban layout and infrastructure in a more reasonable way, like building better public transport infrastructure, encouraging green construction, and improving urban layout; advancing integrated development of construction and transport sectors with the energy system and intelligent technologies, like supporting solar integrated building, intelligent transport, and smart cities. For example, Belgium has successfully tested the world's first ever solar-energy-powered train. The power is solely generated by the 16,000 solar panels installed on the roof of the railway tunnel. The railway is exclusive for TGV high-speed train. The total area of solar panels amounts to 50,000 $m^2$, equivalent to eight basketball courts, and generates 3300 MW of power annually, enough for powering some 1000 households. Power generated by solar energy is mainly used for railway infrastructures such as lighting and signals and power supply. Germany has built up a highly developed smart transport network. Detectors on some urban roads have been replaced by small computers, which are capable of monitoring road conditions around the clock. With the help of auxiliary equipment, drivers could be informed of the duration of traffic lights through the signal system. In this way, they could decide whether to stop the engine or when they should restart the vehicle based on the information. The system effectively alleviates traffic congestion and reduces fuel consumption.

### 7.2.6 *Improving Legislation for Low-Carbon Energy Development*

Besides advancing technological researches and application to mitigate climate changes, developed countries and regions have made efforts to regulate, plan, and guarantee energy mix shift in terms of legislation. The EU provides an exemplary model for other countries and regions by building a complete legal system in this regard.

Laws and regulations regarding carbon dioxide emissions set by the EU cover the three levels. 1. The EU's commitment to the international community. In line with the *United Nations Framework Convention on Climate Change* and the *Kyoto Protocol*, the EU pledged the target for the first commitment period (2008–2012) of reducing carbon dioxide emissions by 8% on the 1990 level; over the second commitment period (2013–2020), it will reduce the emissions by 20% on the 1990 level. Second, the EU emissions' reduction targets are responsibility sharing among member states. Third, it is the EU's mid- and long-term plan for low-carbon development and climate change mitigation. The EU's commitment to the international community is based on its low-carbon development goals and the results of international negotiations.

The responsibility sharing of the EU's commitment among member states is also the outcome of reconciling each member state's development goals with the overall targets set by the EU. Taking account of population size and its growth rate, living standards, economic development level and structure, energy efficiency, energy mix and climate conditions among other national conditions, Phylipsen, and other researchers classified emission sources into power generation sector, industrial sector, and other domestic emitting sectors. The power generation sector is required with permits in coal combustion and a minimum proportion of renewable energy. Industrial sectors required with a minimum annual increase rate in energy use efficiency on average. Other sectors are required with permits of carbon dioxide emissions per capita. Based on these, the accounting quantity of each member state is calculated. In the first commitment period, the EU established a plan for redistributing the 8% target among its 15 member states based on the ideas of Phylipsen and other researchers through negotiations. In the second commitment period, the EU redistributed the emissions target among sectors covered by the European Union Emission Trading Scheme (EU ETS) and those not covered by it. Then, the accounting quantity for each member state is assigned by different principles. Sectors covered by the EU ETS emit about 45% of the total carbon dioxides in the EU. They will reduce emissions by 21% in 2020 on the 2005 level, down at an annual rate of 1.74% between 2013 and 2020. Among them, 88% of emission permits will be allocated based on the 2005 share to sectors covered by the EU ETS. And 10% will be used to subsidize less-developed EU member states, and 2% to reward those which had emissions reduced by 20% on the baseline year level set by the Kyoto Protocol as early as in 2005. For sectors not included in the EU ETS, the emission amount will be reduced by about 10% on the 2005 level. Emission permits of member states for sectors not covered in the EU ETS are linked to GDP per capita. The higher the GDP per capita is, the higher emission allowances will be. The allocation of emission reduction targets among sectors not covered by the EU ETS in member states is shown in Table 7.2. As emission permits for EU ETS sectors are directly allocated to businesses through auction or presentation instead of countries, EU member states have no economy-wide emission reduction targets during the second commitment period. EU member states' total emission allowance is calculated by summing up the emission permits of EU ETS sectors and the assigned amounts of non-EU ETS sectors.

Leaders of EU member states pledged to promote low-carbon development in 2007 and formulated the 2020 low-carbon development plan. Besides reducing GHG emissions by 20% in 2020 on the 1990 level, they agreed to increase the proportion of renewable energy in end energy consumption to 20%, and the energy use efficiency by 20%. These targets are collectively called the EU 20/20/20 plan, which was incorporated into EU laws in 2009, but no binding obligations in terms of energy efficiency improvement were imposed on member states. There are also some requirements in specific sectors. For example, 10% of energy consumption in the transport sector should be from renewable sources before 2020, and carbon contents in fuels should be reduced to 6%. The EU released its *Roadmap for Moving to a Competitive Low-carbon Economy in 2050*, and *The EU 2030 Policy Framework for Climate and Energy* as mid- and long-term plans for low-carbon development. The former aims to shape a competitive and climate-neutral economy featuring low-energy-consuming,

**Table 7.2** The assigned amount of GHGs emission target in non-EU ETS sectors among member states

| Country | 2020 emission reduction target/% | 2012 GDP per capita/(USD10,000, 2005 level/person) | Country | 2020 emission reduction target/% | 2012 GDP per capita/(USD10,000, 2005 level/person) |
|---|---|---|---|---|---|
| Austria | −16 | 3.99 | Latvia | +17 | 0.85 |
| Belgium | −15 | 3.65 | Lithuania | +15 | 1.01 |
| Bulgaria | +20 | 0.46 | Luxembourg | −20 | 7.79 |
| Cyprus | −5 | 1.67 | Malta | +5 | 1.63 |
| Czech Republic | +9 | 1.42 | Netherlands | −16 | 4.06 |
| Denmark | −20 | 4.64 | Poland | +14 | 1.06 |
| Estonia | +11 | 1.18 | Portugal | +1 | 1.79 |
| Finland | −16 | 3.84 | Romania | +19 | 0.56 |
| France | −14 | 3.42 | Slovakia | +13 | 1.49 |
| Germany | −14 | 3.75 | Slovenia | +4 | 1.86 |
| Greece | −4 | 1.86 | Spain | −10 | 2.51 |
| Hungary | +10 | 1.10 | Sweden | −17 | 4.39 |
| Ireland | −20 | 4.61 | United Kingdom | −16 | 3.78 |
| Italy | −13 | 2.84 | EU27 | −10 | 2.87 |

low-emitting, intelligent, and clean lifestyle. It will work with other major economies to keep the global temperature increase to well below 2 °C of the twenty-first century. To achieve the goal, in 2050, the EU needs to cut emissions by 80% compared to the 1990 level, and save energy by 30% on the 2005 level. Correspondingly, it has to cut emissions by 40 and 60%, compared to 1990 levels, by 2030 and 2040, respectively. On October 24, 2014, EU leaders adopted *The EU 2030 Policy Framework for Climate and Energy*, which sets the following key targets for the year 2030: at least 40% cut in greenhouse gas emissions from 1990 levels, including EU ETS sectors cutting by 43%, and non-ETS sectors by 30% compared to 2005, at least 27% share for renewable energy, and at least 27% improvement in energy efficiency.

### 7.2.7 Improving Long-Term Mechanism for Low-Carbon Energy Development

To boost low-carbon energy development, major developed countries and regions have made continuous efforts in improving pricing, taxation, and charging mechanisms for the energy sector. A market system with the "cap and trade system" as the core has been established, playing a constructive role in fostering favorable climate for low-carbon energy development.

The EU Emissions Trading System is one of EU's most important policies to boost low-carbon economy and tackle global warming. In line with provisions of the *Kyoto Protocol*, the EU established the emission trading system through legislation in 2003. Taking effect on January 1, 2005, the trading system has made Europe the most dynamic market for emissions trading worldwide. At the beginning of the trading, the price of emission allowances was about 6 Euros per ton $CO_2$. Since then, it has been climbing all the way up to about 30 Euros per ton in the first 10-day period of April 2006. Too many permits issued in the first phase of EU ETS (2005–2007) and the provision that the surplus allowances are not allowed to extend to the second phase resulted in the price plunging to zero at the end of 2007. Similar problems happened during the second phase (2008–2012). Nevertheless, the EU ETS has become the world's most successful practice to solve environmental issues with the market approach, offering experience for other countries to developing their trading system.

The United Kingdom incorporated its previous emission trading practices into the EU ETS and introduced the Climate Change Levy. The two policies were incorporated into a new policy of carbon floor price, which sets the lowest price for the UK's power generation sector in carbon dioxide emissions. It requires power plants consuming fossil fuels to pay climate change taxes or fuel taxes at a rate of the difference between the future price of EU ETS and the expected carbon floor price. The UK's annual budget act sets the temporary tax rate for the next 2 years. For example, the 2011 budget act set the rate for 2013–2014 as 4.94 lb per ton $CO_2$, the 2012 budget act set the rate for 2014–2015 as 9.55 lb, and the 2013 budget act set the rate for 2015–2016 as 18.08 lb. The policy took effective on April 1, 2013.

The UK's Climate Change Levy, in effect, is a type of energy use taxation. It is levied based on the amount of coal, natural gas, and electricity consumed by businesses and public agencies in lighting, heating, and power. Such taxation has distinct features. First, price is levered to increase energy efficiency and promote adjustment of energy mix as the taxation varies by energy type, and combined heat and power projects and renewable energy are tax-free. Second, it is a fiscal-neutral taxation by principle as it does not increase businesses' tax burden in general. Third, a flexible approach to high-energy-consuming businesses is employed by setting a clear target for carbon dioxide emissions and offering appropriate tax cut to maintain their competitiveness. Fourth, the taxation targets at businesses and public agencies rather than private households, lowering political risks as it has no direct influence on voters. Fifth, it helps raise public awareness of environmental protection. Therefore, the UK's Climate Change Levy is an effective instrument with slightly negative effects.

Germany has long begun to levy energy tax. As early as in 1879, the German government taxed on oil imports. After several modifications, the prevailing energy taxation and administration is validated by the *Energy Sources Act* enacted and took effect in 2006. The act aims to translate the *Proposal for a Council Directive Restructuring the Community Framework for the Taxation of Energy Products* enacted on October 27, 2003 by the EU Council into a domestic law. The purpose is to influence energy price with the lever of taxation and further regulate demand mix in the energy market, stimulate technological improvement on high-energy-consuming products, and increase productivity. Meanwhile, the energy taxation helps the public increase their awareness of the shortage of natural resources and encourage them to lower energy consumption and reduce environmental pollution. Moreover, Germany revised the *Vehicle Tax Act* in 2009 to encourage consumers to buy low-exhaust-emitting vehicles with emissions less or equivalent to 120 g $CO_2$ per km or emission-free vehicles, and actively promoted low carbonization of means of transportation.

## 7.3 Focuses and Directions of China's Low-Carbon Energy Development

China consumes the most energy and emits the most greenhouse gases worldwide, and is the main source of growth of energy consumption and greenhouse gas emission. So far, China's greenhouse gas emission per capita has outnumbered the world's average level, and the figure in some cities has exceeded that of developed countries and regions like Europe and Japan. The pressure of cutting $CO_2$ emissions is extremely severe for China. As the world's largest developing country, China will witness continuously soaring energy demands and greenhouse gas emissions in the following years given the reality that it has to maintain rapid economic growth to complete the building of a moderately prosperous society in all respects and step up efforts in building itself into a modern socialist country. Compared with developed countries, China has a larger population, fewer resources per capita, fewer natural resources, more fragile ecology, and lower degree of economic and social development. As a result, China faces an unprecedentedly daunting challenge and pressure in energy development. Under such domestic and international conditions, China's efforts in exploring new energy development path and accelerating the transition to low-carbon energy and economic development are essential to realize the Chinese Dream, and are of great significance for the international community to cope with climate change and realize sustainable development.

Against the backdrop of active global response to climate change, China strives to achieve low-carbon energy development that is scientific, rational, energy-saving, efficient, green, diverse, safe, and reliable, focusing on advancing revolution of energy production and consumption. On the premise of meeting the requirements for building ecological civilization and a Beautiful China, China should greatly improve energy efficiency, improve the energy mix, and attain the goals of completing the building of a moderately prosperous society in all respects and the modernization drives with $CO_2$ emissions per capita remarkably lower than that of developed nations.

### *7.3.1 Specifying the Strategic Goals of China's Revolution of Energy Production and Consumption*

Strategically, China needs to incorporate revolution of energy production and consumption and low-carbon energy development into the program of building ecological civilization, as well as each of the specific tasks for national economic and social development, urbanization, and industrialization. The revolution of and transition to low-carbon development should run through the entire process of energy production, circulation, consumption, and disposal. It is necessary for China to set specific goals and define steps for the revolution in each field and at each stage, integrate fundamental transformation of energy production with globalization, IT application and independent innovation, give play to its advantages as a latecomer, and improve its overall national strength and competitiveness.

On the premise of achieving the goal of total energy supply and demands, we should follow the principle of satisfying rational energy demands by scientific supply, restrict all relevant actors, ensure joint participation of the government, businesses and society, and build ecological civilization, so as to advance the revolution of energy production and consumption. We should resolutely curb irrational energy use, effectively implement in the policy on giving priority to energy conservation, save energy across the board in all spheres of economic and social activities, adjust the structure of the energy industry, attach great importance to energy conservation during urbanization, foster new thinking on consumption characterized by diligence and thrift, and work faster to build an energy-conserving society. We should resolutely control the total energy consumption and continue to work to achieve a higher goal of energy conservation, which requires us to control the economic growth within a reasonable range, guide the investments to low-carbon sectors like services, and shift the focus of evaluating officials' performance from on growth rate to on development quality. This is in line with the requirements on accelerating the transformation of growth model and modernizing the government's capacity for governance.

First, we should impose a ceiling on total energy consumption, especially the total coal consumption, and strive to reach the peaking of coal consumption by 2020 and scale it down gradually afterward. With the possible ceiling of energy supply, we should support economic growth with lower energy elasticity coefficient, rationally allocate energy supply for industry, constriction and transport sectors, and establish an energy consumption system that is green, low-carbon, and circular. Second, we should rationally control the scale and area of urban construction. We should keep the total building area within 60 billion $m^2$, and the area per capita at the level of Asian developed nations like Japan and South Korea, which measures 40–45 $m^2$ per person, prevent from high environmental expense and maintenance cost like that of the U.S., and avoid unnecessary large-scale demolition and reconstruction projects. The construction speed and level should be well controlled. Efforts should be made to ensure soft landing of building material and construction sectors after urbanization is basically finished. Last, the expansion of vehicle ownerships should be moderately contained. Car owners should be educated to choose more environmentally friendly travels by public transport, and effective allocation of transport resources should be achieved.

### *7.3.2 Significant Improvement of Energy Efficiency*

The core to revolutionize energy consumption is substantial increase in energy efficiency so that it supports economic performance to continuously improve, while the total consumption is effectively controlled. To substantially increase the energy efficiency, on the one hand, economic transformation should be accelerated so that economic growth will rely less on high-energy-consuming and polluting sectors. Instead, the economy will be driven by technological advances, independent innovation, and improvement in performance. On the other hand, energy efficiency should be significantly improved in construction and transport sectors, and the energy conservation and moderate consumption should be encouraged.

As the world's largest energy consumer, we should prepare itself with a development path and top-level design in advancing the long-term endeavor of revolutionizing energy consumption. In line with the requirements for realizing modernization drives by 2050, we should design compatible plans for industrialization, urban and rural development, as well as the development of construction and transport sectors. An energy consumption system compatible with the requirements of building ecological civilization and a Beautiful We should be established, and a social system that is effective, green, low-carbon, and circular should be put in place. For major projects, technologies, and processes associating with energy environment, we should fully consider international standards and future trends, adopt strict management on energy conservation, environmental protection, and strategic environmental assessment, so as to ensure, from the origin, that China's energy efficiency reaches international leading level as soon as possible.

It is necessary to foster a culture and social environment where rational and moderate energy consumption and energy conservation are encouraged. Traditional Chinese virtues like harmony between nature and human and thrift should be integrated with modern notions of building green and low-carbon society to guide people to save energy and rationally consume energy. By implementing rule of law, reducing market distortions, introducing economic incentives, promoting public education, and giving full play to the government's role, we should resolutely prevent extravagance and energy waste and actively foster green and low carbon consumption culture in the whole society. Green consumption, eco-friendly transport services, and classification of domestic wastes should be encouraged, and green and low-carbon lifestyle and consumption pattern should be advocated.

### 7.3.3 *Making Low-Carbon Energy as the Dominating Source of Energy Supply*

The important targets of optimizing energy mix include accelerating low-carbon energy development and substantially reducing the proportion of coal consumption. By improving the market system and policy environment that are conductive to fair and orderly competitions, we should promote technological advances in developing nuclear power, hydropower, and renewable energy and make green and low-carbon energy to be the main source of energy supply. Economically developed regions and cities should be encouraged to take the lead in reducing fossil energy consumption and using clean energy. The newly emerging energy demands should mainly depend on the supply of renewable energy. Efforts should be made to explore multiple paths to use renewable energy based on resource conditions in different regions.

i. Developing nuclear energy in a safe, steady, and efficient manner

Confidence in developing nuclear energy should be boosted. Efforts should be made to guide the public to understand and accept projects and plans in this area. During the 13th Five-Year Plan period, preliminary preparations for new nuclear power projects should be started as soon as possible, especially for those in China's landlocked regions. It is necessary to comprehensively strengthen safety management on nuclear power plants and the emergency response capabilities in case of nuclear accidents. The overarching guideline of putting safety first should be implemented across all associated industries and throughout the entire process of nuclear energy projects from planning, construction, operation, and shutdown. State capacity for governance should be improved to make decisions in a sound manner. The pace of building projects should be well controlled. Nuclear projects in landlocked regions should be launched in a steady and orderly manner. The technological threshold should be improved. At some previous sites, nuclear projects of the second generation should be developed on the basis of passing the comprehensive project review. A three-step progressive path starting from developing thermal reactors to fast reactors and lastly fusion reactors should be followed. Focusing on mega-kilowatt pressurized water reactors, efforts should be made to develop new technologies like high-temperature gas-cooled reactors, commercial fast reactors, and small-scale reactors. Meanwhile, the nuclear fuel supply system should be improved to fulfill the long-term development of nuclear energy. We should fully use the limited time and take advantage of the existing projects to ensure self-reliance in equipment manufacturing. A modern industrial system for nuclear power generation should be established so as to build China into a country with powerful strengths in terms of nuclear power generation. By 2020, the installed capacity of nuclear projects in operation will reach 65 million kW and that of projects under construction will reach 25 million kW.

ii. Accelerating hydropower development

Given the dominating position of hydropower in its clean energy mix, we must accelerate the development of hydroelectric resources in southwest China. First, consensus must be achieved on this, and a national strategy and overarching goal on hydroelectric development should be made as soon as possible. Second, rational pricing mechanism for hydropower should be established to address profit distribution regarding construction of hydroelectric projects, especially issues like environmental protection, relocation of affected people, and sustainable development of local economy. Third, environmental impact assessments on the planning and projects should be connected with an established mechanism so as to improve the environmental impact assessment system for strategic planning of hydropower projects in the watershed. By 2020, the installed capacity of conventional hydropower projects will reach 350 million kW, and that of pumping storage projects will reach 70 million kW, totaling 420 million kW.

iii. Promoting both centralized and distributed wind power and solar projects

We should make more investments in research and development of large-scale wind turbines and pilot the achievements on a trial basis to keep abreast with the international trend. Efforts should be made to strengthen the manufacturing of key components of wind turbines, and gradually build a complete industrial chain for wind turbines manufacturing. Large-scale wind farms should be built. In the third batch of wind power projects approved during the 12th Five-Year Plan period, the number of large-scale wind farms will increase to 10 with a total approved capacity of 16.9 million kW. The supporting power grid should be improved so as to integrate the wind farms into the power grid network as soon as possible. We should also take measures to encourage distributed wind power application. In areas where the wind speed is slow, distributed projects should be developed to supply electricity to nearby power grids. More pilot wind power projects should be established on the sea, related projects should be well planned, and policies on fixed pricing of wind power on the sea should be in place as soon as possible.

In terms of solar projects, technological upgrading should be promoted, including technologies of materials, polysilicon manufacturing in a Siemens process, mass production in silane process, and low-cost processes like physical and chemical metallurgy. Researches on key supporting materials for solar cells should be strengthened. Breakthroughs should be made in manufacturing technologies of key supporting materials for solar cells, such as silver paste, silver aluminum paste, TPT back sheet material, EVA packaging material, and TCO glass substrate for hull cell. One focus of technological upgrading should be on making efficient and low-cost hull cells. Upgrading of the integrated system should be based on grid-connected photovoltaic power generation systems with the capacity over 10 MW, and focus on the key technologies and equipment for grid-connected photovoltaic power generation systems with the capacity over 100 MW. We should solve such essential technological problems as efficiency of grid-tied inverters, waveform control, photovoltaic micro-grid inverter control, and energy management of micro-grid photovoltaic power generation system. We should strive to make breakthroughs in key technologies of advanced independent photovoltaic power generation system, grid-tied photovoltaic power generation system with capacity of megawatt to gigawatt, and micro-grid photovoltaic systems complemented by multiple sources. The photovoltaic industry should be better planned to encourage more distributed solar projects rather than centralized ones.

We should improve the technology and process of harnessing solar thermal energy to improve the energy efficiency. Safe, efficient, all-weather, and intelligently controlled low-temperature thermal systems for heat storage should be developed and promoted. New type of solar water heaters that is easy to integrate with and dismantle from buildings, pressure-bearing, and recyclable should be developed and manufactured. Solar water heaters should be improved to better fit for application in residential sectors. Technologies of high-efficiency flat-plate solar collectors should be developed. Automatic production lines for continuous panel coating and flat-plate collector manufacturing should be developed and constructed so that the production process and equipment of flat-plate solar collectors reach the international level. Efforts should be stepped up to commercialize technologies of medium- and high-temperature thermal systems for heat collection and promote the R&D and manufacturing of related products. R&D and application of heat storage technologies should be strengthened. Solar heating and air conditioning technologies should be promoted to expand application of solar thermal energy in heating and cooling systems for buildings. Technologies of solar thermal energy should be encouraged to be applied in industrial and agricultural activities. The integration of solar energy system with conventional energy system, and thermal measuring instruments and control systems should be better developed. The service system supporting the application of solar thermal energy should also be improved. The quality control system should be enhanced, the market should be well regulated, testing capability and standard should be improved, and a state-level R&D testing platform should be established.

iv. Advancing development and application of biomass, geothermal, and ocean energy

Based on local conditions, we should encourage the utilization of biomass energy for diverse purposes, and support the utilization of biomass energy and recycle of resorted refuse to be strategic emerging industries. Biomass power generation technologies like biomass gasification, cogeneration, and co-combustion power generation should be energetically promoted. Targeting existing problems in refuse resorting for power generation, relating technologies and process and public awareness, efforts should be made to improve the refuse resorting system in urban areas to improve the recycling efficiency. Adopting technologies of grate furnace, fluidization, waste hydrolysis, cryogenic fracturing, gasification, and plasma processing will promote the process of converting garbage to power to be more efficient and consume less energy. Developing biogas should be in line with the requirements for large-scale development of livestock and poultry farming. To adapt to the transition of biogas generation from a distributed pattern to a moderately centralized pattern, the use of biogas should also change from government-funded way to the market-driven way. Industrialized large-scale biogas plants and megawatt-scale cogeneration biogas projects should be established to supply heat and electricity to rural households and public facilities like schools, hospitals, and public venues. It is necessary to develop biogas purification technology and reduce the cost of purification of biogas. Biogas should be integrated with the natural gas pipeline network or used as fuels to power vehicles and for other purposes. Advanced technologies for converting biomass to liquid fuels should be promoted. We should promote the technology of converting plants such as cassava and sweet sorghum to 1.5-folds ethanol fuels as the transitional technology for the near- and mid-term plan of developing ethanol fuels. Continuous efforts should be made to promote the application of the second-generation technology of ethanol fuels converted from cellulose, commercialize related projects and improve the utilization of biomass energy for multiple purposes.

The development and utilization of geothermal and ocean energy should be highlighted. R&D and popularization of geothermal energy development and utilization should be strengthened. Heat pump technology should be adopted to exploit shallow geothermal energy for heating and cooling in areas with suitable conditions. The research and application of thermal water recharge technology under different geological conditions should be encouraged. It is necessary to develop technologies to utilize dry hot rock for power generation and expand the scale of geothermal utilization. Demonstration programs of geothermal utilization should be established in various regions to ensure sustainable development and utilization. In regions where geothermal resources are shallow, demonstration programs adopting heat pump technology, geothermal water recharge technology, geothermal central heating technology, geothermal gradient utilization technology, and the technologies using oil field to exploit geothermal resources and developing productive agriculture with geothermal energy should be launched. New technologies should be harnessed to promote sustainable utilization and efficient development of geothermal resources. The development of ocean energy should be highlighted on the strategic level, and more inputs should be made in this regard. Governments at all levels should formulate development plans for ocean energy, increase investment, and introduce favorable policies. The construction of ocean power stations should be planned in a unified way and carried out step by step. Supportive policies should be formulated to establish a research team for long-term studies on ocean energy development and utilization. Research focuses include high-performance energy conversion and low-cost ocean energy equipment, stability and reliability of such equipment, integrated system for ocean energy development, and materials against seawater corrosion and marine attaching organism for underwater equipment, among others. We should also explore international cooperation in this regard and introduce foreign capital and technologies to accelerate the development of ocean energy in China.

### 7.3.4 *Using Fossil Energies in a Cleaner, More Efficient, and Low-Carbon Manner*

i. Promoting coal utilization in a cleaner and more efficient way and achieving the peaking of coal consumption as early as possible

It is of strategic significance to realize the peaking of coal consumption and use energy in a cleaner and more efficient manner. Given the natural resources' conditions, the proportion of coal to the primary energy supply is as high as 66% and is expected to remain above 50% in the mid- and long-term future. Using coal energy in a rational, clean, and efficient way has a bearing on China's green and low-carbon development. At present, coal production and consumption in China far exceed the environment capacity and the ceiling of coal production. They are the root causes of ecological and environmental crises, worsening air condition, increasing greenhouse gas emission, mercury pollution, and safety in production. Overcapacity in the coal making sector has led to extensive and irrational coal consumption, caused poor performance and low quality of economic growth, hindered the development of clean and low-carbon energy to some extent, and adversely affected efforts for upgrading and optimizing the energy mix.

We should achieve the peaking of coal consumption as early as possible during the 13th Five-Year Plan period and control coal consumption within 4 billion tons in 2020. This is essential for fundamentally solving environment problems like smog, improving the economic structure and energy mix, and increasing the quality of China's low-carbon development. Meanwhile, we should significantly improve the level of cleaner coal consumption and create the policy environment favorable for the application and promotion of advanced technologies by enhancing the energy efficiency and emission standards throughout the entire production and consumption process. We should improve the entire system from coal supply to conversion and consumption in order to change the extensive way of coal consumption through direct combustion of raw coal. Coal efficiency should be increased from the perspective of the whole life cycle and entire industrial chain to reduce various impacts on the environment. Technical thresholds and emission standard for key coal-consuming industries like coal-fired power generation, metallurgy, chemicals, and building materials should be formulated and implemented. Greater efforts shall be made for the promotion of policies on existing technologies for consuming coals in a cleaner and more efficient manner, such as improving coal washing and dressing rate, high-efficiency coal-fired power generation, clean combustion of industrial boilers, and high-efficiency and environment-friendly stoves. The coal chemical industry, especially the emerging coal chemical industry, should pay high attention to restrictive factors like the amount of coal resources and water resources, the bearing capacity of the environment and greenhouse gas emissions, contain the production capacity within a rational scope, formulate strict standards for energy efficiency, water consumption and lifecycle emissions by referencing the latest international standards for energy efficiency and emissions, and update them on a regular basis.

ii. Increasing supply of natural gas and optimizing the way of consumption

China boasts abundant natural gas and unconventional gas and favorable conditions for accelerated development. In 2013, the U.S. EIA released *Technically Recoverable Share Gas and Shale Oil Resources: An Assessment of 137 Shale Formations in 41 Countries outside the United States* and *Overview of Global Shale Gas and Shale Oil Developments—Focusing on the United States, China, Argentina, Australia, Indonesia and Britain*, saying China has the world's largest reserve of technically recoverable shale gas resources as high as 31.6 trillion $m^3$, making it the most promising producer of shale gas. Domestic experts estimate that recoverable shale gas in China amounts to 22 trillion $m^3$, outnumbering the reserve of conventional natural gas. In the near- and mid-term future, natural gas exploration should be expanded and taken as the focus for transiting to low-carbon energy development.

First, production capacity of natural gas in key regions should be expanded based on breakthroughs in exploring conventional natural gas in the Sichuan Basin. The exploration of tight gas should be accelerated by expanding the Sulige model, so as to increase the reserves and output. High attention should be paid to sustainable increase of reserve and production of natural gas in central and western regions and coastal areas. Preliminary exploration and production capacity building projects should be started as soon as possible. Resource exploration in the South China Sea waters should be strengthened by introducing international advanced technologies of deep-sea oil and gas exploration.

Second, exploration of unconventional gas should be enhanced. The focus should be fostering a technological system for exploring shale gas. Production capacity projects should be started in selected regions, and shale gas can be explored on a large scale for commercial uses. Based on Qinshui Basin and Ordos Basin, we should make breakthroughs in exploring coal bed gas, build industrial bases for the exploration, and further expand to Xinjiang and other regions.

Third, we should import more natural gas, explore new global partnerships for exploration under new circumstances, improve product sharing contracts and supporting policies, and encourage technological cooperation with foreign companies.

Fourth, we should accelerate the construction of natural gas pipelines to enhance the gas storage and peak-shaving capacities. Diverse investors should be encouraged to invest in infrastructure projects for long-distance gas transmission and urban pipeline network. The governments should offer fiscal and taxation incentives for such projects, set rational pricing standard for pipeline transmission, and optimize the allocation of resources by gradually introducing third-party operators.

Fifth, we should advance reform of the natural gas pricing system in an orderly manner and adjust policies for using natural gases on a regular basis. With peak-shaving policy on electricity pricing, the use of natural gas can be expanded especially among combined heat, cooling and power plants, natural gas peak-shaving station, and distributed users. Inter-city and inter-provincial transport of liquefied natural gas (LNG) by freight vehicles and ships should be expanded. Progressive pricing of natural gas for residential use should be rolled out as soon as possible. China plans to work to supply 400 billion $m^3$ of natural gas nationwide before 2020. The production of conventional natural gas including tight gas will be further expanded toward the goal of exceeding 180 billion $m^3$. The targeted production of unconventional gas is 60–90 billion $m^3$, including 30 billion $m^3$ of coal bed methane and 30–60 billion $m^3$ of shale gas. In total, China's natural gas production is expected to exceed 240 billion $m^3$. China will import 35 billion $m^3$ of natural gas through pipelines from the second phase of the Central Asia project and another 5 billion $m^3$ from the China–Myanmar project, which total 70 billion $m^3$. The LNC import will amount to 630 billion tons, roughly equivalent to 80 billion $m^3$. We should energetically explore unconventional gas and try to increase the import of natural gas.

iii. Encouraging oil conservation and replacing oil with clean energy

To cut greenhouse gas emission caused by oil consumption, we should save oil and replace it with other energies. Oil is mainly used in industrial and transport sectors. Industries such as chemicals, power generation, and building materials are large consumers of oil. The petroleum industry itself consumes a large amount of oil and has a great potential for oil saving.

Chemical industry: Efforts should be made to optimize the existing production process, thoroughly solve the problems of waste of oil, further expand low-oil-consuming production process, improve product mix, and shut down small-scale chemical companies adopting backward technologies.

Electric power: Technologies like small gasified oil gun, plasma oil-free ignition, and low-load stable combustion should be promoted. Fuel-powered generator sets should be upgraded to be natural-gas-powered ones.

Building materials: Heavy oil should be gradually replaced by natural gas in regions with suitable conditions.

Oil and gas exploration: Oil-saving measures include optimizing oil production system, recycling vented gas and associated petroleum gas, replacing fuel oil with flammable gas, and reducing the rate of self-used oil in oil field.

Transport sector: 1. Promote manufacturing technology of fuel-efficient gasoline and diesel engines and vehicle lightweight technology, promote small-displacement and energy-efficient vehicles, and curb the increase of large-displacement vehicles with necessary supporting policies. 2. Work to address traffic congestion problems and encourage public transport to reduce vehicle fuel consumption. 3. Develop technologies to power vehicles with LNG, electricity, hydrogen fuels, biofuels, and other new energies, speed up R&D and application of aviation biodiesels, and encourage the purchase of LNG- and biofuel-powered vehicles and electric vehicles through subsidies and tax cuts, and hydrogen-fueled cars in the future. 4. Upgrade the fueling energy in the shipping industry by replacing diesel with LNG, trial, and apply compact power reactor technologies in oceangoing freighters. 5. Recycle swill-cooked oil and reduce petroleum consumption.

### 7.3.5 Balancing the Roles of the Market and the Government and Building an Institutional System for Low-Carbon Energy Development

Energy is the foundation of the modern economy and society. Energy exploration, conversion, and consumption bring various externalities. How to give play to the basic role of market in allocating resources and effectively offset market failure have been the focus of policy-making among countries. As global efforts for improving energy security and mitigating climate change are strengthening, major developed countries adopt a combination of market, administrative, and legal means, which include accelerating market-driven development of the power supply sector, setting mandatory emission cut targets, improving energy efficiency and renewable energy consumption, enhancing energy efficiency standards for building, home appliances and vehicles, developing incentives, improving fiscal and taxation policies, and establishing a system of emission ceilings and carbon trading. Thanks to these policies and measures, some developed countries have realized the goal of reducing energy consumption, total greenhouse gas emissions, and the per capita greenhouse gas emission, from which we can draw on experience from.

In terms of the market role, the reform on the pricing mechanism for energy resources has been lagged behind for a long time. The relationship between the pricing, taxation, and surcharges of energy resources has not been adjusted. Pricing of factors associating with energy use, such as capital, land, water, environment, and workforce, is also twisted and distorted. These are the root causes of the extensive energy use in China. During the 13th Five-Year Plan period, we should step up efforts to reform the energy market, adjust resource tax, environment tax and carbon tax, and put in place a pricing mechanism that truthfully reflects the supply mix, scarcity, and environmental externalities of energy products to ensure price signals play a core role in guiding energy exploration and consumption. Meanwhile, reform of other factors such as capital, land, labor, and environment should be furthered so as to eradicate the root cause of the institutional environment that stimulates excess investment and energy and resource consumption.

In terms of the government role, the situation in which energy-saving targets are separate from macro-economic regulation and energy-saving measures heavily depend on technical means should be gradually shifted to a new approach where the government effectively guides the investment, imports, and exports, and builds energy-saving lifestyle and consumption during the 13th Five-Year Plan period. A high-level decision-making and coordinating mechanism should be established to ensure clearly defined powers and responsibilities and creating a strong synergy in terms of management, supervision, and institutional reform. A performance assessment mechanism should be further improved so that the energy-saving targets will be better aligned with pollutant discharge reduction, climate change mitigation, and sustainable energy targets. Factors like resource consumption, environmental damage, and ecological benefits should be incorporated into the assessment system. Random inspection shall be conducted to check the effects of policy implementation. By strengthening media supervision and introducing third-party evaluation, we shall advance official performance evaluation so as to build a service-oriented government. Meanwhile, efforts should be made to improve laws and standard system regarding resources and environment so that green low-carbon development is faithfully acted upon inland function zoning and policies for industries and import and export. A fair and just macro policy system and the mechanism of supervision and management should be improved.

# Chapter 8
# Low-Carbon Mode of Production

With the ongoing improvement of productive forces, the history of the human society is a course in which the mode of production evolves. In the pre-industrialization era, people worked to satisfy basic living necessities. During the industrial revolution, the accelerated advancement of productive forces and technology fundamentally changed people's life and the mode of production in the human society. On the one hand, while fulfilling people's basic needs for food, clothing, housing, and transport, the rapidly expanding production capability spurs diversified material demands. On the other hand, emerging new demands, in turn, boost the progress of productive forces, and the continuous expanding and upgrading of production capabilities. With the ongoing interplay between production and demands, the relationship between man and nature has undergone major changes. While harnessing resources and environment for their well-being, human beings have brought serious damages to the environment and ecology. Against the backdrop of global efforts for climate change mitigation, the essence of transforming the mode of production to a low-carbon one is exploring how to maintain sustainable relationship between man and nature.

## 8.1 Conditions and Characteristics of China's Traditional Mode of Production

### *8.1.1 As the World's Largest Manufacturer, China has Powerful and Fast-Growing Production Capacity*

China's economy has been expanding with an average annual growth rate of nearly 10% over the past year. Now, the country has become the second largest economy in the world. Between 1978 and 2013, China's GDP increased by 25-fold, the fastest worldwide in the period. As a late developing country, China had its GDP growth driven by investment and export. With rising industrial production capacity, China had its industrial output soar by 40-fold and grew to be the world's largest manufacturer in the period. Now, China has built a massive industrial capacity covering a


X. Du et al., *Overview of Low-Carbon Development*,
https://doi.org/10.1007/978-981-13-9250-4_8

full range of products, and became a global manufacturing powerhouse. The global market shares of more than 900 products made in China rank first. China produces the most steel, cement, coal, aluminum oxide, fertilizer, and microcomputers in the world. Half of many industrial products in the global market are produced by China. The market shares of major industrial products have exceeded the record high of developed countries in history as shown in Table 8.1.

**Table 8.1** The output, export, and export/output ratio of major products in China (2008)

| | Output | Export | Export/Output ratio (%) |
|---|---|---|---|
| Shoes/100 million pairs | 113 | 81.7 | 72 |
| Motorcycles/10,000 units | 2508 | 1097 | 44 |
| Bicycles/10,000 units | 7475 | 5923 | 79 |
| Washing machines/10,000 units | 4005 | 1644 | 41 |
| Vacuum cleaners/10,000 units | 6514 | 3800 | 58 |
| Refrigerators/10,000 units | 4397 | 1943 | 44 |
| Electric fans/10,000 units | 15,440 | 9680 | 63 |
| Air conditioners/10,000 units | 8014 | 3295 | 41 |
| Telephones/10,000 sets | 16,516 | 8467 | 51 |
| Fax machines/10,000 sets | 889 | 526 | 59 |
| Mobile phones/100 million sets | 5.48 | 4.8 | 88 |
| Display/100 million sets | 1.44 | 1.3 | 90 |
| Color TV sets/10,000 sets | 8478 | 4788 | 56 |

*Note* Compiled based on data from *China Statistical Yearbooks* and *China Custom Statistical Yearbooks* over the years

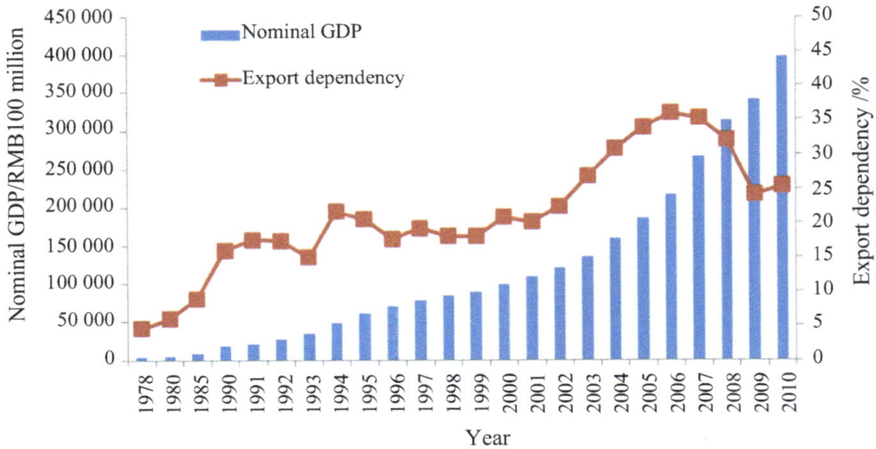


**Fig. 8.1** China's nominal GDP and export dependency

### *8.1.2 Continuous Expanding Export and High Dependency of Production on Export*

Since the founding of the People's Republic of China, we have begun foreign trade exchanges. Export increase was relatively slow under the highly centralized planned economy. Since 1978, China has gradually established a socialist market economy system through domestic reform and opening, and started to get involved in international competitions and division of labor. Foreign trade entered a stage of rapid development especially after China's entry into the World Trade Organization (WTO) in 2001. During the 10th Five-Year Plan period, the average annual growth rate of China's total exports stood at 24.9%, with the total volume tripled in 5 years. In the 11th Five-Year Plan period, influenced by the international financial crisis and global economic downturn, China's export in 2009 decreased by 18.29% from that in 2008, but bounced back in 2010 with a growth rate of 23.16% from the previous year. Compared to the early days of reform and opening up, China's export had leapfrogged and made the country a large, fast-growing, and all-round exporting economy (Fig. 8.1).

### *8.1.3 China's Industrial Structure is Improving, But is Still Dominated by Manufactured Goods*

In terms of the structure of export commodities, China first mainly exported primary products, textiles, and other light industrial goods, and then changed to manufactured goods and mechanical and electronic goods in the 1980s. The proportion of high-tech

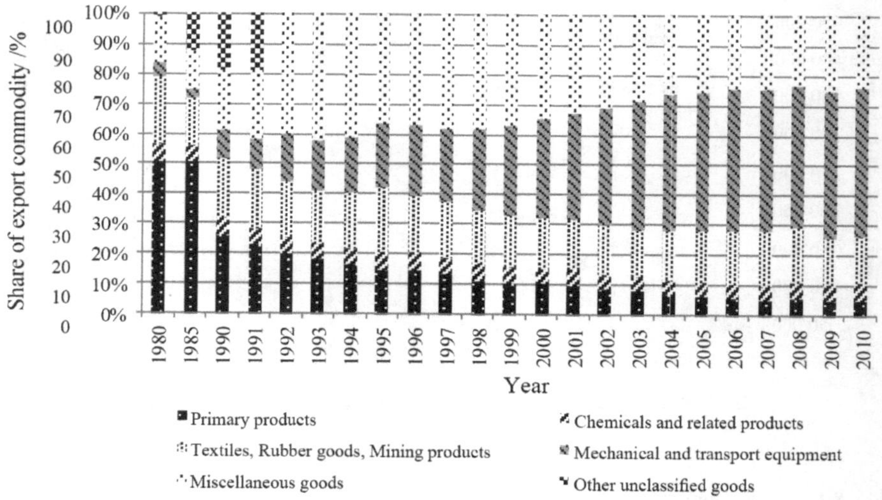


**Fig. 8.2** The commodity structure of China's export and its changing tendency

products represented by electronic and information technology has been expanding in the total export (Fig. 8.2). From the perspective of global competitiveness, both the export of labor-intensive manufactured goods and that of the value-added products represented by electronic and mechanical equipment and high-tech products is on the rise. In 2013, China exported mechanical and electronic goods valuing USD1.27 trillion, making the country the largest exporter of such products for 19 consecutive years. The export of high-tech products totaled USD 660.3 billion, and its proportion to the total exported commodity continued to grow. While maintaining the advantages of manufacturing and exporting labor-intensive products, China is gaining competitiveness in terms of products with high technology and added value. The quality and efficiency of its export have been significantly enhanced.

## 8.2 Problems of China's Traditional Mode of Production

### *8.2.1 Unsustainable Extensive High-Carbon Development Path*

Over a long period of time, China's manufacturing scale has been expanding. From the perspective of supply, it is driven by inputs of factors like labor force, land, capital, environment, and resource. In essence, this model is identified as the continuation of driving the economy with massive input and production in the traditional industrial era. With the rapid development over the last three decades, China's productive factors have changed profoundly. China can no longer depend on growing the economy by expanding manufacturing with more factor inputs.

In terms of labor forces, China's population structure has also changed. The working-age population has reached its peak, and population aging has become an increasingly prominent issue. With the advent of the Lewis Turning Point, China has grown old before it gets wealthy. The comparative advantage of low labor cost that drove fast-growing economy is further undermined. Estimates show that labor forces contributed 0.5% points to GDP growth in 2001, and the figure decreased to 0.2% points in 2010, showing China's demographic dividend is diminishing.

In the coming years, China will undergo the most dramatic demographic changes in history, and face the dual pressure of reducing working-age population and increasing dependency ratio (Fig. 8.3). Chinese Academy of Social Sciences predicted that the supply of and demand for new labors in China would basically balance around 2015, and since then, the economic growth driven by scale expansion would face widespread labor shortage. The 20-year period when China has demographic dividend is coming to an end. Labor costs will be on rapid rise, and the aging population will become an ever-increasing burden to the economy. Although some scholars maintain that China still has a large rural surplus population, labor shortage in manufacturing sectors in some regions is expected to come early given high housing, education, transport, and living expenses in urban areas and regional disparities.

From the perspective of capital, China's economic potential depending on expanding manufacturing capacity driven by investment has reached the limit. Poor investment efficiency and excess production capacity are very common and have spread from traditional industries and infrastructure sector to emerging industries. Statistics by the World Bank show that the investment rate of China is much higher than that of other countries, and even outnumbers the historic high of developed nations since 1960 (Fig. 8.4). In 2010, the investment rate in many regions had exceeded 70%, and the figure in some regions was even as high as 80–90%. In terms of sustainability, with the new investments generating manufacturing capabilities, the problem of overcapacity in China will be more prominent in the future.

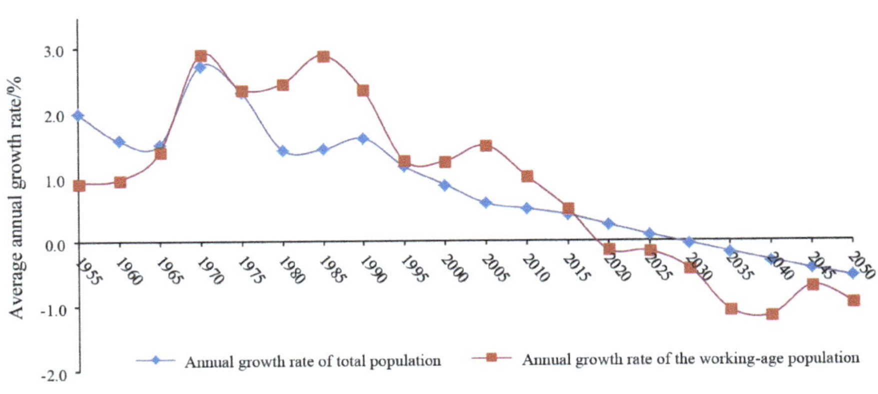


**Fig. 8.3** Changes of China's total population and working-age population (age between 15 and 64)

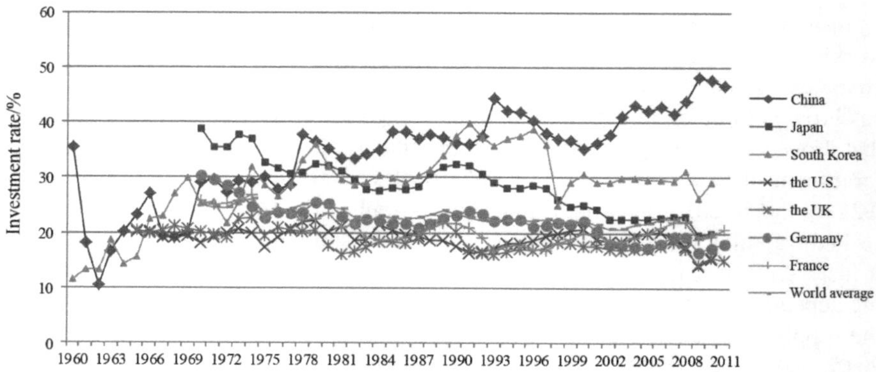


**Fig. 8.4** Historical investment rates in major countries

### 8.2.2 *Continuous Expansion of Manufacturing Capacity is Restricted by Demand Saturation*

In terms of external demands, China faces major changes in external conditions to expand export and daunting challenge in accelerating transformation of export structure under the influence of the 2008 global financial turmoil and the widespread contraction of demands in developed nations. Between 2010 and 2013, the annual growth rate of China's export was 8.6% on average, far lower than the level of 24.9% during the 10th Five-Year Plan period. Despite recent economic recovery in many developed nations, the trend that China's export will grow at lower rate is unlikely to change. In 2013, the export volume only increased by 7.6% over the previous year, lower than the actual GDP growth, and the processing trade even experienced negative growth. As the world's largest exporter, China will face contracting demands, increasingly stronger competition, and aggravating trade frictions in the long period of future. The traditional growth model in which China exports manufactured goods for the consumers of developed economy will no longer continue.

In terms of domestic demands, overcapacity is serious in high energy-consuming sectors. In the cement-making sector, the output reached 2.21 billion tons in 2012, accounting for 48% of the world total. The aggregate production capacity of the sector has amounted to 3.3 billion tons. Taking more than 200 production lines under construction into account, the aggregate capacity will continue to increase. In the steel-making sector, in 2012, China's crude steel yield was 720 million tons, more than 46.3% of the world's total, outnumbering the aggregated output of all steel-making countries from the second place to the 20th place. Structural problems like large-scale backward capacity, poor product quality, and loose industrial structure are prominent. The profit margin of the industry is meager, being the lowest among all domestic industrial sectors, and many enterprises operate at a loss. In the chemical industry, China has over 20 refinery bases with the capacity above 10 million tons and dozens of ethylene production bases with the capacity above one million tons under construction. It was estimated that the refining capacity would reach 750 million tons, signifying a serious problem of overcapacity.

Urban building scale is also saturated. Conservative estimates show that the total area of existing buildings in urban and rural areas has been close to 50 million $m^2$, and the housing conditions for both rural and urban residents have been fundamentally improved. In 2012, the area of buildings under construction reached an astonishing high of 11.67 billion $m^2$, and the area of completed building was even higher, standing at 3.36 billion $m^2$. The calculation based on the number of permanent residents shows that housing area per capita in urban areas has exceeded 35 $m^2$, and is still rising at an annual average growth rate of about 3 $m^2$. The calculation based on the sales area of commercial residential buildings shows that housing area per capita in urban areas increased by about 1.5 $m^2$ on average annually. Even if the current building scale was not expanded, China would catch up with Japan, France, and Singapore at the end of the 12th Five-Year Plan period in terms of housing area per capita (about 40 $m^2$ per person), and reach the level of European countries around 2020. Provided the scale of completed buildings no longer expanded on the 2011 level, the area declined by 1.5% year on year, and half of the existing buildings were demolished, China would still have buildings with the total area reaching 132.2 billion $m^2$ by 2050. Assuming that there is a population of 1.45 billion, housing area per capita would be as high as 91 $m^2$, far higher than the European level and paralleling the level in the U.S.

The infrastructure sector is also in a state of saturation. In recent years, the completed road and railway mileage hit new records in the world. In 2010, the total length of China's existing highways exceeded 4 million km, the second longest in the world. According to incomplete statistics, the total planned highway mileage was 165,000 km during the 12th Five-Year Plan period, including 85,000 km in the national highway network. The figure was about twice that in the U.S., far exceeding the reasonable level, and leading to idle resources and waste of investment. In terms of port construction, some coastal and riverside cities proposed to build themselves into port cities or international ports, and even county-level ports rolled out ambitious plans. Excessive ports brought about vicious regional competitions. Estimates show that the actual handling capacities of these ports were excessive by 30–40%. Even for those ports in the Yangtze River Delta region, more than 20% of the handling capacities were idle. The figure of ports in the Pearl River Delta region was 30%. In terms of airport construction, 80% of airports nationwide suffered from losses. Even though, the impulse of building more airports is strong. According to associated plans, 70 new airports would be built during the 12th Five-Year Plan period, more than twofold over that in the 11th Five-Year Plan period.

### 8.2.3 *Traditional Mode of Production Cost Huge Environmental and Economic Expenses*

China's demand for energy has risen rapidly as the manufacturing capacity continues to grow. While fueling economic and social progress, energy consumption has brought a series of severe challenges such as soaring greenhouse gas emissions, worsening environmental quality, deteriorating ecological conditions, and imminent risks of energy security. At present, the emissions of major pollutants have far exceeded the capacity of environment. Smog has become large-scale regular air pollution in China, directly threatening the living conditions of the majority of the population, and thus becoming the most urgent issue affecting people's livelihood. Water shortage, overexploitation, and damage to the water environment have all been far beyond the ceiling of its bearing capacity. The land has been over farmed, causing significant shrinking of the area of the arable land and reduction of its fertility. Over 10% of arable land has been heavily polluted. Generally, China's ecological environment has been overloaded, with no room to accommodate more pollution and damaging activities. Whether the issues can be solved as soon as possible has become a matter of maintaining social stability, testing the government's governance capability, and fulfilling China's international responsibilities. In the future, China's existing development path will be restrained by the red lines of ecological and environmental capacity. On the one hand, traditional industrialization and urbanization continue to stress the environment capacity by discharging more pollutants and increasing the use of environmental resources. Even with intensifying efforts for end-of-pipe treatment, the recovery and self-purification of the environment are weakening. On the other hand, to keep environment from further deteriorating, it is urgent to relax the pressure on the environment. We need to significantly reduce artificial occupation of environmental resources, reduce environmental loads, and pollutant discharges. Given that the energy demands are likely to continue to rise, how to cope with both domestic ecological challenges and global climate change has become a hard constraint on China's economic and social development.

Overemphasis on GDP growth and export-oriented economy, in essence, is a development model built on intentionally lowering the costs of production factors like labor, land, capital, resource, and environment, which cost China a huge economic expense. Compared with developed countries, China remained at the lower end of international industrial division. High energy-consuming, high-polluting and resource-based products, primary raw materials, low-end manufactured goods accounted for a large proportion of China's total exports, while the share of high-tech and high-value-added products and those with recognizable brands or independent property rights remains relatively low. The export of services grows slowly and its share in total exports is lower than that of developed countries and regions like the U.S. and Europe, and even lower than that of developing economies like India. From the enterprise level, there is still a huge gap with those in developed countries in terms of independent innovation, standard setting, marketing network, and capacity of integrating resources. The overall technological strength, innovation capability,

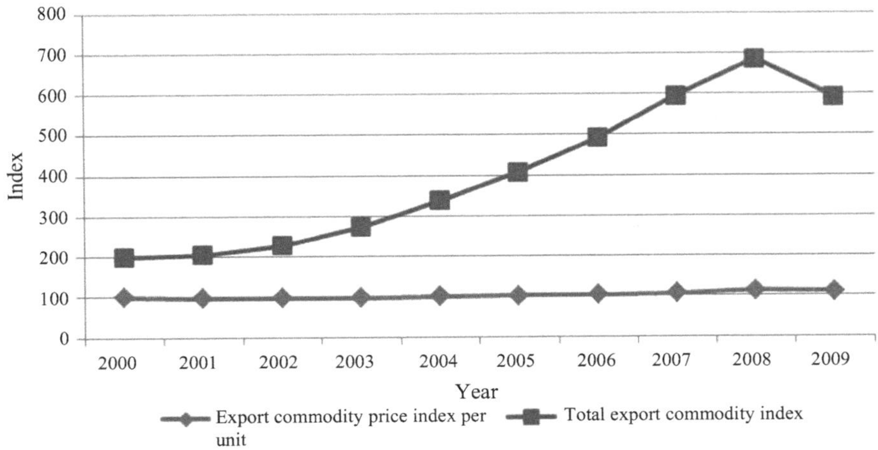


**Fig. 8.5** China's export commodity price index per unit and total export commodity index

and profitability of the export sector are disproportionate with China's status as a large exporting economy. In 2013, the profit margin of some 1000 exporting enterprises supported by the Ministry of Commerce was less than 3%. Between 2000 and 2009, in terms of constant prices, China's export more than doubled, but the commodity price per unit had no significant increase (Fig. 8.5).

### 8.2.4 *Limited Participation in the Value Distribution of Global Industrial Division*

With the accelerating globalization and refining industrial division of labor, the proportion of direct investment by multinational companies and foreign businesses is on the rise. Intra-industry trade, vertical division of labor, and outsourcing have become the main forms of international trade. The complexity of one country participating in global division of production factor input and distribution of value have brought challenges to the efforts for transforming and upgrading traditional mode of production. Under the new global situation of industrial division, global resources are being allocated among the entire world instead within one single factory, region, or country. Expanding export or developing technology advocated by traditional theories for economic development can no longer faithfully reflect the comparative trade interests of a country. How to constantly reduce production factor inputs of domestic resources while raising the value share in the international industrial chain when intensively participating in global industrial division of labor is a common issue faced by all countries across the world. Especially, as a late developing country, China has its overall economic strength at a lower level despite large export volume. It has few

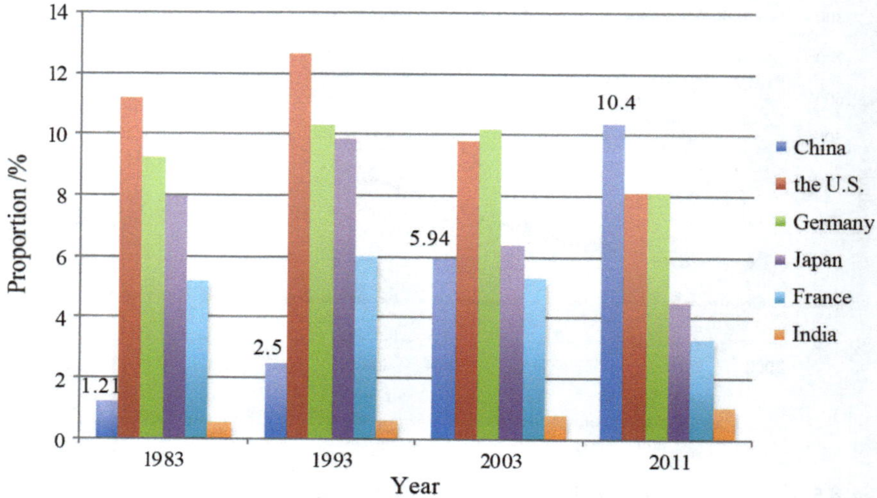


**Fig. 8.6** Proportion of commodity export of major countries to the world's total

exporting companies with global competitiveness. In this sense, China is far from being a strong exporting country. Under new circumstances where global industrial division is undergoing profound changes, and where export volume is approaching its ceiling, China faces ever-greater challenge in transforming and upgrading the mode of production (Fig. 8.6).

## 8.3 Focuses and Direction for Low-Carbon Mode of Production

The traditional mode of production overemphasizes fulfilling demands by increasing supply, but overlooks reasonable guidance on supply mix, constraints of environment and resources, and the sustainable development of the economic and social system. In the context of tackling global climate change, realizing low-carbon mode of production requires one country or one region not only to take local environmental issues and global warming serious but also fundamentally transforms the purpose of production and organizational methods in order to satisfy reasonable demands and push forward the shift of production system toward the direction of low carbon and sustainability.

For China, low-carbon mode of production means fundamentally changing the overemphasis on GDP growth and investment- and export-driven development. Instead, it should satisfy rational domestic demands, participate in international division of labor at a higher level, and drive the economy with investment, consumption, and import and export in a balanced manner. The mode of production should be focused on improving the quality and performance rather than expanding in quantity. It should significantly improve the quality and performance of economic growth, which is driven by technology and innovation. The production structure should be dominated by the services sector instead of industrial expansion, and efforts should be made to foster modern services, strategic emerging industries, and high-end manufacturing as major industries. Economic momentums should be shifted from investment of resource and investment to inputs of innovation and technology. Efficiency of various production factors including capital, labor, energy, land, and ecology should be improved.

### *8.3.1 Changing Export-Driven Economic Growth for the Purpose of Satisfying Rational Domestic Demands*

Low-carbon mode of production not only means decarbonization of the supply side, but, more importantly, integration between low carbon production and low carbon consumption. The traditional mode of production in which demands were blindly satisfied should be changed. Instead, an energy-saving, moderate, and rational approach should be adopted to satisfy demands, and the traditional mode of production should be improved constantly. Rational demands include various low-carbon technologies, products, and services, such as high-efficiency home appliances, green buildings and energy-saving vehicles, and also low-carbon public goods and services, such as low-carbon urban infrastructure, public transport system, and rural–urban system.

For China, it is a long-term development trend to step up efforts for integrating in globalization and global division of labor more intensively. Climate change mitigation and energy efficiency improvement are long-term strategic tasks. In line with the requirement on low-carbon development, export-driven development should be changed to a new economic growth model in which the focus should be on satisfying rational domestic demands, increasing public services, and enhancing the environmental quality. Instead of quantity expansion relying on cost advantage, we should foster global competitiveness with technology, brand, quality, and service at the core.

### *8.3.2 Optimizing Organizational Structure and Planning for Production and Making Full Use of Resources and Markets at Home and Abroad*

There are distinct gaps between rural and urban areas and different regions in China in terms of resource endowment and development level. China has favorable conditions for differential and progressive development. The living, production, and ecological functions should be appropriately planned in line with the national land zoning and based on resource conditions, environmental restrictions, ecological capacity, industrial basis, technological level, and market demands of different regions. The idea of focusing on local balanced development encompassing a complete range of industries should be discarded; instead, we should think in the big picture and promote regional coordinated development to ensure rational layout and optimized system between producing regions and consuming ones and throughout the entire industrial chain. In terms of organizational structure, we should promote mergers and acquisitions to increase industrial concentration and the level of technological equipment, and encourage the enterprises to become bigger and stronger.

Resources and markets at home and abroad should be fully used to participate in global division of labor at a higher level. On the premise of internalizing the external cost of resource and environment, we should give full play to China's comparative advantages. A more proactive trade strategy should be adopted in import and export. More resources, energy, and raw materials with strategic significance should be imported. The strategy of going global should be further implemented to expand international cooperation. Resource supply bases and manufacturing bases should be built in areas with sufficient conditions. Enterprises in industries such as steel and ethylene making should be encouraged to upgrade and relocate their businesses to address domestic overcapacity, poor supply of raw materials, and limited market demands. The model by which the country imports raw materials and exports primary products should be upgraded to a new model by which raw materials are imported and high value-added ones are exported to form new development advantages.

### *8.3.3 Promoting Low-Carbon Efficient Industrial Development with Modern Services as the Focus*

As a late developing country, China will see the manufacturing sector play a supporting role in its economy for quite a long period. In light of China's national conditions, we should maintain moderate expansion of its manufacturing sector and significantly improve its quality and efficiency in the process of accelerating economic development. With modern service sectors as the focus, efforts should be made to open the service sectors to both domestic and foreign businesses. Reform of financing, cultural, and medical sectors should be accelerated to create a healthy environment of market competitions for the development of the service industry. Efforts should be made to boost knowledge-based and innovation economy, deepen reform of cultural system, promote the export of cultural industry, and improve the development level and international competitiveness of the industry with modern services as the key.

It is necessary to significantly improve the production efficiency of present industrial activities, not only the efficiency of energy use but also the productivity; capital efficiency; and the efficiency of utilization of land, water, and other resources. With the progress of human economy, natural resources become more and more scarce. Countries and businesses that could fully and efficiently use the resources can stand out in future economic competitions. Improving energy efficiency is not only an important approach to the transformation of production mode but also represents the inevitable requirements of future economic development. We should work to reach international advanced levels in terms of industrial technology by 2020 and reach the international leading level by 2030.

### 8.3.4 Enhancing the Position in the International Chain and the Global Competitiveness of the Industrial System

From trend of international industrial division of labor, we can see the modern industrial process includes more inputs of intermediate products (and services), which differ significantly in terms of intensiveness. For example, R&D, design, innovation, and advanced material manufacturing are high-tech-intensive, whereas assembling, transport, and simple parts processing are low-skill-intensive. Especially with the rise of trade in services and the extension of industrial value chains to both ends, information- and management-intensive processes, such as marketing, brand operation, specialized services, financial services, and logistics management, keep increasing their positions in the industrial value chain. In this sense, the global industrial value chain is like a U-shaped curve or smiling curve (Fig. 8.7), in which different countries and businesses acquire different value distributed by involving in different sections of the industrial chain.

As far as China is concerned, the essence of improving China's position in the industrial value chain and upgrading the trade is to acquire more added values by investing in as few domestic resource factors as possible. With rapid expanding

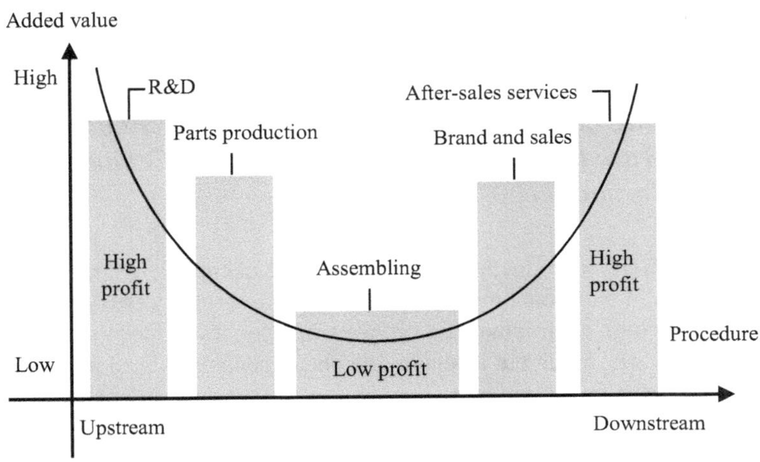


**Fig. 8.7** Global industrial value chain (smiling curve)

foreign direct investment (FDI), vertical division of labor, inter-product trade, and processing trade in international division of labor, the gains of China's industrial system are not purely what the country has acquired, but the trade gains out of configuration of international factors in a specific country, including FDI, imported raw materials, and intermediate products. From the perspective of added-value creation and value distribution, we should discard the ideas focusing on blindly expanding production scale, but should actively participate in global industrial division of labor and upgrade the position and value distribution in the system, which is the key to fundamentally enhance China's comprehensive competitiveness in the world.

### *8.3.5 Follow the Principle of Development of Circular Economy and Establishing a System for Sustainable Production and Utilization*

It is necessary to follow the from-cradle-to-cradle design principle, change the traditional from-cradle-to-grave model of product production, use and discard, advocate recycling and minimizing the use of materials (wiping out the concept of waste), and realize the radical transformation of mode of production. As the life cycle of a product ends, it transforms to harmless substance and returns to water or soil as ecological nutrients, or to useful raw materials for industrial production as process nutrients. In urban planning and the construction of transport infrastructure, systematic optimizing and recycling development should be integrated from the design stage to reduce energy intensity, pollutant discharge, and emissions to the utmost extent. By minimizing and recycling the resources, the service life of products will be extended, and energy consumption and carbon emissions will be cut from the source.

### *8.3.6 Adjusting the Market System and Policy Mechanism to Improve the Institutional Guarantee for Economic Transformation*

From the perspective of production factors, the inputs of capital, labor, land, energy, and environment in traditional mode of production are seriously twisted, which is a major reason for high environmental cost, overcapacity, low added value, and poor competitiveness. We should put an end to the phenomena of twisted pricing system in terms of domestic factor inputs, and create a business environment that truthfully reflects the external cost of environment to encourage enterprises to improve energy efficiency in a fair business environment, and increase the contribution of factors like human capital, technology, brand, and intellectual property. In this way, a win-win outcome for the economy, society, and environment can be achieved.

# Chapter 9
# Direction and Focus of Guiding Low-Carbon Consumption Mode

Low-carbon consumption is an important part of low-carbon development. As countries in the world pursue green and low-carbon development and low-carbon transformation, they attach greater importance to low-carbon consumption. In the process of reducing GHG emissions and developing low-carbon economy, at first people emphasized carbon emission reduction by technical approaches, implemented energy conservation and emission reduction in the energy and industrial fields, improved the energy efficiency of buildings and vehicles, and actively developed low-carbon energies including renewable ones. Nevertheless, the GHG emission is still increasing worldwide, and economic growth and the rising consumption level of the traditional mode are creating new demands for fossil energies and industrial GHG emission. People realized that if they do not resolve the final consumer demand that can effectively reduce carbon emission, it is hard to meet the low-carbon goal merely through technical approaches and energy substitution. The topic of low-carbon consumption has received extensive attention and lots of theoretical discussions, explorations, and practices have been carried out in many countries. It has evolved to be a deep-level theoretical and practical issue involving multiple aspects and fields, including consumption mode, morality of conduct, philosophy of life, outlook on life, values, economic and social systems, the goal and mode of human development, and even faith and religion.

The current per capita GHG emission in China is about 7t $CO_2$, which is the general level in European countries. Of GHG emission in China, direct emission from production activities takes up a large proportion, while direct emission from consumption activities takes up a low percentage. Given the export structure, part of the production emission is to serve foreign consumers. However, the ultimate purpose of production is meeting consumer needs, so all production emissions should be categorized as indirect emission from consumption. Whether Chinese people can shift to the low-carbon consuming mode and contents will be fundamental and critical for China's low-carbon development.



X. Du et al., *Overview of Low-Carbon Development*,
https://doi.org/10.1007/978-981-13-9250-4_9

## 9.1 Current Situation and Characteristics of Consumption Mode in China

China has basically met the basic needs of its people and is working hard to build a moderately prosperous society in all respects and further improve the consumption quality. Although about 100 million Chinese are still living under the poverty line and our consumption level remains low, a great many people have strong buying power and have even boosted the huge luxury goods market. As far as material consumption is concerned, China has moved from the stage of survival needs to the stage of improvement and selective needs. In this stage, a pressing task of low-carbon development is to decide which consumption mode to be chosen as the development goal.

Housing consumption in urban and rural areas has obviously improved with significantly increased construction area, and energy guarantee reaches a high level. Housing consumption in China began to pick up speed after the 1990s and became the primary drive of consumption growth. The floor space under construction has kept rising nationwide and reached 7.26 billion $m^2$ at the end of 2014, including 5.15 billion $m^2$ residential areas. The annual sales area of commercial housing exceeded 1 billion $m^2$ for years and was more than 1.2 billion $m^2$ in 2014. According to urban population, the per capita housing area increased by nearly 2 $m^2$ every year. Housing construction in rural areas was the main consumption for local residents. At present, per capita housing area in rural and urban areas is more than 37 $m^2$ and 35 $m^2$, respectively, and the area of public buildings in urban areas has increased by a large margin. According to surveys and statistics, the total area of civil construction in China is well above 50 billion $m^2$ and close to 55 billion, averaging more than 40 $m^2$ per person. While heating is continuously improved in northern China, civil heating facilities become popular in the Yangtze River basin. In urban areas, the promotion of gas use progresses rapidly, power supply is sufficient, and electric appliances both for household and office use have reached a high penetration rate. All rural areas have basically had access to electric power, and residents in the suburban villages of some large cities have higher per capita energy consumption than city dwellers (Tables 9.1, 9.2 and 9.3).

### *9.1.1 Urban Traffic System Quickly Shifts to Motorized Travel*

Chinese cities are getting bigger, but they only focus on the land economy and the showcasing function, and make China quickly shift from the bike-based traffic mode to motorized urban traffic mode, while the development of public transit and rail transit obviously lags behind that of private cars. China had more than 126.7 million cars in 2013, and the car sales maintained a growth rate of more than 25 million every year. The number of private cars reached 91.98 million in 2013, and car sales quickly

**Table 9.1** Ownership of main consumer durables per 100 urban households at year-end

| Indicators | 1990 | 1995 | 2000 | 2005 | 2010 | 2011 | 2012 |
|---|---|---|---|---|---|---|---|
| Motorcycle | 1.94 | 6.29 | 18.80 | 25.00 | 22.51 | 20.13 | 20.27 |
| Washing machine/set | 78.41 | 88.97 | 90.50 | 95.51 | 96.92 | 97.05 | 98.02 |
| Fridge/set | 42.33 | 66.22 | 80.10 | 90.72 | 96.61 | 97.23 | 98.48 |
| Color TV/set | 59.04 | 89.79 | 116.60 | 134.80 | 137.43 | 135.15 | 136.07 |
| Audio system/set | – | 10.52 | 22.20 | 28.79 | 28.08 | 23.97 | 23.63 |
| Camera/set | 19.22 | 30.56 | 38.40 | 46.94 | 43.70 | 44.48 | 46.42 |
| Air conditioner/set | 0.34 | 8.09 | 30.80 | 80.67 | 112.07 | 122.00 | 126.81 |
| Water heater (shower)/set | – | 30.05 | 49.10 | 72.65 | 84.82 | 89.14 | 91.02 |
| Computer/set | – | – | 9.70 | 41.52 | 71.16 | 81.88 | 87.03 |
| Video camera/set | – | – | 1.30 | 4.32 | 8.20 | 9.42 | 10.00 |
| Microwave/set | – | – | 17.60 | 47.61 | 59.00 | 60.65 | 62.24 |
| Fitness equipment/set | – | – | 3.50 | 4.68 | 4.24 | 4.09 | 4.27 |
| Mobile phone/set | – | – | 19.50 | 137.00 | 188.86 | 205.25 | 212.64 |
| Fixed-line telephone/set | – | – | – | 94.40 | 80.94 | 69.58 | 68.41 |
| Family car | – | – | 0.50 | 3.37 | 13.07 | 18.58 | 21.54 |

**Table 9.2** Ownership of main consumer durables per 100 rural households at year-end

| Indicators | 1990 | 1995 | 2000 | 2005 | 2010 | 2011 | 2012 |
|---|---|---|---|---|---|---|---|
| Washing machine/set | 9.12 | 16.90 | 28.58 | 40.20 | 57.32 | 62.57 | 67.22 |
| Fridge/set | 1.22 | 5.15 | 12.31 | 20.10 | 45.19 | 61.54 | 67.32 |
| Air conditioner/set | – | 0.18 | 1.32 | 6.40 | 16.00 | 22.58 | 25.36 |
| Kitchen ventilator/set | – | 0.61 | 2.75 | 5.98 | 11.11 | 13.23 | 14.69 |
| Bike | 118.33 | 147.02 | 120.48 | 98.37 | 95.98 | 77.11 | 78.97 |
| Motorcycle | 0.89 | 4.91 | 21.94 | 40.70 | 59.02 | 60.85 | 62.20 |
| Fixed-line telephone/set | – | – | 26.38 | 58.37 | 60.76 | 43.11 | 42.24 |
| Mobile phone/set | – | – | 4.32 | 50.24 | 136.54 | 179.74 | 197.80 |
| Black-and-white TV/set | 39.72 | 63.81 | 52.97 | 21.77 | 6.38 | 1.66 | 1.44 |
| Color TV/set | 4.72 | 16.92 | 48.74 | 84.08 | 111.79 | 115.46 | 116.90 |
| Camera/set | 0.70 | 1.42 | 3.12 | 4.05 | 5.17 | 4.55 | 5.18 |
| Computer/set | – | – | 0.47 | 2.10 | 10.37 | 17.96 | 21.36 |

**Table 9.3** Newly built residential area and civil housing in urban and rural areas

| Years | Newly built urban residential area/100 mil m$^2$ | Newly built rural residential area/100 mil m$^2$ | Per capita housing area of urban residents/m$^2$ | Per capita housing area of rural residents/m$^2$ |
|---|---|---|---|---|
| 1978 | 0.38 | 1.00 | – | 8.1 |
| 1980 | 0.92 | 5.00 | – | 9.4 |
| 1985 | 1.88 | 7.22 | – | 14.7 |
| 1990 | 1.73 | 6.91 | – | 17.8 |
| 1995 | 3.75 | 6.99 | – | 21.0 |
| 2000 | 5.49 | 7.97 | – | 24.8 |
| 2001 | 5.75 | 7.29 | – | 25.7 |
| 2002 | 5.98 | 7.42 | 24.5 | 26.5 |
| 2003 | 5.50 | 7.52 | 25.3 | 27.2 |
| 2004 | 5.69 | 6.80 | 26.4 | 27.9 |
| 2005 | 6.61 | 6.67 | 27.8 | 29.7 |
| 2006 | 6.30 | 6.84 | 28.5 | 30.7 |
| 2007 | 6.88 | 7.75 | 30.1 | 31.6 |
| 2008 | 7.60 | 8.34 | 30.6 | 32.4 |
| 2009 | 8.21 | 10.21 | 31.3 | 33.6 |
| 2010 | 8.69 | 9.63 | 31.6 | 34.1 |
| 2011 | 9.49 | 10.26 | 32.7 | 36.2 |
| 2012 | 10.00 | 9.51 | 32.9 | 37.1 |

*Note* Per capita housing area of urban residents is based on the sample survey of urban residents (excluding collective households)

expanded to rural areas. For car-owning families, vehicle energy consumption is their largest source of GHG emission. China has the world's second largest car population and the largest car market.

### 9.1.2 *Fast Popularization of Household Appliances Boosts Drastic Increase of Household Electricity Consumption*

China is a major manufacturer of all kinds of household and commercial electric appliances. Main household electric appliances are popularized rapidly with a huge quantity. Common home appliances like TV set, fridge, washing machine, and microwave are already popularized; air conditioner is having a high penetration rate; and large-power appliances are entering the households continuously. With the greater variety and longer service life of household appliances, household electricity consumption increased from 159.4 kWh per person in 1999 to 506 kWh per person in 2014, the fastest-growing part of household energy consumption.

## 9.2 Problems in China's Consumption Mode

Before the reform and opening-up, China had a very low consumption level and many of its basic consumer needs could not be met. As a result, the main purpose of economic development was continuously improving the supply capability to meet the people's material and cultural needs. At that time, China lagged far behind developed countries in material consumption, so it took them as the target and was trying to catch up with them in consumption, while developed countries also strongly advocated their consumption mode and contents through economic, cultural, media, advertising, and publicity approaches. To date, China's consumption mode is mainly influenced by the lifestyle advocated by the US and, while pursuing it, China is moving quickly toward the traditional high-carbon consumption mode. In particular, during the urbanization process and the transformation of consumption culture, China lacked the necessary independent studies and rational orientation, and did not compare the differences in per capita carbon emission among developed countries caused by different consumption modes. These made China's efforts quite blind. Although China is undergoing industrialization, urbanization, and modernization and its consumption level remains low for many people, the high-carbon tendency in everyday consumption, covering clothing, food, housing, and travel, has become prominent.

### *9.2.1 Uncoordinated Dietary Structure and Nutritional Goal*

China has made tremendous achievements in developing agriculture, ensuring food supply, and improving people's nutrition, and invested immensely in maintaining the quantity of farmland and guaranteeing the conditions for agricultural development, while also paying a high eco-environmental price for it. However, the phenomenon of food waste is increasingly conspicuous in this process. It is reported that in 2010, meat consumption was only 66% of output, vegetables 27%, fruits 52%, egg 80%, dairy products 50%, and aquatic products 63%, considering the factor of dining out. The primary reason for the output-consumption gap was food waste. China has huge food losses in the process of storage and distribution, with the grain loss rate of about 5% and the loss rate of agricultural and sideline products such as fruits and vegetables being as high as 25–30%. Moreover, food waste on the dining table becomes increasingly serious in China. In the catering industry, at least 10% of food was wasted on the dining table, which amounted to RMB60 billion in 2004 and increased to RMB80 billion in 2005 and RMB200 billion in 2010. This phenomenon also spread from cities to rural areas.

### 9.2.2 *Urban Residents have More Unused Clothes as They have a Shorter Life Span*

China is the largest textile and clothing producer in the world. In 2010, urban and rural residents bought 8.43 and 2.7 pieces of clothing per person on average. While it is necessary to make the clothing more diverse and beautiful, massive waste does not conform to the requirement for low-carbon consumption and the resource conditions in China. Some Chinese people already have a high clothing consumption. 12% of Chinese wear a piece of clothing only once or several times on average, 25% wear it for three months to a year, and only 35% wear it for more than two years on average. Clothing waste has become a common phenomenon.

### 9.2.3 *Construction Area Increases Too Fast, Energy-Saving Buildings Develop Too Slowly*

China has a large population, but its territory suitable for living is very limited, and national conditions have to be taken into account when increasing per capita housing area. After nearly one or two decades of high-speed development, the per capita housing area in China has approached the level in developed countries and regions such as Europe and Japan, and the substantive housing shortage is replaced by structural imbalance caused by unreasonable housing distribution. People with self-owned housing in China take up a larger proportion than in most developed countries, but housing construction is still in full swing and per capita housing area increases by nearly 2 $m^2$ every year in urban regions. According to many studies on low-carbon development, the per capita housing area of 30 $m^2$ is a reasonable target of housing development given China's national conditions. If China's construction scale is not reduced soon, its per capita construction area will probably surpass that in most developed countries, including the US and Canada, by a large margin. Nevertheless, the current policies still take real estate development as an important way of maintaining steady economic growth, and per capita construction area keeps increasing at a high speed.

Regarding the ratio of vacant commercial housing in all commercial housing up for sale and lease during the reporting period, China had a housing vacancy rate of 20–30% in recent years. Regarding the part of housing that is sold but not used, CCTV-2 aired two sessions of "vacant housing" report in May and August 2010, according to which the vacancy rate in some popular commercial properties in Beijing and Tianjin was up to 40%, and this situation has not obviously improved in recent years.

Besides, the energy-saving standards in the construction industry remain low in China. Even in Beijing that has the highest standards, the energy efficiency standards for new buildings are lower than those in many European countries, while the standards adopted by most Chinese cities are 50% lower than those adopted by Beijing.

Another major problem in the development of residential buildings is that newly built apartments are too large. Because of the irrational competitions in housing consumption and the once excessive housing price hike, there is a general trend of building apartments larger than actually needed.

Building larger housing than needed is also a common phenomenon in rural areas. Rural China has the tradition of taking housing as a basic form of wealth, and the blind competition in large housing is common in many economically developed rural areas. As rural population moves to cities, the actual utilization rate of rural buildings falls significantly, causing serious waste.

Public buildings have developed rapidly in China. In most places of the country, government buildings are super large and more than necessarily luxurious. In some places, large commercial public buildings are blindly developed, ending up with an appallingly high vacancy rate. As to business layout, it is common to copy the American model and there are in many places large supermarkets, central business districts, and retail centers that are far from the residential area, forcing the people to drive for large-scale shopping.

According to relevant studies, energy consumption for producing building materials used for housing construction, such as cement, iron and steel, and glass, is about 131.6 kg standard coal/$m^2$ completed area when converted to construction area, while energy consumption during construction is about 6.9 kg standard coal/$m^2$ construction area according to statistics. Assume the building's life cycle of 50 years, the annual energy consumption for building materials and construction is about 2.8 kg standard coal/$m^2$. In a sense, housing vacancy wastes the energy used for building materials and construction. For buildings in northern China that use central heating, the annual energy consumption for heating is 10–20 kg standard coal/$m^2$ even if they are vacant, also causing immense energy waste.

Energy waste is also common in building operation. Many hotels in China demonstrate their star-rated superiority through luxurious and impractical premier services. For instance, they offer eiderdown bedding in summer, so the customers have no choice but set the air conditioner to a very low temperature. In the northern region, heating cannot be measured on a household basis for a long time, so some households have to open the window for ventilation and cooling as it is too hot in the room. It is also common to see office buildings with the air conditioning on and the windows open at the same time.

### *9.2.4 City Planning Lacks Low-Carbon Guidance, Fast Development of Motorized Traffic Leads to High Carbon Emission*

Traffic congestion is a serious problem in most cities of China, both big ones and medium-sized ones. When planning for their development, many cities base the design of urban functional zones on motorized traffic. They first build super-wide and super-long roads, then build production and business zones far from the residential zones, so that driving for work or consumption is a necessity. Even if people want to walk or not travel by car, it is hard to have a second choice.

### *9.2.5 Traditional Advocacy for Austerity Shifts to Western Consumerism*

The western idea of consumerism has penetrated every aspect of the Chinese society. For a long time, the Chinese government neither conducted proactive studies on our consumer concept, culture, and contents nor set a green, low-carbon target suitable for our national conditions, not to mention effective policy guidance. It did not conduct any specific study as to what kind of average consumption level the people can reach, so its actions are obviously blind. Capitalism lured people to make luxury consumption and constantly stimulated various kinds of wasteful consumptions. Due to the partial understanding of the idea that "we allow a few people to get rich first," a social trend of wealth flaunting has been formed in China, and a group of billionaires is created in a short span of 10 or 20 years. As a result, consumerism is even more prominent in the country, and the consumption inclination of the few rich people is regarded as representing the general consuming wishes and actual consuming capability, leading to the wrong consumption policy guidance and the wrong idea that big housing and luxury car are the symbols of success. SUV (Sport Utility Vehicle) and off-road cars, which are criticized in many developed countries, are quite a vogue in China, where the annual SUV sales hit more than 5 million units, accounting for more than 20% of all auto sales and being the fastest-growing segment. China has very limited land and road resources, but it has the largest cars in the world. Almost all auto brands have lengthened their products specifically for the Chinese market, which is the largest market for luxury cars with large displacement, whereas the oil price policy adopted by developed countries to curb excessive oil consumption can hardly be put into practice in China. Some people are calling for developing private airports and docks, private jets, and luxury yachts, all luxury products that only serve a small number of rich people.

## 9.3 Strategic Thoughts on and Direction of Promoting Low-Carbon Consumption Mode

### *9.3.1 Change the Mindset, Re-choose the Low-Carbon Consumption Mode, Targets, and Contents*

To tackle global climate change, we have to reduce the global carbon emission as soon as possible and realize zero carbon emission in the second half of the twenty-first century. Nowadays, fossil energy still takes up more than 85% of global energy supply and consumption, and world economic development will continue to depend on fossil energies for a long time. For most developing countries, a better life and rising consumption imply the increased use of fossil energies. The global efforts for GHG emission reduction lag far behind the needs to control the temperature rise. Such a dilemma fully proves that the traditional path of industrial civilization that depended on the massive consumption of fossil energies will not sustain, nor will the high-carbon consumption mode that constituted the high-carbon industrial civilization, and we have to make a change soon. We have to consider how to achieve low-carbon consumption in order to develop ecological civilization in China.

The American-style high-carbon consumption mode should not be our target. There are several reasons why the US can keep the high-carbon lifestyle and consumption mode for so long. It has abundant resources, occupies global resources through hegemony and power, and keeps stimulating consumption in such aspects as housing, transport, and material consumption. America has a much higher per capita energy consumption and carbon emission than European countries and lower energy efficiency in all aspects, but this high consumption is based on the consistently low consumption level in developing countries that account for most of the world population. Even without the GHG limitation, China cannot copy America's consumption concept and mode. As the US is incapable of criticizing its own development model, it cannot make up its mind to make the low-carbon transformation and contentions on how to tackle climate change have never stopped in the country. European countries, however, have long chosen a different path of energy consumption owing to the long-term restrictions of resource security and environmental capacity. They paid close attention to energy conservation in housing, transport, and the business model of social services, and to greater social balance. The per capita energy consumption and carbon emission in European countries is only half of that in the US, their per capita housing area is around 40 $m^2$, and they have an obviously smaller car ownership than the US, but more developed public transit, more intensive cities, and more convenient conditions for slow travel. But even the carbon emission in Europe is far from the required low-carbon level. European countries are making active efforts toward the low-carbon transformation, in hopes of occupying the new high ground of low-carbon development.

As a developing country with a population of nearly 1.4 billion people, China, based on its national conditions, must see the general trend of world development and explore a new low-carbon development path, including the mode and contents of low-carbon consumption. We should integrate low-carbon consumption with the new model of industrialization and urbanization, and realize the shift to low-carbon consumption as soon as possible. We should explore the new consumption mode along with other countries in the world, and strive to continue improving the quality of Chinese people's material and cultural consumption and the general consumption level with per capita energy consumption obviously lower than that in developed countries and much lower carbon emission.

### 9.3.2 Promote Low-Carbon Consumption Through Accelerated Low-Carbon Transformation and Development

To achieve low-carbon development, we need first of all significantly raise the utilization efficiency of energy and other natural resources. Various efficient energy and resource technologies can largely reduce carbon emissions. For instance, the new type of ultralow-energy buildings can skip the traditional heating system and keep warm indoors in cold winter through insulation and efficient air–heat exchange. Electric vehicles can not only drastically raise the energy efficiency but also realize low-carbon motorized traffic in the case of low-carbon power generation. There is vast room to improve the efficiency of all kinds of energy-using equipment, so we can get the same energy services with less energy input. Nonfossil energies may eventually substitute fossil energies and help us realize zero carbon emission in the energy system. What energy development strategy we choose is an important part of low-carbon consumption. Many low-carbon energy technologies will be more costly than traditional fossil energies, at least in the early stage, and require more investment. The users may also have to pay more for them while having to change their old consuming habits and make new choices.

At present, the progress of low-carbon technologies, especially in low-carbon energies, is not enough to meet the aggressive expansion of the traditional consumption mode, and the input of fossil energies is inevitable in the traditional production of many raw materials. Infrastructure construction will have to depend on fossil energy consumption for a very long time to come, and it is a long-term necessity to save resources, cut unreasonable energy consumption, and restrict or abolish luxury consumption. While building infrastructures and planning for the city and housing construction, we should change, as soon as possible, the traditional rule of taking a few people's luxury consumption as the guide for general social consumption, and put necessary limitations on luxury consumption. China is a socialist country, so we can determine the market orientation in accordance with the consumer needs of the majority of the people. For example, we should insist on reasonable apartment type in housing construction; keep city planning and economic development from being "hijacked" by real estate economy, restore the social property of housing, and keep the scale and aggregate of housing construction within a reasonable range. We should also promote the rational distribution of housing resources and change the situation that a few people own too much housing while most people find it ever more difficult to improve their living conditions.

The choice of consumption mode is primarily a social choice. On the condition of fixed city plans, infrastructure construction, and inflexible market conditions, individuals actually have very limited low-carbon choices; on the condition of unreasonable urban layout and insufficient public traffic services, it is hard for individuals to stick to slow travel and non-motorized travel too. The individual choice of low-carbon consumption should first and foremost support the social development of low-carbon technologies, low-carbon energies, and low-carbon utilities and infrastructure. It should also support the construction of low-carbon cities, and the adjustment of price and tax policies aimed to encourage thrift and curb high-carbon consuming behaviors. Of course, the change of personal behaviors is also important and useful, including the advocacy for thrift, objection to all forms of extravagance and waste, and proactive choice of low-carbon products and behaviors.

### 9.3.3 *Speed Up Policy Design and Guidance to Establish Low-Carbon Consumption Mode*

The current price and value system in the market economy neither tilts resource allocation to low-carbon development nor guides the establishment of low-carbon consumption mode. Low-carbon consumption not only means the choice of low-carbon goods and services but also the restriction on luxury high-carbon consumption. Traditional market economy depends on the continued increase of end consumption to drive the economy. The capitalist market economy also protects and enlarges the rich-poor gap and depends even more on the rich people's consumption to drive new demand, which usually involves high material input. While exploring the low-carbon transformation, developed countries have to regulate the market by formulating compulsory measures, such as the emission quota, and adjust market signals by economic means such as carbon tax. Besides, steering the public opinions, criticizing high-carbon consumption behaviors, and advocating low-carbon consumption are also important approaches to establish the new mode of low-carbon consumption.

To establish the low-carbon consumption mode in China, we need to effectively steer the market. On the one hand, we should reflect the carbon emission cost in the goods price system by adjusting the prices and taxes; on the other hand, we need to adopt necessary regulatory measures to systematically put in place the laws, regulations, and standards to promote low-carbon consumption. We have to prevent the situation that the market continues to operate in a high-carbon way despite the theoretical low-carbon goal, foster the idea of low-carbon consumption in the whole society, and guide the people to adopt low-carbon-consuming behaviors.

## 9.4 Low-Carbon Consumption Mode and Contents in Key Areas

China has a generally low consumption level at present. The people who got rich first have lived a traditional high-carbon affluent life, while the nation is under tremendous resource and environmental pressure, and the discharge of main pollutants has exceeded environmental capacity in many places. We have to establish the new low-carbon consumption mode as soon as possible and advocate low-carbon development.

### 9.4.1 Green and Low-Carbon Food Consumption

Food is the paramount necessity for the people. China has a fine and long dietary culture, and its dietary diversity is a wealth for the Chinese people. The traditional dietary habits in China dominated by grain, vegetables, and fruits have a natural low-carbon nature. Instead of blindly following the western meat-rich dietary habits, we can further improve our dietary structure, balance the nutrition from the health perspective, and reduce unnecessary meat consumption. In this way, we can meet the goal of reducing the consumption of water, energy, and other resources and cutting GHG emission while maintaining health.

What's more important is the insistence on food conservation, objection to food waste and upholding the fine tradition of being industrious and thrifty. We should further advocate the "clean plate campaign" and reduce food waste in the dining table; make overall arrangements for urban–rural common development when pursuing the new model of urbanization, and protect the rural food supply capability and "vegetable basket" around cities. Efforts should be made to reduce losses during long-distance transport and distribution and protect the diversity of agricultural products and supply. Basic guarantee measures must be strengthened to minimize food losses during distribution. The agricultural produce loss rate during logistics is less than 3% in some developed countries, and the loss rate of fruits and vegetables is 1–5%. We must accelerate the development of cold-chain transportation of perishable food. Cold-chain transportation accounts for more than 90% of total food transportation in western countries, but it is only 20% in China. Going forward, we will control the food loss during logistics within 5%.

### 9.4.2 Green and Low-Carbon Clothing Consumption

A 200 g cotton T-shirt emits nearly 7 kg $CO_2$ from cotton planting to T-shirt making to end sales, and to the constant washing, drying, and ironing after it is bought by the consumer. Clothes made of chemical fiber have higher carbon emission. A pair of trousers made of 100% polyester fiber emits about 47 kg $CO_2$, including during the assumed 2-year life cycle and the production and consumption, and that is 117 times its own weight. If every Chinese person buys one less piece of clothing a year, they will reduce the $CO_2$ emission by more than 10 million tons.

China should avoid unnecessary waste of clothing, call for higher use ratio of it, and extend its life cycle appropriately. By reducing unnecessary clothing consumption, we can meet the goal of reducing resource consumption and protecting the environment while saving personal spending. Chinese consumers should choose, as much as possible, clothes made of environment-friendly materials like cotton and linen, which consume less energy during production. In the use of clothes, we should use washing machine more efficiently and practice natural drying. We should control the market of household clothing dryer.

### 9.4.3 Green and Low-Carbon Housing Consumption

Housing consumption in a green and low-carbon society should have the following characteristics:

(1) Total construction area is controlled within a reasonable range: based on the actual conditions in China, the reasonable per capita housing area can be set at 30 $m^2$ with 10–15 $m^2$ residential pool area, and the national housing construction area (excluding plant construction) should be controlled around 63 billion $m^2$, not exceeding 68 billion $m^2$. Construction land plans and urban–rural development plans should involve beforehand planning and management for achieving reasonable development scale of the construction industry.
(2) Rational regional plans: buildings should have a life cycle of more than 50 years and short-lived buildings must be eliminated. Rational city plans should be made and intensive residences should be dominant, while massive demolition and construction should be stopped. The area of rural construction should be kept within a reasonable range and the quality of rural buildings should be improved according to population change, and energy efficiency standards should be introduced as soon as possible.
(3) A reasonable and natural home lifestyle should be maintained. There is no need to create the extreme high-carbon living environment featuring "constant temperature, constant humidity, and fixed ventilation." Rather, natural approaches should be prioritized in architectural design and operation to create the indoor environment, so that room temperature will rise and fall along with different seasons, and natural ventilation and lighting are people's top choice.
(4) Buildings of ultralow energy consumption should be promoted across the board as soon as possible, including passive buildings. Energy efficiency standards should be significantly raised in the construction industry, and the new system of building energy efficiency standards should be promoted that is based on the actual operating energy consumption. Efforts should be made to develop and promote various types of renewable energies for buildings and to realize building carbon balance to the greatest extent. To build a new countryside, energy-saving building technologies and standards should be adopted as soon as possible.
(5) Technical progress in household electric appliances is advanced and efficient household appliances are popularized. For example, semiconductor lighting, high-efficiency air conditioner, and fridge are promoted on a large scale.

According to a relevant research group at the School of Public Policy & Management of Tsinghua University, on the condition of the aforementioned low-carbon construction development, energy consumption in China's construction industry will peak in 2025 (about 2.1 billion ton standard coal) and fall afterward until it stays around 1.4 billion ton standard coal in 2050 (Fig. 9.1).

The research group held that if "household-based measurement and measurement-based charge" is realized in northern China where collective heating is adopted, 90% residents will adjust their heating approach and reasonably lower the over-high

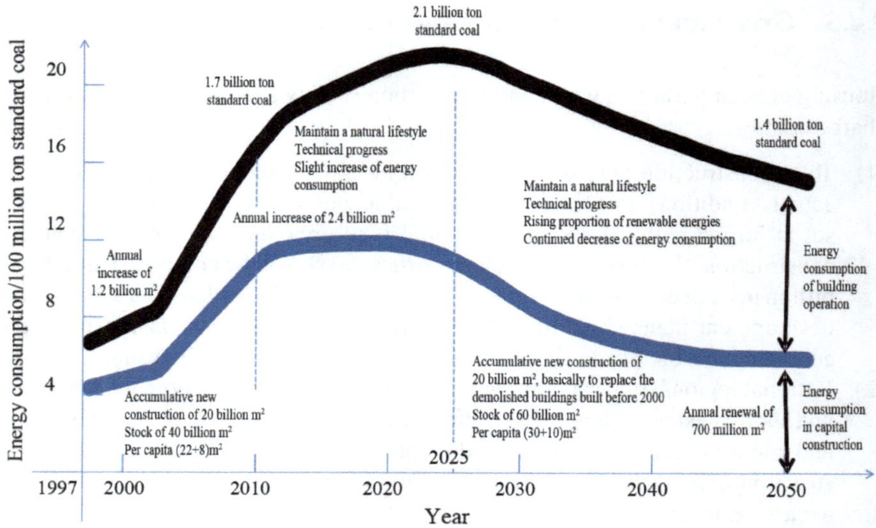


**Fig. 9.1** Construction energy consumption in a green and low-carbon scenario

heating temperature. Meanwhile, heating suppliers will actively adopt energy-saving operation and management and use efficient heating sources, thus greatly reducing the energy waste in the whole collective heating system and further bringing down the per capita heating energy consumption by 40% from the 2010 level.

### *9.4.4 Green and Low-Carbon Travel*

To lower the energy consumption in travel, we need to make plans and provide guidance in such aspects as the amount of travel, travel mode, and low-carbon vehicle. We should make reasonable city plans so that people can reach the main living functional zones either by walk or through non-motorized travel, thus reducing the amount of travel to the largest extent. We should improve the urban traffic environment and advocate the travel mode dominated by slow traffic and public transit. Rail transit should be given priority, the fuel economy of motorized vehicles should be improved, and great efforts should be made to strongly develop new-energy vehicles, so as to minimize the impact of motorized travel on resources and the environment (Fig. 9.2).

The formation of a low-carbon traffic system requires the joint efforts of all walks of life. For example, given the existing traffic and road conditions, giving priority to public transit may imply further restricting the right of road of private cars and allocating more special lanes for buses. To curb the high-carbon development of cars, we need to raise the consumption tax on refined oil products by a large margin, or set more limitations on the purchase of traditional cars, levy more taxes on

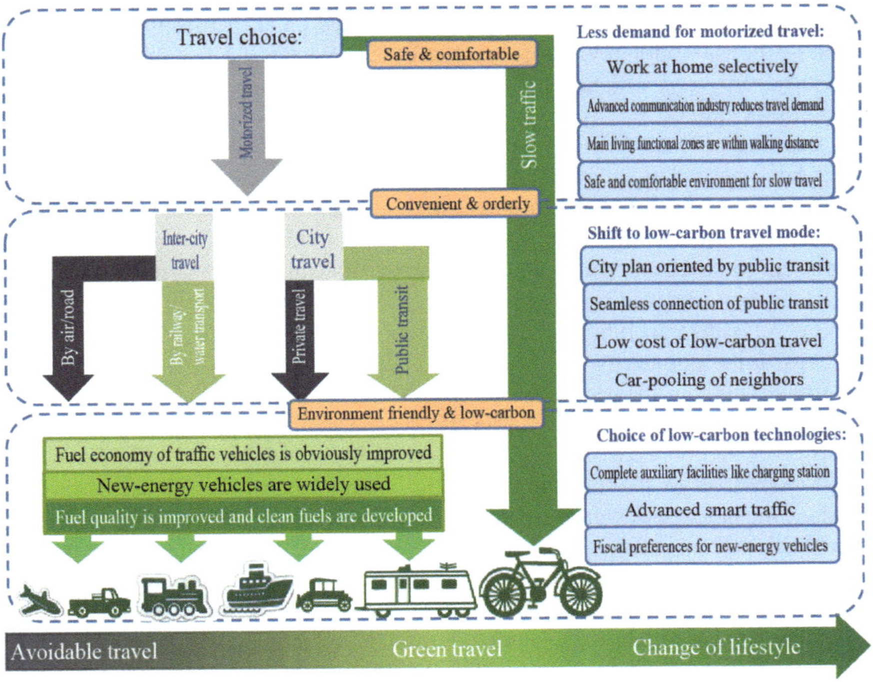


**Fig. 9.2** Outlook of green and low-carbon travel

large-displacement cars, and implement differentiated parking fees based on different car types. These economic measures have been adopted in many countries and taken obvious effects. At the current stage, consumption tax on refined oil products in China is much lower than that in most oil importers, which is one of the reasons for the large displacement and high oil consumption in China's automobile development.

## 9.4.5 *Establish Green and Low-Carbon Consumption Culture*

China always upholds hard work and frugality in running the country and the household alike and opposes waste and extravagance. This has something to do with our low level of productive forces in the past, but it also summarizes why and how China has been able to feed the largest population in the world for so long with its limited territory and natural resources. This basic national condition—large population and limited resources—remains today. Although China has basically got ride of the age of shortage, it has to carry on the fine traditions. China proposes the basic state policy of building a resource-saving nation and advocates the energy strategy that puts frugality first; moreover, the central government has taken strong measures in recent years to fight against extravagance, and called on all parties to push forward the energy consumption revolution, restrict irrational energy consumption, and control total energy consumption. All these efforts are in essence calling for a good consumption culture.

A new fad of "neo-frugality" featured by "reasonable consumption, shame on waste" has emerged in the international community over recent years. The LOHAS, for instance, emphasizes the Lifestyles of Health and Sustainability. According to them, taking mass transit, eating green and organic food, wearing linen fabrics, and using used goods is the future lifestyle. From exercising in the gym or stadium to doing morning and evening exercises in the neighborhood park; from self-driving or taking a taxi to traveling by public transit or by bike; from pursuing brand-name handbags to the popularity of LOHAS and cotton "I'm Not A Plastic Bag"...The Neo-frugality guides the people to shift from the luxury-oriented consumption mode to one that places equal emphasis on health and environmental protection while meeting their living needs.

In comparison to the consumption mode under "consumerism", neo-frugality abandons the practice of signifying social status with money and spending and advocates health and environmental protection. Consumption mode under such a cultural influence is sure to largely reduce carbon emission and energy and resource consumption. This is not parsimony, but a public virtue and a reflection of the progress of human civilization. This trend initiated by the elite group in the society is bound to lead the development of social culture for being low-carbon and environment-friendly while making the people healthier, more optimistic, and giving them a stronger sense of happiness. It will be the basic principle of low-carbon lifestyle in the future.

### 9.4.6 Institution, Mechanisms, and Guarantee Measures

(I) Green and low-carbon consumption mode needs a rational urban–rural morphology

As an important social process in human socioeconomic development, urbanization is not only an inevitable way of achieving socioeconomic development but also exerts immense pressure on the resources and the environment on earth, most of which comes from energy production and consumption, as well as the consequent environmental impacts beyond the urban boundary.

(1) People's consumption mode is affected by urban morphology.

    (a) The size and spatial layout of city decide the travel needs in everyday work, life, and recreation.

(b) The population, density, and spatial layout, and transportation infrastructure in a city decide people's choice of traffic mode.
(c) The density of residential buildings, standards for building energy efficiency, and consumption mode decide the building's water, electricity, heating, and gas demands.
(d) Consumption mode has direct effects on a city's supply system, water treatment system, and garbage generation and disposal system.

(2) Build urban infrastructures suitable for low-carbon consumption mode

(a) Adhere to the high-density and compact urban development.
(b) Rationally arrange and organically integrate the living zone and functional zones.
(c) Implement the most rigorous and advanced building energy efficiency standards.
(d) Give priority to public transit across the board.
(e) Guide the reasonable ownership and use of cars.

There is no conclusion yet what urbanization rate is suitable for China, but to keep the agricultural diversity and necessary intensive farming, we should keep a large rural population. It is possible that 30% of Chinese people still live in the countryside in 2050. In the process of urbanization, it is also important to step up the building of a new countryside and lay a solid infrastructure foundation for low-carbon consumption in rural areas.

(II) Adjust market signal system, encourage low-carbon life, and consumption

We should speed up the energy price reform and let price guide reasonable and low-carbon consumption. China still offers multiple general subsidies for civil energy consumption, including electric power, natural gas, and heating, which neither conforms to the actual cost nor contributes to energy conservation in the civil sector, and the few people who use more energy even receive more subsidies. These general subsidies should be canceled as soon as possible, and we shall further implement the differentiated, tiered price system, so that those who use more energy will have to pay a higher price and green and low-carbon consumption is promoted. Take energy-saving air conditioner, for example, it is several hundred yuan or over one thousand yuan more expensive than general air conditioner, but the policy of tiered power tariff gives a strong impetus to the market share of large-power household electric appliances, including high-efficiency air conditioner. Consumption tax on refined oil products needs to be raised, so that the price of oil products will be at least on a par with that in developed countries and regions like the EU, so as to encourage the development of high-efficiency and new-energy cars and prioritize public transit. The heating fee collection mechanism should be adjusted. A heating charge system based on heat measurement should be established and promoted as soon as possible so as to replace the current area-based charge mechanism with household-based measurement.

(III) More advanced energy efficiency standards, certification, and labeling should be adopted more extensively to guide market consumption

Energy efficiency standards and labeling system is an important means of guiding economical market consumption under market economy conditions. The minimal compulsory energy efficiency standards for products (including buildings) can be used as the "lowest threshold" to keep increment from entering the market. Meanwhile, energy efficiency label can also provide the consumers with energy efficiency information on relevant products.

The scope of equipment in China that involves energy efficiency standards, certification, and labeling system is far from enough for building a resource-saving society, and is on a low level compared with that of developed countries. We have to expand the scope of equipment covered by minimal energy efficiency standards, certification, and labels, and revise and update relevant standards in a timely manner.

In the implementation of energy efficiency standards, the minimal compulsory standards are usually not put into practice due to weak supervision. We have to foster and improve the capability of supervising and managing energy efficiency standards, certification, and labeling, give full play to the government's function in market regulation and supervision, and establish an effective market access system for energy-saving products. In this way, we will make sure that all products that enter the market meet the energy efficiency standards, and there should be the conditions and environment for consumers to choose energy-saving products.

(IV) Improve relevant infrastructure and push consumers to choose a reasonable consumption mode

The choice of consumption mode is either active or passive, and it is urgent to build efficient, convenient, and user-friendly infrastructure to adapt to the passively accepted lifestyle. For instance, many consumers choose to drive by themselves because the current public transit infrastructure is not well developed enough, not to mention "seamlessly connected." The household-based heating adjustment and measurement facilities are also a necessary technical condition to reform the heating charge system. We suggest allocating more public finance for infrastructure that has a strong public nature and is good for choosing the energy-saving consumption mode, and guiding private investors to invest in the construction of energy-saving infrastructure. In this way, we will gradually create the sound social infrastructure and market environment that is conducive to sustainable development and public interests.

(V) Make intensified efforts for publicity and education to raise the public awareness of conservation

The low-carbon consumption mode eventually has to be realized through the market, with the consumers being the main entities in this process and public engagement being the foundation for building a resource-saving society. Chinese people's awareness of resource conservation has to be further increased, and practices against this principle are commonplace in work and life. Making intensified efforts for publicity and training to raise the public awareness of energy conservation is a long-term, important task for promoting the low-carbon consumption mode and building a resource-saving society. We suggest establishing a long-term publicity mechanism of energy conservation as soon as possible, adopt a wide range of publicity and educational approaches at multiple levels and in various forms, and give full play to the role of media, so as to lay a solid public foundation for building a resource-saving society.

# Chapter 10
# Technical Support for Low-Carbon Development

## 10.1 Great Importance of Technical Support for Low-Carbon Development

### *10.1.1 Technical Progress is the Driving Force of Human Development*

Science and technology are the primary productive forces and the concentrated reflection and main symbol of advanced productive forces. In the human history, every substantial reform in the economic or social sector always started with technical innovation. In the twenty-first century, the new technological revolution developed rapidly, and a series of new and major technological breakthroughs have profoundly changed the economic and social outlook, industrial layout, and organizational form of all countries around the world. Nowadays, the way of interpersonal connection has undergone essential changes because of the Internet and information revolution. The convenient communication, smooth exchange, and interconnectivity of knowledge have significantly enhanced the human capability of technological creation, development, and application. Technological innovation has entered a "golden age" never seen before, while technological progress is changing the way of human production and life at an unprecedented speed, with its influences penetrating every corner of the society.

### *10.1.2 Low-Carbon Technology is the Starting Point and Goal of China's Economy in Its Shift from High Carbon to Low Carbon*

The idea of low-carbon development essentially originates in the human reflection on the traditional industrialized development mode. The shift from the traditional


X. Du et al., *Overview of Low-Carbon Development*,
https://doi.org/10.1007/978-981-13-9250-4_10

high-carbon development mode to the low-carbon mode not only stems from the mankind's endless pursuit for development quality and benefits but also symbolizes the evolution of human development concept, thoughts, and civilization. Establishing an intensive, clean, and efficient production system and consumption mode is the integral requirement of low-carbon development. On the one hand, low-carbon energy is the basic feature of energy supply in the low-carbon economic system, and low-carbon economy would not have been possible without low-carbon energy supply. On the other hand, a better energy consumption mode and intensive and efficient energy utilization is the essential requirement of low-carbon development and an important way of transforming the economic development mode. To achieve the historical transformation toward low-carbon development, we can not do without a solid low-carbon technological foundation. The shift from high-carbon to low-carbon economy needs the strong support of numerous low-carbon technologies in the field of energy production, supply, and consumption. These technologies, as important tools to reform the energy system, industrial system, and production mode, will play an ever bigger role in the future.

### *10.1.3 Low-Carbon Technology is the High Ground of Future Global Competition and a Comprehensive Demonstration of National Competitiveness*

Under the pressure of energy security, ecosystem degradation, and climate change, the world economy is going through massive adjustment and reform that has never been seen before, and green economy that is characterized by clean energies, low-carbon industries, and environmental technologies represents the future trend. The new round of technological revolution will reshape the international division of industries. Developed countries and regions, including the US, Japan, and EU, are all making huge investments in hopes of taking a preemptive step in this round of revolution and obtain the leading strategic position in future competitions. They even set trade barriers such as carbon tariffs and green standards to undermine China's competitiveness in traditional industries. Against such a background, China, more than any time before, has to firmly rely on technological innovation to materialize substantive progress in productive forces and push the comprehensive, coordinated, and sustainable socioeconomic development. If China cannot follow the trend, seize the opportunities, and develop the speed up its green economy, not only we will lose the chance of catching up with and surpassing other countries but our competitive advantages in traditional industries will be further weakened too.

### *10.1.4 Low-Carbon Technology Will Comprehensively Reshape the Energy System of China*

Low-carbon development will exert fundamental impacts on China's energy system. Since the Industrial Revolution, fossil energies like coal, oil, and natural gas have dominated the energy production system, and low energy efficiency and energy waste is only too common due to the existence of cheap energies. Problems such as climate change and environmental pollution caused by the massive consumption of fossil energies are now common challenges faced by the whole mankind. To deal with them, China has to follow the global trend and make changes in three aspects. First, we have to fundamentally change the energy supply system dominated by fossil energies; second, we have to change the consumption mode of low energy efficiency; third, we have to change the energy supply–demand model that meets unreasonable energy demand without limit, with the final goal of establishing a clean energy supply system and efficient energy consumption system. To accomplish this task, we must rely on powerful low-carbon technologies as our basic guarantee to make systematic innovations in the whole production chain, including energy exploitation, processing and conversion, storage and transport, consumption, and end-of-pipe treatment.

## 10.2 Current Development of Low-Carbon Technologies in China

Tackling climate change, improving energy security, and addressing environmental pollution are high on the agenda of all countries. The Chinese government is attaching growing importance to low-carbon technologies, which have received even more input and taken effects gradually. In recent years, China has made remarkable progress in low-carbon technologies, with significantly enhanced capability of basic R&D, technological innovation, and equipment manufacturing.

China is making all-round progress in the application of low-carbon technologies. As the world's largest coal consumer, China takes a leading position in efficient coal exploitation and clean utilization. Especially regarding the water-coal-slurry technology, clean coal-burning technology (e.g., IGCC, CFB, and USC thermal unit), coal gasification and liquefaction technology, and modern coal chemistry, China is an international leader with a large amount of independent IPRs, and advanced thermal power generator set is an important part of its energy equipment export and foreign exchange creation. In terms of the exploration and development technology of unconventional oil and gas, China has carried out international cooperation actively and made great breakthroughs. The shale gas field in Chongqing's Fuling had established the annual production capacity of 2 billion $m^3$ and produced 1136 million $m^3$ shale gas accumulatively by the end of 2014; national shale gas output is expected to exceed 1.3 billion $m^3$; and surface exploitation of coal bed methane (CBM) is likely to exceed 4 billion $m^3$. In regard to renewable energies, China, while maintaining its traditional advantages in hydropower development technology, has made rapid advances in wind power and PV. As a result, the situation of "imported raw materials and exported products" has been greatly mitigated, and the domestic industry has a greater capability of independent innovation. In the field of energy utilization, a range of advanced energy-saving and emission-reducing technologies have been widely applied, such as TRT in the iron and steel industry and heat recovery technology in the cement industry, with the industrial energy efficiency significantly improved. As to power transmission, China is dedicated to the construction of UHV grid and is likely to be a pioneer and proponent of the global energy network.

At present, all countries are preparing for a new round of energy technology revolution characterized by cleanness, high efficiency and low carbon, various low-carbon technological and theoretical innovations keep emerging, and new industries and technologies are upgrading quickly. Exactly because of this, China cannot be contented with its past achievements and sit on the laurels. We must be clear that compared with developed countries, China still lags far behind in the overall capability of low-carbon technology, including basic R&D and innovation, and we are still dependent on others when it comes to many high-end and key technologies. For instance, electric vehicle is a key part in our future auto development, but we are a long way behind developed countries regarding the electric motor and electric control systems, the most critical systems for electric vehicle. As a major developing country, China is faced with multiple tasks and challenges, such as energy supply security, eco-environmental governance, and climate change. Under such circumstances, we have to develop low-carbon technologies at a faster pace and with a firmer determination, narrow the distance from developed countries, and break the technical bottlenecks on low-carbon development, thus ensuring the technical support for our low-carbon transformation.

## 10.3 Key Areas and Directions of Technological R&D

### *10.3.1 Low-Carbon Technology on Energy Production End*

The energy production end mainly refers to energy development, processing, and conversion. Energy development means the production of primary energies, including fossil and nonfossil energies, whereas energy processing and conversion means the process of converting primary energies to secondary ones, including power generation, heat supply, oil refining, and coal chemistry. In a narrow sense, low-carbon technology on the energy production end should focus on low-carbon energy development, but the fact is that the traditional process of fossil energy development, processing, and conversion has vast room for energy conservation and carbon reduction, so it should also involve these areas.

## 10.3.2 Low-Carbon Technology for Energy Development

1. Safe, efficient, and environmental-friendly coal exploitation technology

China is the largest coal producer and consumer in the world, but its coal exploitation approach is quite backward. The utilitarian "predatory" exploitation and the exploitation of rich and main coal mines and abandonment of lean and subsidiary ones lead to serious waste of resources. In comparison with large coal producers like the US and Australia, China lags far behind regarding the automated level of coal exploitation, resource recovery efficiency, safe operation, and geological and water protection. In general, China has to get on the track of safe, efficient, and environmental-friendly coal development in the future.

(1) We should build digitalized and automated mines. Through digitalized and automated mining, we can realize the automated control of coal production, make it possible to monitor frontline production data and equipment status at the ground dispatching center, and ensure safe production at the mine (Fig. 10.1). We should make full use of electronic information technology and communication technology, improve the mechanized level of coal exploitation, realize remote and automatic control during mining, digging, transport, washing, measuring, ventilation, drainage, air compression and power supply, and push for centralized and unmanned control of mine production and equipment operation. In this way, we can gradually achieve the automated and informatized safe mine production and consequently raise the labor productivity and economic benefits. More importantly, through digitalized and automated coal mining, we can significantly reduce the personnel input in dangerous underground locations and enhance the overall safety of mine production.

**Fig. 10.1** Automated coal mine

(2) We should carry out demonstration project of underground coal gasification. Underground coal gasification means the process of burning underground coal in a controlled way and generating flammable gas through the thermal and chemical effects of coal. The purpose is to extract the energy-containing components of coal and replace physical coal mining with chemical mining. It is reputed as the second-generation coal mining approach for its good safety performance, small investment, high efficiency, and little pollution. Underground coal gasification technology can not only recover the coal resources discarded in the well but also can be applied to thin and deep coal bed that is hard to exploit or has poor economic and safety performance, to coal mining under buildings, underwater and under railway, and to coal bed with high sulfur, ash, and gas content. The resulting underground coal gas can be used as gas directly for civil use and power generation, or be used to extract pure hydrogen, or as the raw gas of synthetic oil, DME, ammonia, and methanol. The slag of underground coal gasification will be left underground, and filling technology can be adopted to reduce ground depression without solid waste discharge, thus causing less destruction of the geological environment. Therefore, underground coal gasification technology has great economic and environmental benefits and considerably raises the utilization rate and level of coal resources. It is an important direction of clean coal technology R&D and development in China.

(3) We should develop green exploiting technologies such as symbiotic and associated mining, water protection mining, and filling mining. China has complicated coal reserve conditions and abundant symbiotic or associated mineral resources with high potential economic value. Therefore, strengthening the overall development and utilization of these resources will not only fully utilize and effectively protect the mineral resources and realize a sustainable mining economy but also help protect the environment and maintain ecological balance. This is of great importance for promoting the Chinese coal industry to shift from the extensive to the intensive approach, optimizing resource allocation and achieving sustainable development. Water protection mining protects the groundwater resources and recycles mine drainage, with a view to conscientiously protecting water resources while preventing water inrush, and minimizing the water temperature disturbance in the mining area. By setting up waterproof columns, adopting filling mining, and grouting reinforcement, we can suppress the upgrowth of diversion fissure zone and the abscission and rupture of key water-resisting strata, as well as the seepage through rock stratum.

2. Technologies to stabilize and increase oil and gas output

We should adhere to the strategy of "stabilizing the east, accelerating the west, making breakthroughs on the sea." Following the two centerlines of large scale, efficient development of new oil and gas fields and improving the recovery ratio of old ones, we encourage the development of low-quality resources in order to steadily increase the crude oil output and natural gas output. In the east, we should tap the potential of main oil-producing areas, step up the renovation of old oil fields, and explore resources in surrounding basins; in the west, we should promote risky prospecting and the prospecting of low-quality resources, work hard to tackle technical difficulties, and carry out large-scale, effective development; on the sea, we should continue to accelerate oil exploration in the South China Sea, highlight the strategic position of offshore oil and gas, and reach to deeper waters.

(1) We should develop new oil and gas prospecting technologies. Technologies that are likely to significantly improve the prospecting efficiency include seismic technology, CSEM, interpretation technology, earth system simulation, and underground measurement technology, while auxiliary technologies include drilling technology, nanomaterial technology, and computer control and data collection technology. Furthermore, basic research on hydrocarbon accumulation mechanism and their distribution and upgrowth pattern should be conducted. Studies on the hydrocarbon accumulation mechanism, process, and effect in typical superimposed basins in the west need to be carried out, and theories on the oil and gas formation and distribution forecast in those basins should be proposed, so as to provide the scientific basis for oil and gas prospecting there. We should study the storage type of oil reservoir in fracture-vug-type carbonate rocks and the distribution features of reservoir fluids, reveal their flowing and seeping pattern, and lay the theoretical foundation for the exploitation of this type of oil reservoir.
(2) We should develop unconventional oil production processes. As it takes longer to exploit oilfields, oil output has come to a decline period and some old oilfields are in the middle or late period of their lifespan. Going forward, China should make more R&D input in unconventional oil production processes, overcome key technical "bottlenecks" regarding the development of oilfields with extra high water content, low or extra-low percolation and of thick oil, improve auxiliary processes and equipment, raise the recovery ratio, and lower the cost. A range of new processes and methods have been widely applied in various oilfields, including polymer flooding, microbial enhanced oil recovery, hydraulic vibration, and high energy gas fracture, effectively increasing the output.

3. Technologies for exploiting unconventional oil and gas resources

(1) We should develop shale gas exploitation technology. According to the estimates published by the Ministry of Land and Resources, China has about 25 trillion $m^3$ of recoverable shale gas reserves, the largest in the world and with immense potential (Fig. 10.2). Based on America's experiences, the most fundamental technology of "shale gas revolution" is horizontal well fracturing, which fractures the shale bed with high-pressure water and forces out the natural gas contained therein. To be more specific, high-pressure water is injected into the oil and gas well to fracture the borehole and get oil and gas. To make the water pressure penetrate deeper, some additives would be added in the high-pressure water and grits or ceramic granules would be put in the well to help open up the fractures. To make progress in our future shale gas exploitation, we have

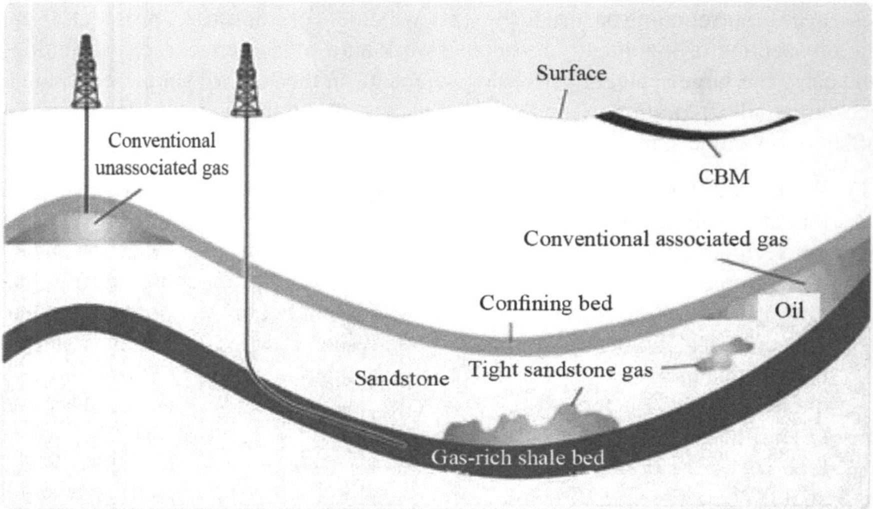


**Fig. 10.2** Unconventional oil and gas reserves

to make substantial breakthroughs in the horizontal well fracturing technology. Other technologies that also need breakthrough include horizontal well technology and multilayer fracturing, hydraulic fracturing, re-fracturing, and the latest synchronous fracturing, as well as auxiliary technologies such as shale gas evaluation, seismic treatment and interpretation, and fracturing monitoring.

(2) We should develop CBM (coal bed methane) exploitation technology. CBM refers to the flammable gas contained in underground coal bed comprising methane and a small amount of CO, $CO_2$, and nitrogen. As a kind of unconventional natural gas, CBM is a quality energy and chemical raw material. According to estimation, China has more than 30 trillion $m^3$ of geological resources buried less than 2,000 m underground, basically equivalent to the conventional natural gas reserves. By the end of 2012, the proved geological reserves of CBM nationwide reached 540 billion $m^3$ accumulatively, and over 10,000 CBM wells were drilled with more than 2.5 billion $m^3$ output. Given such huge CBM resources and developing potential, China has the resource foundation to form the CBM industry, which will focus on the following aspects: integrated coal and gas development, reservoir engineering and dynamic assessment, underbalanced drilling, fracturing stimulation and production planning, and the safe collection, transmission, and utilization of low-concentration CBM.

4. Wind power: raise the utilization efficiency and capability, and promote its on-grid consumption

Wind energy is the kinetic energy created by the massive airflow on the earth surface. It depends on the wind energy density and the annual cumulative hours of usable wind energy. According to data from the State Meteorological Administration, China has about 3226 GW wind energy reserves 10 m above ground and about 1000 GW exploitable reserves, including around 700 GW onshore wind energy reserves (10 m above ground) and 300 GW offshore exploitable and usable reserves. At present, wind power is a renewable energy with the highest value of large-scale and commercial development in China and will be a critical pillar for our low-carbon development. In the future, China's wind power technology will focus on the following aspects:

(1) Onshore wind power technology. This kind of technology is getting more mature in China and basic technology and equipment is no longer a bottleneck for onshore wind power development. However, with its large-scale development, how to consume the massive amount of intermittent and uncontrollable wind power is a major problem. Accordingly, intelligent management and high-performance materials will be the main contents of future technological development, including refined wind energy evaluation, multi-MW wind power unit manufacturing, disaster relief of wind power units (against low temperature, typhoon, lightning stroke, and earthquake), and composite materials of wind power blades and recovery thereof. Special attention will be paid to the R&D of large-scale wind power grid connection and cross-regional wind power dispatching and management (Fig. 10.3).
(2) Offshore wind power technology. Offshore wind power generation, which is characterized by high wind speed, relatively stable airflow, and proximity to power load, is a key area of China's future wind resource development. As this area is in the starting stage in China and considering our deficiency in offshore wind power technology, priority should be given to the development of materials and technologies of offshore wind power units in order to enable them to better resist typhoon and erosion. Efforts should also be made to improve the technology of offshore wind farm construction, testing and maintenance of offshore wind power equipment, complementary power generation from wind energy, wave energy and ocean current energy, and HVDC power transmission based on voltage source converter (Fig. 10.4).

**Fig. 10.3** Onshore wind power generation

**Fig. 10.4** Offshore wind power generation

5. Solar power: expand the scope of solar power utilization and develop new battery materials

China has a vast territory and abundant solar resources. It is estimated that the ground surface of China receives about $50 \times 10^{18}$ kJ solar radiation every year, and the annual solar radiation across the country reaches 335–837 kJ/(cm$^2\cdot$ a), with the median of 586 kJ/(cm$^2 \cdot$ a). There are currently two ways of solar energy utilization in China—photovoltaic conversion and solar thermal utilization (CSP). The basic principle of photovoltaic conversion is converting solar radiation energy directly into electric power through the photovoltaic effect, and solar battery is its main component; while the basic principle of CSP is collecting the solar radiation energy, converting it into thermal energy through reaction with other substances and using it.

(1) Photovoltaic conversion. Photovoltaic conversion (PV) means using the solar battery to directly convert solar radiation energy into electric power in accordance with the photovoltaic effect. The PV system mainly consists of the solar panel (component), control unit, inverter, and conventional power metering and delivery system, which are basically all electronic parts without mechanical ones. Therefore, PV equipment is highly compact, reliable, and stable with a long life cycle and easy installation and maintenance. At present, the polysilicon and monosilicon solar battery technologies are rather well developed (Fig. 10.5), so the future PV R&D will focus on new battery technologies so as to raise the battery's conversion efficiency, make silicon wafers thinner, and develop non-silicon batteries. Thin-film silicon solar cell, CIGS, organic solar cell, and concentrated photovoltaic cell are the future directions of PV battery R&D, among which CIGS thin-film solar cell has received extensive attention for its high conversion efficiency (21%) and relatively low material cost (Fig. 10.6).
(2) CSP. Apart from traditional solar thermal technologies like solar hot water, CSP is also an important area of solar energy utilization. It collects solar radiation energy, converts it into thermal energy and then into mechanical energy to motivate the power generator. Given the different approaches of solar radiation

**Fig. 10.5** Polysilicon PV battery component

**Fig. 10.6** CIGS thin-film solar cell

collection, there are mainly four types of CSP technologies—tower, parabolic trough, disk, and Fresnel (Fig. 10.7, 10.8 and 10.9). In the future, CSP technology will feature higher efficiency and lower cost. To be more specific, the direction of concentrating system is lowering the cost of current products and improving their reliability; the direction of heat-absorbing system is raising the solar thermal conversion efficiency and reducing heat loss; while the direction of heat-removing system is developing a new medium for heat conduction and storage.

6. Nuclear energy: safe, orderly, and efficient development

Nuclear energy is a clean, efficient, and quality modern energy. The development of nuclear energy is of great importance to improving China's energy structure and ensuring our energy security. In future nuclear development, we should adhere to the concept of scientific and rational nuclear safety, specify the technological routes, and implement the principle of "safety first" in the whole process ranging from planning, site selection, R&D, design, construction, operation all the way to decommissioning, with the aim of developing nuclear energy in a safe, orderly, and efficient manner.

Fig. 10.7 Tower CSP

Fig. 10.8 Trough CSP

Fig. 10.9 Disk CSP

**Fig. 10.10** Gen-IV nuclear power units

(1) Strategy of nuclear technology development. We should follow the "three-step" path of thermal reactor, fast reactor, and fusion reactor. Focusing on 1000 MW class advanced pressurized water reactor (PWR), we will actively develop new technologies such as high-temperature gas-cooled reactor, commercial fast reactor, and small reactor.
(2) Gen-IV nuclear power technology. The Gen-IV nuclear power system is an advanced system with better safety performance and economic competitiveness and little nuclear waste that can effectively prevent nuclear proliferation. It represents the development trend and technology frontier of advanced nuclear power system (Fig. 10.10). There are six types of Gen-IV conceptual reactors, namely, gas-cooled fast reactor (GFR), lead-cooled fast reactor (LFR), molten salt reactor (MSR), sodium-cooled fast reactor (SFR), very-high-temperature gas-cooled reactor (VHTR), and supercritical water-cooled reactor (SCWR).
(3) High-level radioactive nuclear waste treatment technology. Nuclear waste is usually highly radioactive with its half-life being as long as thousands, tens of thousands, or even hundreds of thousands of years. The rays emitted by nuclear waste go through ionization and excitation when passing through objects and cause radioactive damages in living organisms. Nuclear radiation might trigger cancer or other diseases in a person. Two methods for disposal of nuclear waste are generally adopted in the world today—oceanic and land disposal. Usually, the nuclear wastes are cooled and stored in a dry fashion, and then metallic tanks containing the wastes are plunged more than 4000 m undersea in the designated area, or buried in nuclear waste depositories built deep down in underground rock beds. The land burial approach is adopted in countries like the US, Russia, Canada, and Australia for their vast territory and large areas of wasteland. In the future, the researches on new technologies of nuclear waste disposal will focus on "reprocessing" and "direct disposal," the former is to recycle nuclear raw materials from the wastes, including plutonium that can be used to manufacture nuclear weapons, and the latter is to bury high-level radioactive wastes underground. The second approach usually consists of three stages—cooling, dry storage, and final disposal, and it is very demanding on the site of burial, deep drilling, and nuclear waste isolation.

7. Geothermal energy: diversified utilization based on local conditions

China has rich geothermal resources that are distributed extensively across the country. According to incomplete statistics, the 12 main sedimentary basins contain geothermal reserves of up to 853.2 billion ton standard coal, and dry hot rock reserves at the depth of 3000–10,000 m underground amount to 860 trillion ton standard coal. As far as quantity is concerned, China is in a leading position regarding the utilization of geothermal resources, but they are mainly utilized in simple and direct ways, such as hot spring bath. An inevitable choice in the future is to widely apply geothermal energy in such production sectors as power generation, heat supply, agriculture, and industry.

(1) Geothermal power generation technology. Geothermal power generation is converting geothermal energy into electric power. Current technologies that generate electric power from regular hydrothermal geothermal resources mainly include dry steam, flash, and binary cycle. Dry steam power generation draws high-temperature geothermal dry steam from the well, clears it of various impurities through the decontaminating separator, and injects it into the steam turbine to drive the generator. It is mainly suitable for high-temperature steam-type geothermal resources. Flash power generation first lowers the pressure and expands the volume of geothermal water through the flash evaporator to produce steam, and then uses the flash-induced steam to drive the steam turbine generator set. It is mainly suitable for high/medium-temperature hot water or mixed geothermal resources. Binary cycle power generation draws geothermal water from the well and transmits the heat through heat exchanger to organic working media with a low boiling point (e.g., n-butane, chlorohexane, and $CO_2$) to produce steam, which then drives the turbine generator set. This technology is mainly suitable for medium/low-temperature hot water geothermal resources (Fig. 10.11).

**Fig. 10.11** Yangbajing geothermal power station in Tibet

**Case study 1: Yangbajing geothermal power station in Tibet**
The Yangbajing geothermal power station in Tibet was completed and generated electricity on September 23, 1975. The Yangbajing geothermal power station started the trial run in September 1977, and it is the only one of its kind that has been consistently producing electricity so far. Its first 1 MW testing unit successfully generated power in 1977 and was decommissioned in 1991 when the other eight 3 MW units were successively completed. Since then, its total installed capacity has been kept at 24.18 MW, accounting for 41.5% of all capacity of the Lhasa grid. In winter, which is a dry season, the geothermal power generation accounts for 60.0% in the Lhasa grid and is one of the main producers. Yangbajing geothermal power station generates about 100 GWh electricity per year, but its potential has been tapped in recent years and the output has been increasing and hit records frequently, exceeding 140 GWh in 2009.

The geothermal development and utilization at Yangbajing set an international precedent of using medium/low-temperature geothermal energy for power generation and occupied an important position in new energy development and utilization in the world. Yangbajing is not only the largest geothermal power station in China but also the only power plant in the world that uses geothermal reservoirs for industrial power generation. Of all geothermal stations in more than 20 countries and regions, Yangbajing geothermal station is the only one that can use medium-temperature reservoir heat storage less than 200 m underground and lower than 150 °C to generate electricity. In 1997, a piece of exciting news came from Yangbajing: a high-temperature well of 205 °C, high pressure, and dense energy was dug on the Nyainqêntanglha Range north of the geothermal city. This made Yangbajing one of the few high-enthalpy geothermal fields in the world with single-well power-generating potential exceeding 10,000 kW, and promised great prospects of developing the deep-level high-enthalpy geothermal resources.

(2) GSHP (Ground Source Heat Pump) technology is an efficient, energy-saving, and environmental-friendly system that takes the rock–soil body, groundwater, or surface water as the heat sink or heat source for heating or refrigeration. Given the different forms of geothermal energy exchange, there are several kinds of GSHP systems—buried pipe, groundwater, and surface water. As a mature energy supply technology, GSHP has enormous potential. First of all, it has a low demand on resources. The shallow geothermal energy in most places in China can be utilized. Second, geothermal resources have little fluctuation and distinct patterns, so they can serve as stable energy supply sources. At last, GSHP is a mature technology with good economic efficiency. Generally speaking, GSHP is of high practical value in regions that have both heating and cooling demands, such as the Yangtze River and Yellow River basins, the northeast, northwest, and north China.

### *10.3.3 Low-Carbon Technologies for Energy Processing and Conversion*

1. Clean and efficient coal power generation

The coal-dominated energy endowments of China will change fundamentally for a long time, and the proportion of coal-fired installed capacity in the total installed capacity will stay at a high level. At present, more than half of the coal in China is used to generate electricity, and how to develop coal power in a clean and efficient way is an important topic in our low-carbon development.

(1) Supercritical (SC) and ultra-supercritical (USC) coal power units. SC units refer to those whose main steam pressure is higher than water's critical pressure 22.12 MPa. They are usually divided into two types: regular SC units whose main steam pressure is about 24 MPa and main steam and reheat steam temperature is 540–560 °C, and USC units whose main steam pressure is 25–35 MPa or higher and main steam and reheat steam temperature is above 580 °C. Compared with traditional coal-fired units, SC and USC units have an obvious effect of energy conservation and environmental protection. It is estimated that the thermal efficiency of a single SC unit can reach 50% and the minimal coal consumption per KWh is only 255 g (BWE of Denmark), lower than the 327 g minimal coal consumption per KWh of subcritical unit. Meanwhile, the adoption of low-$NO_x$ technology reduces 65% of $NO_x$ and other hazardous substances during burning with the de-$SO_x$ rate of over 98%, meeting the goal of energy conservation, consumption reduction, and environmental protection.

**Case study 2: USC thermal units at Waigaoqiao No. 3 Power Plant, Shanghai**
The two 1000 MW USC units at Waigaoqiao No. 3 Power Generation Co., Ltd. (hereinafter referred to as Waigaoqiao No. 3) started construction in July 2005 and were put into operation in June 2008, creating a world record of coal consumption that very year and significantly reducing it every year since then. In 2013, the two units (including de-$SO_X$ and de-$NO_X$ facilities), with 78% load rate, reached the net coal consumption rate of 276.82 g/kW · h, much lower than the 286.08 g/kW · h (2009) of Unit 3 of Denmark's NORDJYLLAND power plant, the former world record holder, and the 303.7 g/kW · h of Unit 1 of Isogo power plant, which had the lowest coal consumption in Japan. The efficiency of Waigaoqiao No. 3 was already equivalent to the expected efficiency of the next-generation high-efficiency USC unit with the steam temperature of 700 °C that was still being developed.

Waigaoqiao No. 3 has made remarkable innovations in low-carbon coal power technology, but it never stopped the steps of progress. After the two units were put into operation, the plant carried out a series of technical improvements, including the world's first "de-$SO_X$ with zero energy use," "energy-saving all-weather de-$NO_X$," and "efficient life extension of de-$NO_X$ catalyst." These innovations solved several world-class difficulties, such as sulfur erosion during low-temperature heat recovery and low-load exit of de-$NO_x$ device, not only greatly improving the unit efficiency but also considerably improving the environmental performance. Waigaoqiao No. 3 was highly recognized by the Ministry of Environmental Protection for its win-win results in energy conservation and environmental protection.

In 2013, Waigaoqiao No. 3 had the average dust emission concentration at chimney exit of 11.10 mg/m$^3$, much lower than the standard 20 mg/m$^3$ in strictly controlled areas; $SO_2$ emission concentration was 35.03 mg/m$^3$, much lower than the national standard of 200 mg/m$^3$; and $NO_x$ emission concentration was 27.25 mg/m$^3$, much lower than 100 mg/m$^3$, the most rigorous national standard ever that came into force on July 1, 2014. The power plant ranked first in Shanghai in overall aspects, with much better performance than the new national standards and even the standards for gas turbines.

(2) Integrated coal gasification combined cycle (IGCC). IGCC is an advanced power system that integrates the coal gasification technology with efficient combined cycle. It consists of two parts—coal gasification and purification and power generation through gas–steam combined cycle. The process of IGCC is as follows: coal is gasified to produce coal gas of medium or low heat value and then purified to strip such pollutants as $SO_x$, $NO_x$, and dust and become clean gas fuel. Then, it is injected into the combustion chamber of gas turbine for burning, where it heats up the gas to drive the gas turbine, and the exhaust gas from the gas turbine goes to the heat recovery boiler (HRSG) to heat up the water, producing overheated steam to drive the steam turbine (Fig. 10.12). By combining clean coal gasification technology with efficient gas–steam combined cycle system, IGCC features both high generating efficiency and excellent environmental performance, being a clean coal power technology with bright prospects. With the current technology, its net generating efficiency can reach 43–45% while its pollutant discharge is only 1/10 that of regular coal-fired power stations, de-$SO_x$ efficiency is 99%, $SO_2$ emission is around 25 mg/m$^3$, and $NO_x$ emission and water consumption are only 15–20% and 1/3–1/2 that of regular power stations. IGCC is of great significance for environmental protection.

(3) Utilization of low-heat-value coal resources such as gangue, coal slime, and middling for power generation. Gangue and coal slime are impurities generated during the exploitation of raw coal that are usually discarded by coal developers, but these so-called impurities, if reused through special processes, can be the most effective approach to realize social, environmental, and economic benefits. For instance, gangue power generation is a key technology in the field of overall resource utilization in China, and it takes up a large proportion of large-scale non-coal projects operated by coal developers. At present, gangue-fired power plants in the country have the total installed capacity of more than 5GW, and more than 50 million ton gangue is burned to generate electricity every year, accounting for over 60% of its total use.

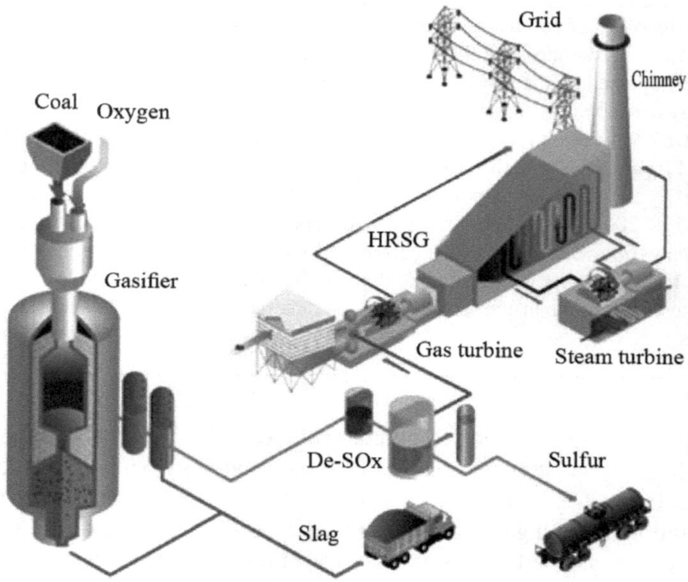


**Fig. 10.12** IGCC plant

2. Intensive development of oil refining and processing industry

The market demand for oil products and petrochemicals has increased rapidly in recent years, which has driven the flourishing of China's oil refining industry and the improvement of its oil processing capability. However, due to the limited domestic crude oil supply and the price fluctuation and geopolitical impacts on our crude oil import, China's oil refining industry has to take the path of intensive development. We will develop and apply more advanced crude oil deep processing technologies and adopt cleaner refining processes, so as to raise the light oil yield, reduce pollutant emission, and lower the refining cost.

(1) Hydrocracking technology. In terms of heavy oil and poor-quality oil processing technology, large refineries mainly adopt the hydrocracking and hydroprocessing approaches as they are more beneficial than the coking process and can make full use of raw materials. One of the steps of hydrocracking refining is making the heavy oil undergo cracking reaction and convert into gas, gasoline, jet fuel, and diesel under the condition of heating, high hydrogen pressure, and addition of catalyst. The advantage of this approach is that it can change poor-quality oil fraction into products with high added value and produce quality catalytic-reformed naphtha and jet fuel that cannot be produced through catalytic cracking, thus making up for its deficiency. The liquid product yield of hydrocracking is higher than 98% with a much better quality than catalytic

cracking. But despite its multiple advantages, hydrocracking is not as generally applied as catalytic cracking because it takes place in high-pressure and rigorous conditions and requires a lot of alloy steel, hydrogen, and investment.

(2) Residual oil and heavy oil processing technology. Cutting the heavy fuel oil production is the general trend of world refining industry today. Technologies of residual oil and heavy oil processing will focus on the hydrogenation process and catalytic cracking of heavy oil will also be developed appropriately with a view to improving the light oil yield and meet the demand for transport fuel and chemical raw materials to the largest extent. Meanwhile, coking process still has some potential thanks to its wide applicability to raw materials, mature technology, and low investment.

(3) High-octane gasoline production technology. About 90% of gasoline $SO_x$ comes from catalytic cracking, so lowering the sulfur and olefin mass fraction in catalytically cracked gasoline is critical for meeting China's gasoline quality standard and improving its gasoline quality. To meet the refineries' requirement for high-octane gasoline, alkylation and olefin modification technologies are being developed, including alkylate oil substitution, alkylation of green solid acid catalyst, olefin modification, and olefin-to-gasoline technology.

(4) New technologies of gasoline and diesel desulfurization. The hydrofining of gasoline is an effective method for lowering the sulfur content in catalytically cracked gasoline, but the usual way of hydrodesulfurization is accompanied by olefin saturation that causes the loss of octane in gasoline. Selective hydrodesulfurization technology, through the effect of selective hydrogenation catalyst, reduces olefin saturation and consequently octane loss. As to diesel, to meet domestic refineries' needs to produce clean diesel with high cetane, low density, and low sulfur content, we developed the medium-pressure hydrogenation technology to produce high-quality diesel and designed a special catalyst for poor-quality catalytic diesel. The new catalyst is capable of hydrodesulfurization, hydro-denitrification, olefin and aromatic saturation, and selective ring-opening.

3. Heat supply system of appropriate temperature and gradient utilization

(1) Popularize cogeneration. Cogeneration is a process that produces electricity and heat at the same time while saving fuel in comparison to separate production. The power plant not only produces electric power but also uses the steam from the steam turbine generator set to supply heating for households. Cogeneration is a typical example of the gradient utilization of temperature—while high-level thermal energy is used for power generation, low-level thermal energy is used for heating. At present, cogeneration in developed countries has reached a high level, and its purpose has expanded from heating supply alone to cooling and heating as well as seawater desalination (Fig. 10.13). Rough statistics show that the installed capacity of cogeneration power plants in developed countries accounts for 30% of the total electric installed capacity, and 54% of heating in Denmark is provided through cogeneration. Cogeneration serves two equal

**Fig. 10.13** Guodian's cogeneration project in Tai'an

purposes—industrial production and zoned heating, the former's main users include papermaking, iron and steel, and chemical industry. As its steam has almost zero cold source loss, cogeneration can reach the thermal efficiency of 85%, twice as high as the thermal efficiency of large condensing generation unit (40%) and 30% higher than separate heat and electricity production, while concentrated heating is 40% more efficient than heating supply from small boilers. Therefore, cogeneration is one of the strategic directions of China's electricity and heating system development in the near future.

(2) Promote the natural-gas-based combined cooling heating and power (CCHP) technology. Natural-gas-based CCHP uses natural gas as the main fuel to drive gas turbine, micro-gas turbine, or engine generator. The resulting electricity is provided to meet the power demand and the exhaust heat from the power generation system is recovered (by HRSG or direct-fired unit) for heating or cooling. This approach significantly improves the primary energy utilization rate of the whole system, realizes gradient energy use, and provides complementary grid-connected electricity, increasing the overall economic benefits and efficiency. As a kind of distributed energy, CCHP is capable of saving energy, improving the environment and increasing power supply, and is one of the necessary means of urban governance of air pollution and improvement of overall energy utilization rate. It conforms to the national strategy of sustainable development. Moreover, it is easier to integrate natural-gas-based CCHP with renewable energy as the system can either be independent of the grid or be integrated with it to form a power supply system, thus making power supply much more secure and reliable for the users.

### 10.3.4 Low-Carbon Technology on Energy Consumption End

The energy consumption end includes industrial, construction, transportation and living sectors. Low-carbon technologies on the energy consumption end are focused on energy efficiency while involving low-carbon energy substitution as well. Considering the complexity and diversity of energy efficiency technologies, this book only introduces those low-carbon technologies in various industries and fields that have great potential for energy conservation and a high level of maturity.

(I) Low-carbon technology in industrial sector

In this section, the introduction to energy-saving low-carbon technologies in the industrial sector and the analysis on their potential are in reference to the *Research on the Potential and Cost of $CO_2$ Emission Reduction Technologies in China* and *Research on the Potential of Energy Conservation in Chinese Industrial Sector 2020 and Technical Roadmap* published by the Energy Research Institute of NDRC.

1. Iron and steel industry

(1) Oxygen blast furnace iron-making technology. On the basis of existing blast furnaces, oxygen (with over 90% oxygen content) is used as a substitute for air blow. Part of the top gas, after $CO_2$ separation and heating, is blown into the furnace through the furnace stack and hearth in order to increase coal injection and decrease coke ratio.
(2) Oxygen-rich injection technology. On the basis of basic research on reducing the gas properties via hydrogen injection and on the in-furnace reaction mechanisms of blast furnace injection of coke oven gas, we carried out process design and the R&D on key technology and equipment concerning the purification, pressurization, reforming, heating, and injection of coke oven gas. Industrial experiments on coke oven gas injection were also conducted, with the injection quantity larger than 100 $m^3$/t iron, replacement ratio larger than 0.45 kg (coke)/$m^3$ (coke oven gas), fuel ratio reduced by 10%, $CO_2$ emission reduced by 10–20%, and furnace efficiency raised by 10%.
(3) New-generation Thermal Mechanical Control Processing (TMCP). This includes the controllable stepless steel cooling technology centered on extra-fast cooling, cooling path control based on phase change and separate-out, overall reinforcing technology covering fine crystallization, separate-out and phase change, and online/offline thermal treatment. These technologies can save 30% of alloy steel, improve steel strength by 100–200 MPa, save 5–10% steel, raise the production efficiency by over 35%, and save 10–15% energy.
(4) Non-blast furnace iron-making technology. This mainly includes the direct reduction and smelting reduction, the former being the iron-making process in which sponge iron is reduced at a temperature lower than the smelting temperature, while the latter refers to all methods that do not use a blast furnace to smelt liquid pig iron.

Besides, a series of cutting-edge technologies should be strongly promoted in the iron and steel industry, including dry coke quenching, dry furnace top differential pressure power generation, power generation with gas recovered from the blast furnace and converter, power generation with residual heat from the sintering machine, high wind temperature, oxygen-rich coal injection, continuous casting and hot transfer and hot charging, coal moisture control, and regenerative heating furnace.

2. Building material industry

(1) Pure low-temperature waste heat power generation. This technology installs HRSG at the kiln head and tail (respectively, called AQC and SP) of cement production line where a great deal of waste gas is discharged. The water and hot flue gas in the HRSG exchange heat and generate overheated steam with a certain level of temperature and pressure, which enters the steam turbine generator set to produce electricity. When a five-stage preheater is installed on the clinker production line, one ton of clinker can produce 30–35 kW · h electricity; if a four-stage preheater is installed, one ton of clinker can produce 36–40 kW · h electricity. The HRSG power output takes up 25–30% of the power used for clinker production, the average power supply cost is about RMB16/kW · h, average investment per kW installed capacity is RMB6,600, and 6200 kg $CO_2$ emission is reduced for 7200 kW · h power output.
(2) Cement kiln is used to help dispose of domestic garbage and other wastes. Compared with other disposal approaches, the coordinated disposal by cement kilns has comparative advantages in energy conservation, environmental protection, and economy, and the wastes can substitute for part of the fuels and raw materials for cement production. As a result, we can conserve natural mineral resources, promote the low-carbon development of cement industry, and save land resources. Therefore, the coordinated disposal of domestic garbage and other wastes by cement kilns is an internationally acknowledged approach, and it has been implemented safely for more than 30 years in developed countries.
(3) Grinding energy conservation technology. Vertical mill, roller-cylinder mill, and roller press are widely used overseas as the finish grinding system in cement grinding, but the application of roller mill as the finish grinding system is somewhat limited in China as domestic cement makers are worried about product performance. For cement grinding, roller mill is 1.6–1.8 times more efficient than ball mill and saves over 30% electricity. Clinker temperature, feed granularity, and degree of wear all have major effects on output and power consumption. It is crucial to adjust the grinding pressure, height of retention ring and wind speed and volume, control outlet temperature, and use high-performance screening device to improve the cement's grain composition and ensure product performance.

There are many other low-carbon technologies that have reduced the energy consumption of cement production by a large margin, including efficient clinker sintering, six-stage preheater, two-gear support kiln, efficient pulverized coal burner, fourth-generation cooling machine, high-performance chrome-free refractory lining, frequency conversion retrofit, and roller press finish grinding.

3. Electrolytic aluminum industry

(1) New energy-saving diversion structure of aluminum reduction cell. By modifying the aluminum reduction cell cathode and lining structure and by way of advanced process and control, this technology sharply shortens the electrode distance, keeps aluminum electrolysis in efficient and stable operation, and creates an insulated energy-saving aluminum reduction cell.
(2) Energy-saving aluminum reduction cell with new cathode structure. This technology replaces the current flat-base cathode structure with one that has embossments on the cathode surface, thus reducing the surface speed and fluctuation of aluminum liquid on the cathode and further stabilizing the liquid in the aluminum reduction cell. In this way, we can lower the aluminum reduction cell voltage by shortening the electrode distance and consequently reduce power consumption.
(3) Technology to enhance current, improve efficiency, and save energy of pre-bake aluminum reduction cell. Targeting at low anode current density and extensive sources of alumina in China, this process features low temperature, low superheat degree, low alumina concentration, low cell voltage, low anode effect coefficient, narrow working area of material balance and thermal balance, narrow regulating area for magnetohydrodynamic stability, and high anode current density. It has realized the goal of efficient and energy-saving aluminum electrolysis.
(4) Bottom oxygen blowing copper smelting technology. In this process, mixed mineral aggregates are continuously fed into the high-temperature molten pool in the furnace through the charging hole on top, while oxygen and air are injected into the copper matte layer with the oxygen lance at bottom. Oxygen floats on the fondant in the form of large quantities of small bubbles, which creates large gas–liquid contact areas and the smelting process is completed.
(5) Direct reduction of liquid high-lead residue. The hermetic reduction furnace structure and high-efficiency thermal reduction process reduce the heat loss of flue gas and solve the flying powder during the storage and transport of high-lead ingot bar. As a result, the loss of valuable metals is reduced, resources are utilized to the largest extent, and the environment is improved, creating a favorable production environment.

4. Ethylene industry

(1) High-temperature radioactive coating of cracking furnace. This technology applies a far-infrared radioactive coating that is resistant to high temperature and is highly emissive to the refractory lining bricks in the cracking furnace hearth, so as to effectively raise the radiance of the furnace wall and the penetration of thermal radiation, and enhance the overall hearth insulation. This kind of coating is able to resist the scouring and erosion of high-temperature

airflow and extend the service life of in-furnace insulating materials, and the technology is suitable for high-temperature steam cracking furnace. Statistics show that this technology saves about 30.8 kg standard coal/t ethylene on average, for which the usual investment is RMB1,000–2,500/t standard coal, and the payback period is usually 1 year.

(2) Recovery of low-heat-value residual energy for air preheating and burning. By adding an air preheater on the burner at the bottom of cracking furnace, the large amount of low-temperature heat source in the ethylene device is used to preheat the combustion air in the hearth to increase the sensible heat of air, thus saving the fuel consumption of cracking furnace and the use of cooling water. This technology is suitable for the situation that the bottom burner in newly built or retrofitted cracking furnace has a lot of residual heat, especially in low-temperature regions. Statistics show that this technology saves about 17.2 kg standard coal/t ethylene on average, for which the usual investment is RMB1,500–3,000/t standard coal, and the payback period is usually 1 year.

(3) Gas pulse soot-blowing technology. Ignite a special mixture of air and flammable gas in the pulse tank and deflagration would happen. The resulting high-temperature, high-pressure gas is ejected from the nozzle at an extremely high speed and in the form of shock wave, blowing soot on the convection section of the furnace tube off the heating surface. Statistics show that this technology saves about 2.5 kg standard coal/t ethylene on average, for which the usual investment is RMB1,000–2,500/t standard coal, and the payback period is usually 1 or 2 years.

(4) Optimized control of turbine compressor set. The control of turbine compressor set is optimized to realize parallel connection of the units, improve and dynamically distribute the load, lower unit energy consumption while improving flexible operation, and achieve process stability. Statistics show that this technology saves about 44.5 kg standard coal/t ethylene on average, for which the usual investment is RMB3,500–5,000/t standard coal, and the payback period is usually 1 to 3 years.

5. Caustic soda industry

(1) Membrane polar distance technology of ion film electrolyzer. Based on the bipolar ion film electrolyzer of high current density and natural circulation, this technology integrates the advantages of various technologies, including zero or small polar distance ion film electrolyzers, solves difficulties in raw material, processed parts, coating technology and assembly technology, and creates the membrane polar distance ion film electrolyzer of natural circulation that suits Chinese users. On the condition of 12.15 and 13.8 kA, cell temperature of 88 °C, and alkali concentration of 32%, membrane polar distance cell has a voltage 140 and 180 mV lower than the original cell, and DC power consumption/t caustic soda is reduced by about 100 kW · h and 120 kW · h, respectively.

(2) Residual heat utilization through hydrogen chloride synthesis. When chloride and hydrogen react to generate hydrogen chloride, a great deal of heat is released that can be used to produce steam. The medium-pressure steam synthesizer uses steel water wall in the high-temperature section and uses graphite on top and at the bottom of the synthesizing section where steel is prone to erosion. This not only overcomes the low strength and limited temperature of graphite tube but also solves the problem of easy erosion on top and at the bottom of the synthesizing section, raising the thermal efficiency of hydrogen chloride synthesis to 70%. Meanwhile, the steam byproduct has a flexible pressure range of 0.2–1.4 MPa, and it can be connected to the medium/low-pressure steam network to make the fullest use of thermal energy. This technology has been applied in some chemical enterprises, where it increases the thermal efficiency of hydrogen chloride synthesis to 70%, with considerable energy-saving effects.

(II) Low-carbon technology in construction

1. Popularize mature energy-saving construction technology

The focus of construction energy conservation lies in reducing the energy loss of the envelope enclosures, which comes from three parts—exterior wall, door and window, and roof. Developing efficient, economic thermal insulation materials is the future direction of construction energy conservation. Regarding energy conservation of exterior wall, composite wall materials will be used for thermal insulation, with rock wool, slag wool, glass wool, and polystyrene foam being ideal new composite materials. As to energy conservation of doors and windows, a number of energy-saving products with high technical content have been developed in recent years, such as aluminum alloy insulating profile, aluminum–wood composites profile, and steel–plastic extrudate, all with excellent energy conservation effects. As for roof energy conservation, the usual method is using light materials with a small thermal conductivity coefficient under the roof waterproof layer for insulation, such as expanded perlite and glass wool, or placing polystyrene foam on the waterproof layer.

2. Strongly develop passive building

Passive building is a kind of energy-saving building that reduces energy consumption through the interference of nonmechanical-electric equipment. To be more specific, it reduces the energy consumption of heating, air conditioning and ventilation of the building by way of rational layout of the building direction, sunshade arrangement, insulation of envelope enclosures and construction openings conducive to natural ventilation. Compared with traditional buildings, passive building makes better use of natural light, heat, and ventilation and uses less energy for internal lighting, heating, and cooling. It not only saves living cost but also solves the conflicts between energy conservation, emission reduction, and living comfort, and is especially practical in regions that use coal-fired or electric heating or cooling. Materials show that certified passive building is over 80% more energy efficient than regular buildings.

3. Promote green building

Green buildings refer to those buildings that save energy, land, water, and materials, protect the environment, and reduce pollution during the whole life cycle, and that are healthy and efficient and in harmony with nature. Green building originates in the construction industry's response to environmental problems. Also known as sustainable building, ecological building, natural building, or energy-saving building, it is harmless to the environment and makes full use of natural resources without disrupting the ecological balance.

Green building integrates multiple green and low-carbon technologies, including new type of envelope enclosure, efficient and independent temperature and humidity control of the air conditioning system, efficient photo-thermal conversion equipment, water reclamation, rainwater collection and recycling, and the use of "hardware" products such as cyclic building materials, high-performance concrete, and high-strength rebar. It also involves "software" applications such as the itemized, real-time and intelligent energy control system, industrial construction design, and one-time decoration.

**Case study 3: Best practice of green building—The IBR building**

The IBR building is the research and office base of Shenzhen Institute of Building Research. Located in Mei'aosan Road in Futian district of Shenzhen, the building covers 3000 $m^2$ with a total construction area of 18,170 $m^2$, and has two floors underground and 12 floors above ground. It was completed and put into use in April 2009 and was granted the three-star rating of National Green Construction Design Evaluation and the three-star rating of construction energy efficiency.

Planned and designed by Shenzhen Institute of Building Research itself, the IBR building upholds the green concept of low cost, low energy consumption, and openness and the principle of "passive priority, active complementarity," namely, it fully utilizes natural conditions to create a pleasant environment and only uses mechanical equipment and systems as a complement when natural conditions cannot meet the demand.

More than 40 technological measures are adopted across the building, including the following:

- Land-saving technology, such as the utilization of underground space and 3D parking lot.

- Energy-saving technology, such as insulating wall (LBG board on exterior wall), Low-E insulating glass, photoelectric curtain wall, sunshade (exterior sunshade and self-sunshade), natural ventilation and lighting, light pipe, and use of renewable energies.
- Water-saving technology, such as rainwater collection and utilization of reclaimed water.
- Material-saving technology, such as environmental-friendly wood–plastic flooring, carpet, simple decoration, and concrete recycling.
- Environmental-friendly technology, such as radon testing and protection, indoor air monitoring, customized air supply, and sound-insulating glass.

According to monitoring data, the annual power use of the IBR building is $52.9\,\mathrm{kW}\cdot\mathrm{h/(m^2}\cdot\mathrm{a)}$, 60% that of government office buildings, and 55% that of commercial office buildings in the same area. Its monthly power use is 40–70% of the average level of office buildings in Shenzhen, lighting power is about 75%, and air conditioning power is about 40%. More than 50% of the water use in the building is reclaimed water, nontraditional water use rate reaches 52%, and the average per capita water use is 31 L per day across the building, lower than the top-level quota provided in the *Standard for Water Saving Design in Civil Building* (GB 50555-2010). What's more praiseworthy is that the building makes rational use of the construction capital and spends most of the money on green technologies while cutting unnecessary cost for indoor decoration. As a result, the construction cost is RMB5,000/m$^2$, more cost-effective than normal office buildings.

The IBR building is a successful exploration made by Shenzhen Institute of Building Research for the feasible solution to green building in consideration of the high cost and high threshold of energy-saving building at present. It gives a good interpretation of green buildings in specific regions and conditions.

### 10.3.5 Low-Carbon Technology in Transportation

1. New energy vehicle (NEV)

(1) Pure electric vehicle (EV). Pure EV refers to cars completely driven by rechargeable batteries (such as lead–acid cell, nickel–cadmium cell, nickel–metal hydride cell or lithium–ion cell) (Fig. 10.14). Pure EV consists of the electric motor and control system, power transmission system, and working devices that carry out specific tasks. The electric motor and control system comprises the motor, power source, and motor control device, while other devices of pure EV are basically the same as those of internal combustion engine vehicles (ICEV). As a kind of vehicle without ICE, EV does not generate exhaust pollution when at work, and its almost "zero pollution" is helpful for environmental protection and air quality control. Its motor also makes much less noise than ICE.

(2) Hybrid vehicle. Hybrid vehicle refers to cars whose drive system consists of two or more single-drive systems that can operate simultaneously, and the vehicle

**Fig. 10.14** Pure EV

is either driven by a single-drive system independently or by several systems jointly according to its actual driving status. There are different types of hybrid vehicle due to different components, layouts, and control strategies. It usually adopts traditional ICE and electric motor as the power source and drives the car with both the thermal energy and electric systems. The ICE can burn diesel, gasoline, or other substitute fuels like compressed natural gas, propane, and ethanol fuel, while the electric drive system consists of efficient motor, power generator, and battery. Commonly used batteries include lead–acid cell, NiMH cell, and lithium cell, and probably hydrogen fuel cell in the future.

(3) Fuel cell electric vehicle (FCEV). The working principle of FCEV is as follows: the hydrogen fuel has redox reaction with atmospheric oxygen in the car-mounted fuel cell, the resulting electric power drives the motor, which then drives the mechanical transmission structure and consequently the traveling mechanisms like front axle or rear axle, and the vehicle moves forward. Fuel cell is the core part. There are several types of FCEV—pure FCEV, mixture of fuel cell and battery, and the combination of fuel cell, battery, and supercapacitor—considering the different configurations of multiple power sources. The latter two are the main configurations of FCEV. The auxiliary power source provides startup current and recovers electric energy from braking feedback. Fuel cell reaction generates very little $CO_2$ and $NO_x$ and its main byproduct is water, so it is called a new type of green vehicle.

(4) Dual-fuel vehicle. Dual-fuel vehicle has two fuel supply systems, one feeding natural gas or liquefied petroleum gas (LPG) while the other feeding other fuels. The two systems feed fuels to the combustion chamber according to the preset ratio, and the fuels are burned in the cylinder together, such as diesel + compressed natural gas, or diesel + LPG. The biggest advantage of dual-fuel vehicle is that it is able to effectively reduce pollutant emission, cut traditional gas/diesel consumption, and lengthen the engine service life without modifying the engine or weakening the power output.

2. Intelligent Transport System

Intelligent transport system, or intelligent transportation system (ITS), is a real-time, accurate, and efficient comprehensive transport and management system that serves a full range of functions on a large scale by effectively integrating advanced information technology, communication technology, sensing technology, control technology, and computer technology in the whole transport management system. Being a complicated overall system, ITS consists of the vehicle control system, traffic monitoring system, operating vehicle height management system, and travel information system. From the perspective of system composition, it can also be subdivided into advanced traffic information service system, advanced traffic management system, advanced public transit system, advanced vehicle control system, goods transport management system, electronic charging system, and emergency aid system (Fig. 10.15).

3. High-speed railway

According to UIC's definition, high-speed railway refers to old railway systems whose operating speed is above 200 km/h due to modification (linearization and track gauge standardization) or newly built ones whose operating speed is above 250 km/h. Compared with other transport equipment, high-speed railway has a series of advantages such as large passenger capacity, strong transport capability, high speed, high level of safety, high punctuality rate, comfort, and convenience. According to a rough comparison of energy consumption on the unit of "person/km", the unit energy consumption of high-speed railway is 1 person/km, car 5 persons/km, coach 2 persons/km, and plane 7 persons/km.

High-speed railway is considered a significant contribution made by China to the world in the low-carbon transportation field. By the end of 2013, China had a total operating high-speed railway mileage of 11,028 km, with another 12,000 km

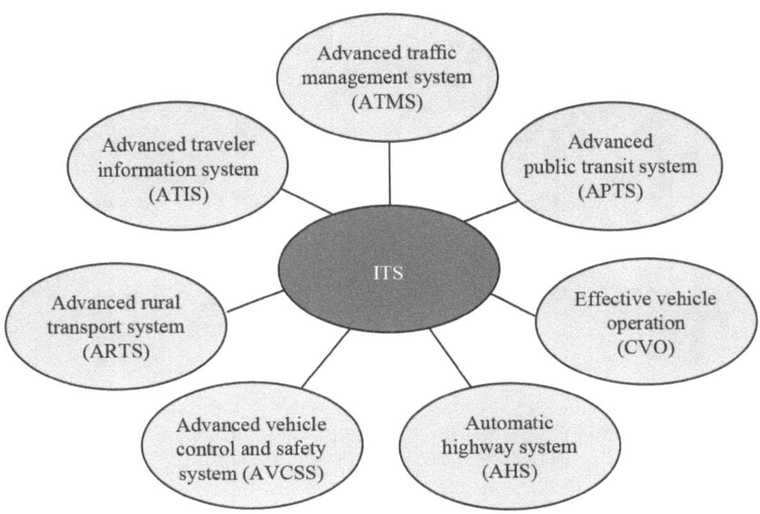


**Fig. 10.15** Connotation of ITS

under construction. With its high-speed railway network primarily taking shape, China boasted the longest operating mileage and the largest under construction scale of high-speed railway in the world. According to the *Medium and Long-term Plan for Railway Network*, China's high-speed railway development will focus on "four vertical and four lateral" trunks in the future. It will set up a general skeleton of fast passenger transport network, build fast, convenient passenger railways with large capacity, and gradually designate separate lines for passenger and cargo trains. High-speed railway is in a golden period and will embrace greater development for a long time to come.

### *10.3.6 General Equipment*

(1) Promote efficient pulverized coal boiler. Pulverized coal boiler is a boiler with pulverized coal as fuel, which enters the combustion chamber along with air and is burned in suspension (Fig. 10.16). As the burning coal is pulverized, its contact area with air is significantly increased, giving the boiler a series of advantages such as fast and complete burning, large capacity, high efficiency, wide coal applicability, and easy control and adjustment. It is estimated that

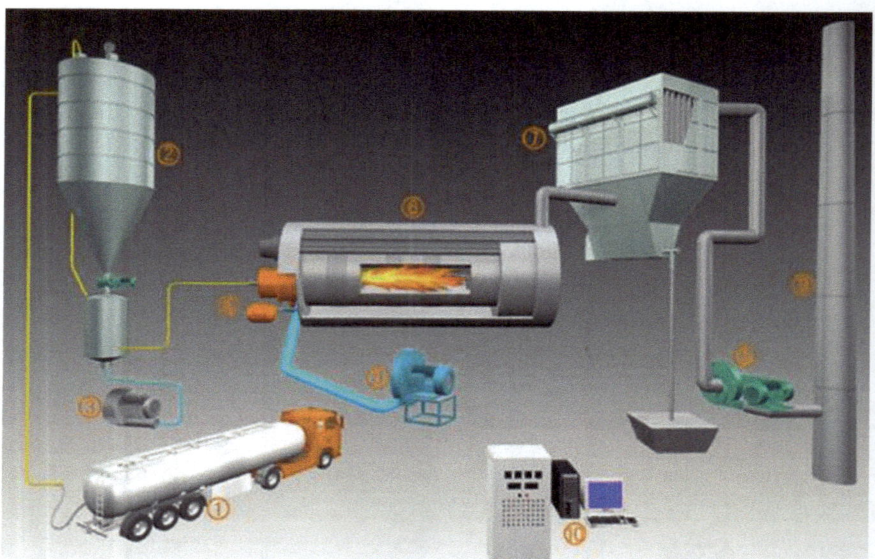

1. Pulverized coal tank; 2. Pulverized coal tower assembly; 3. Primary blower; 4. Burner assembly; 5. Secondary blower; 6. Boiler; 7. Bag-type dust collector; 8. Induced draft fan; 9. Chimney; 10. Automatic control unit

**Fig. 10.16** Working process of efficient pulverized coal boiler pulverized coal boiler can reach the thermal efficiency of 90% or higher, much better than the average 70% of existing industrial boilers in China.

(2) Recovery and recycling of medium/low-heat-value residual heat. China is in serious energy shortage, but the utilization of primary energies and various residual heat resources is at a low level, and a lot of low-heat-value industrial residual heat is wasted, which causes environmental pollution. We use high-temperature water-source heat pump unit to recover residual heat resources of medium/low heat value, and this kind of heating system will be an effective supplement to the urban heating and cogeneration system.

(3) Semiconductor lighting. Also called solid-state lighting or LED, it uses solid-state luminescent devices (solid semiconductor chip) as the light-emitting material. When forward voltage is applied on both ends and charge carriers in the semiconductor are combined, excess energy is emitted to trigger photon emission and consequently generate light. As a new type of lighting device, LED features small size, little power consumption, high efficiency, long service life, rich colors, shock endurance, and easy control, and is recognized in the world as an important lighting material that saves energy and protects the environment. Also called the fourth-generation lighting source or green light source, LED marks another illumination revolution after incandescent lamp and fluorescent lamp.

(4) Rare earth permanent magnet coreless motor. It is a new type of special motor representing the future development direction of the electric motor industry. Being without core, brush and magnetic damping and adopting the rare earth permanent magnet generation technology, this kind of motor changes the traditional motor structure comprising silicon steel sheet and wound stator and adopts China's own electronic intelligent frequency conversion technology to raise the systematic efficiency to be above 95%. Primary estimation shows that if one-third of the incremental motors in China are rare earth permanent magnet coreless motors, we can save nearly 50 billion kWh electricity, 500,000t silicon steel sheets, and 20,000t copper a year and create almost 10 billion yuan of output value, generating remarkable economic and social benefits.

(5) High-voltage variable-frequency governor. Frequency changer is an electric power control device that changes power frequency to another frequency by switching the power semiconductor device. With the rapid development of modern electric, electronic, and microelectronic technologies, high-voltage large-power variable-frequency governor is getting more mature, and the high voltage that was hard to resolve in the past has been well addressed in recent years through series connection of devices or units. This kind of governor is widely used in various blowers, water pumps, compressors, and rolling mills in such industries as large-scale mining, petroleum chemical, municipal water supply, iron and steel smelting, and electricity and energy. Using high-voltage frequency converter to control the speed of pump load is not only good for improving the process and product quality but also meets the requirement for energy conservation and economical equipment operation, and conforms to the trend of sustainable development. Speed control of pump load has many benefits. Most

practical cases have yielded great results (saving energy by 30–40%), significantly lowered the water-making cost of waterworks and raised their automation level, facilitated the operation of pumps and pipe network at reduced pressure, reduced leakage and tube burst, and lengthened the equipment's service life.

(6) Regenerative combustion technology. It is a new combustion technology that is widely promoted in developed countries in recent years. It preheats combustion air from room temperature to 800 °C with efficient heat storage material while drastically reducing $NO_x$ emission, controls flue gas temperature in the 0–150 °C range, and recovers flue gas heat to the largest extent to keep the combustion temperature more even (Fig. 10.17). With multiple advantages, including efficient flue gas heat recovery, high-temperature air preheating, and low emission, this technology can be applied in various fuel-fired heating furnaces (except solid fuel) of the metallurgical industry, and various types of furnaces in the mechanical, petroleum, and chemical industries, as well as combustion, automatic control, refractory materials, and air/gas preheating.

(7) High-efficiency heat exchanger. High-efficiency intelligent heat exchanger is a full-automatic energy-saving product integrating the heat exchange, heat control, heat regulation, and heat metering systems. Based on the working conditions and the change of meteorological conditions, it intelligently controls the primary and secondary heat supply network through the central controller and strikes a balance between heat supply and actual heat load in the end. This kind of heat exchanger is the best type for urban central heat supply (heating, air conditioning, and hot water for domestic use). It consists of the plate heat

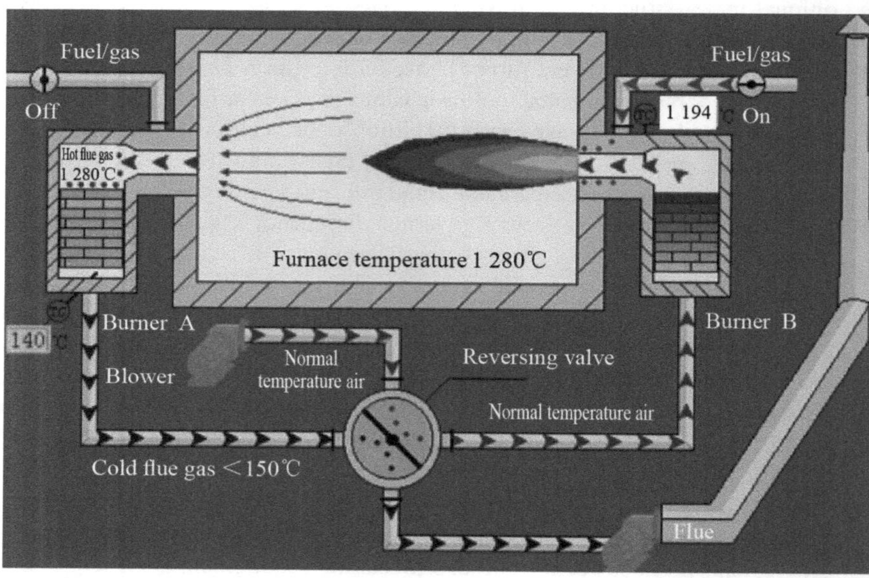


**Fig. 10.17** Working principle of regenerative combustion technology

exchanger, circulating pump, makeup water pump, dirt separator, pipes, valves, instruments, and frequency conversion control system, and can be matched with electronic scale cleaner and full-automatic programming system according to user needs. High-efficiency intelligent heat exchanger is the top choice energy-saving product for its high heat conversion efficiency, small resistance, compact structure, reliable operation, and simple operation.

### 10.3.7 Innovative Low-Carbon Technology in Energy System

(I) Smart grid

The power grid is the bond that connects power plants with end users and the intermediary that keeps the power market running and optimizes power distribution. Although traditional power grid has mature operation management, the traditional mode featuring "large power plants, large grid, and one-way transmission" cannot deal with the load shock on the grid generated by the connection of large amount of renewable-based electricity. Smart grid came into being under such circumstances. It is a new type of power grid based on the physical grid and integrating modern sensing, communication, information, and control technologies. Being a comprehensive economic and technological outcome deriving from the global energy revolution, it demonstrates the efficient utilization of information technology. Also called "Grid 2.0", smart grid is the inevitable choice to realize reliable, safe, cost-effective, efficient, and environmental-friendly grid operation.

By adopting advanced sensing, metering, information and communication, automatic control technologies and new materials, and by making full use of the highly developed information and control technologies, smart grid is a new-generation grid that is highly intelligent, information-based, and interactive with enhanced power distribution network. It not only guarantees the safe and reliable operation of power grid but also meets the end users' various power demands, coordinates the demand and supply of all power generators, power grid, end users, and power market stakeholders, and ensures the effective operation of all components of the system to the largest extent. It minimizes cost and environmental impacts while maximizing system reliability, flexibility, and stability. In sum, the extensive application of smart grid can ensure safe and stable grid operation, reduce the risk of large-scale blackout, utilize intermittent renewable energies and distributed power sources effectively, raise the utilization rate of grid assets, and improve the power efficiency, reliability, and quality for users. It will push the power production and consumption mode to shift toward high efficiency and energy conservation, and increase the power system's economic and social benefits by a large margin.

Different countries have different grid demands. Some have to consume power generated from renewable energies, while others want to share power use information. They are all in the starting or exploratory stage of smart grid development and unified modes, and standards have not been formed yet. Therefore, we should accelerate the strategic deployments of smart grid, specify the development direction and technical routes, and solve the institutional and mechanism obstacles to its development. The current focus is on lifting the intelligent level of grid connection, power transmission, transformation and distribution, power use and dispatching in a coordinated way, boosting the construction of the general information platform, and setting up the intelligent public service platform.

(II) Distributed energy system

Distributed energy is an important direction of energy technology development in the future world, boasting high efficiency, small environmental impacts, and good economic benefits. Distributed energy system refers to the energy production mode on the user end that features small scale, small capacity, and distributed production, which can realize multichannel energy supply and multilayered energy development. This technology has advantages in the exploitation and utilization of renewable energies, improvement of energy utilization rate and supply security, and on-demand energy supply that cannot be matched by traditional centralized energy supply system.

(1) Building Integrated Photovoltaic (BIPV). This kind of PV station installs solar components on the buildings. The PV components can be well combined with the buildings and even become part of them to play the construction functions, hence the name BIPV (Figs. 10.18 and 10.19). It expands the application scope of renewable energies and produces PV electricity power without occupying extra land. As a conjunction point of the vast construction market and the PV market that has a huge potential, BIPV has a promising great prospect. It is foreseeable that PV-building combination will be one of the most important PV applications in the future, with vast prospects and huge market potential.
(2) Distributed natural gas system. Distributed natural gas has advantages in several aspects, including improving energy efficiency and promoting energy conservation and emission reduction. First, it has high energy utilization rate and

**Fig. 10.18** Roof-top PV on civil residence

**Fig. 10.19** Building integrated photovoltaic at industrial plant

tremendous economic benefits. This system can realize gradient energy utilization, make full use of residual heat from power generation, and has an overall energy utilization rate of more than 80%, more than twice that of large coal-fired unit. Second, it can provide heating and power in place and consequently reduce the losses over long-distance transmission. Third, as natural gas can be stored (LNG, CNG, surface or underground gas storage), it can serve as the complement for power grid and heating network at times of emergency. Fourth, featuring quick startup and shutdown, it can participate in grid peak load regulation. As a result, it is able to effectively mitigate the natural gas peak-valley difference in winter and summer, raise the efficiency of gas facilities in summer, and enhance the security of gas supply system. Fifth, natural gas also generates environmental benefits and can reduce pollutant emission by a large margin.

(III) Energy storage

Energy storage is an important guarantee for China's renewable energy development. Although China is already a global leader in low-carbon energies, the intermittence, discontinuity, and uncontrollability of wind and solar power is an impediment to the future development of renewable energies on a large scale and within a large scope. Renewable energies developed by leaps and bounds in China in the past few years. While the installed capacity and output of wind power and PV power keeps rising, the rate of wind and PV abandonment has increased and reached the high level of 15–20% in 2014. The fundamental reason for the abandonment is the timing and structural conflict between energy supply and demand. To solve the grid connection of power from renewable energies, energy storage technology is sure to receive more attention. With China's renewable energy development, electric energy storage will play a bigger role in the economic, efficient, and stable operation of the power system.

Electric energy storage is of great importance for improving the stability of grid operation, bettering electric quality, and perfecting peak-valley regulation. Especially after the grid consumes renewable energy on a large scale, the latter's intermittence will challenge the safe and stable power supply, so a certain ratio of energy storage equipment must be in place as the grid "stabilizer". Based on their working principles, electric energy storage can be divided into several types—mechanical storage (e.g., pumped storage, compressed air storage, and flywheel energy storage), chemical storage (e.g., sodium–sulfur battery, flow battery, lead–acid battery, and lithium battery), electromagnetic storage (e.g., superconducting energy storage), and phase change storage (e.g., ice storage). In view of their developing trend, pumped storage is a mature technology and can be fully applied in areas with the right conditions. Sodium–sulfur battery and flow battery are basically ready for industrialization and can coordinate with the development of renewable energies as advanced energy storage technologies.

(IV) Demand response

Demand response (DR) is short for power demand response. It means when power price in the wholesale market goes up or when the system reliability is threatened, power users receive the suppliers' notice of direct compensation in order to induce them to reduce the load, or when the users get the signal of power price increase, they, for a short term, change their established habits of power use and reduce or postpone the power load within a certain time frame in order to respond to power supply and consequently ensure grid stability and curb the power price hike. DR is one of the solutions for demand-side management (DSM).

DR technology includes the following contents: segmented power meter capable of two-way communication; multilayered customer-friendly communication channels to notice the consumers when they are needed to lower the load; energy information tools that can provide almost real-time segmented load data, analyze load reduction in comparison to the base scenario, and provide operation personnel with energy diagnosis so as to arrive at the possible load reduction goal; the best demand reduction strategy in case of high power price or power system emergency; load controller or embedded energy management and control system that can carry out optimal automatic load reduction on the consumption end; and on-site power generation equipment that can meet emergency load demand either as standby or main power source.

The functions mentioned above can be roughly divided into the following categories. ① Metering technology. Advanced metering technology can support more metering functions than previous metering devices, including storing consumption information at different times and recording and displaying instant load information on active and reactive power, apparent power, voltage, current and load rate, etc. ② Remote communication equipment. It is mainly used for remote metering and control of power suppliers, market operators, intermediary businesses, and consumers, including direct communication device that notifies the consumers through multiple communication channels or adjusts electric installations independently. ③ Control equipment, relevant software, and diverse intelligent electric installations. Advanced metering technology and remote communication technology constitute the Advanced Metering Infrastructure (AMI), while intelligent control technology and remote communication technology constitute the intelligent control system. These technologies can be applied directly on the final user end, including civil users. It makes it possible for different users to make real-time responses according to their price preference and displays immense influence on flexible power demand.

### 10.3.8 $CO_2$ Emission Reduction Technology

(I) Carbon capture and storage

Carbon capture and storage (CCS) is an effective technology of end treatment and reduction of $CO_2$ emission. In practice, it consists of two steps—carbon capture and carbon sequestration.

There are three main technical paths of carbon capture: post-combustion, pre-combustion, and oxyfuel combustion. The first path separates $CO_2$ from the waste gas resulting from fossil fuel burning with the help of liquid solvent and heating, while the second path converts fossil fuel into the mixture of $H_2$ and $CO_2$ first, then absorbs $CO_2$ with liquid solvent or solid absorbent, and finally releases and collects it through heating or depressurization. Compared with post-combustion capture, pre-combustion capture has a higher $CO_2$ pressure and concentration, making it easier to separate carbon and providing the possibility of applying new carbon capture technologies. Oxyfuel combustion is in the stage of lab development and pilot application. Like the other two paths, it also focuses on the process of fuel burning, but what's different is that oxyfuel combustion uses oxygen rather than air as combustion improver, and the post-burning waste gas mainly consists of vapor and high-concentration $CO_2$.

The real difficulty of CCS lies in carbon sequestration. When developing carbon sequestration technology, the focus is on capturing and separating $CO_2$ and then injecting it down into the ocean or deep geological layers. This is not easy. Cost aside, how to find suitable geological layers that can store $CO_2$ and isolate it completely from the atmosphere is a technical challenge. From a geological point of view, three types of geological layers can serve that purpose, the most appealing option being existing oil and gas fields. Based on our deep understanding of the geological profile of pay beds and gas pays, oil and gas fields have proven able to contain hydrocarbon. More importantly, injecting $CO_2$ into oilfields can improve the recovery ratio by 5–15% and correspondingly lengthen the oil well's production cycle. This "ruins-to-riches" approach has helped many oilfields with declining daily output to increase production and extend the life cycle. Another type of geological layer is hydrocarbon-free trap (a location that prevents oil and gas from moving and concentrates it therein), but it has a similar structure to oil-bearing and gas-bearing formations and coal bed. The third type is bottom water-deep aquifer, which is recommended as a long-term carbon sequestration solution for its extensive distribution.

(II) Carbon capture, utilization, and sequestration

Carbon capture, utilization, and sequestration (CCUS) is one of the key technologies to deal with global climate change and a new developing trend of $CO_2$ end treatment technology, capturing close attention around the world. All countries have stepped up R&D and made some headway in $CO_2$-enhanced oil recovery (EOR), but there are technical and economic difficulties in the industrialization of CCUS. Actually, this technology has broad prospects in China. First of all, as the largest GHG emitter in the world, China has the responsibility and obligation to make greater contributions to $CO_2$ emission reduction. Second, as the world's largest industrial country, the complete range of industries, especially developed food processing, petrochemical, and chemical industries, offers vast room for the reuse of $CO_2$.

There are a wide range of technologies for $CO_2$ recycling, such as synthesizing high-purity CO, tobacco expansion, fertilizer production, supercritical $CO_2$ extraction, beverage additive, food preservation and storage, protective gases for welding, fire extinguisher, pulverized coal transmission, synthesizing degradable plastics, improving the quality of saline–alkaline water, cultivating seaweed, and EOR. Among them, the technologies of synthesization of degradable plastics and EOR have broad commercial prospects. At present, more and more power plants, chemical, oil and natural gas enterprises, and equipment service providers view CCUS as a great opportunity for boosting their development.

According to incomplete statistics, by the end of 2013, China had carried out more than 15 CCUS demonstration projects and accumulated some technical and economic data as well as engineering experiences. These CCUS demonstration projects were mainly carried out by large enterprises, such as Shenhua, PetroChina, Sinopec, Yanchang Petroleum, and Huaneng. The $CO_2$ captured by operating demonstration projects is reused in various ways, including EOR. For instance, the integrated project of $CO_2$ capture, EOR, and storage carried out by PetroChina at Jilin oilfield, which has come to the stage of industrial promotion, stores 630,000t $CO_2$, and the annual production capacity of $CO_2$-based EOR is 200,000t. The 100,000t/a full-process CCUS demonstration project carried out by Shenhua in Ordos started construction in 2010 and began $CO_2$ injection in May 2011. So far more than 200,000t $CO_2$ has been sequestered accumulatively, accumulating valuable experiences for other projects.

### *10.3.9 Garbage Recycling and Utilization Technology*

China is in a period of accelerated urbanization. As urban population increases quickly and people's income keeps increasing, the amount of wastes also increases day by day, and domestic garbage in cities is becoming an ever bigger problem for sustainable development. The phenomenon of "garbage siege" is serious in China. It is estimated that over 1/3 of cities nationwide are sieged by garbage, and urban garbage stockpiling has encroached upon 750,000 mu of land (1 mu equals about 666.67 $m^2$). But China's garbage-treating capability is far from enough. Take Beijing for example. The total design capacity of existing garbage treatment facilities is about 10,300t/day, with a daily shortfall of over 8,000t. Going forward, urban garbage generation in China will increase at a speed of around 10% a year. Further breakthrough in garbage recycling is an urgent task.

(I) Garbage sorting and treatment

Garbage sorting means separating different types of wastes based on their components, nature, value of use, and environmental impact (Fig. 10.20). It is the precondition for the scientific treatment of garbage and lays the foundation for the reduction,

**Fig. 10.20** Garbage sorting and recovery system

recycling, and harmless treatment of wastes. In face of the mounting garbage and worsening environment, to maximize garbage recycling by means of garbage sorting and management is one of the pressing issues that all countries in the world are concerned about.

Urban domestic garbage is of very complicated composition, including kitchen waste as well as clothing, glass, and metal, among others. If the garbage is taken to the disposal plant directly and manually sorted there, it is a huge amount of work. Therefore, if a sorting platform is designed and manufactured for pre-disposal garbage sorting, it will not only lower workers' labor intensity and raise work efficiency but also improve the sorting quality. Usual automatic garbage sorting systems include crusher, winnower, magnetic separator, sieving machine, aluminum separator, electrostatic separator, and popping machine. According to the physical or chemical properties of garbage components, these systems, being computer-controlled, sort the garbage by means of crushing, sieving, wind separation, shaking, floating, magnetic separation, and electrostatic separation. In this way, recyclable objects are recycled and treated and the rest is landfilled, incinerated, or made into compost.

(II) Waste incineration

Waste incineration is a new technology that has developed in the recent 50 years and one of the best measures of the "recycling, harmless treatment and reduction" of waste. It is estimated that we can generate 300–400 kWh electricity by burning 1t urban domestic garbage, and the post-burning slag is neutral without smell and will not cause secondary pollution. Furthermore, the bulk of garbage is reduced by 90% and weight by more than 75% after burning. Germany and France used garbage for power generation earliest, but the US and Japan developed rapidly in this field in the past few decades and are in a world leading position now.

Garbage incineration power generation means burning the flammable substances in domestic garbage in incinerators, the resulting high-temperature flue gas enters the HRSG to produce steam, which then drives the gas turbine generator set. Garbage incineration power generation has successful experiences overseas, and long years of practices at the garbage-fired power plant in China's Shenzhen also prove the feasibility of this technology. High-temperature burning can get rid of the massive volumes of harmful substances in garbage and meet the goal of harmless treatment and reduction of waste, while the recovered thermal energy can be used to provide heating and electricity. However, with the promotion of this technology and the construction of more garbage-fired power plants, the post-burning derivatives such as waste gas, wastewater, and slag have caused mounting environmental pollution. A particular concern is that dioxin, a highly toxic substance in waste gas, cannot be effectively controlled and treated.

At present, "3T + E" is a dioxin control measure generally adopted in the international community, which means guaranteeing sufficient glue gas temperature from the incinerator, sufficient stay time in the combustion chamber of flue gas, and appropriate turbulence and excess air during burning. Besides, as dioxin in the incinerator glue gas is attached to fly ash, efficient dedusting can significantly reduce dioxin emission from the incinerating device.

(III) Garbage landfill

There are two types of garbage landfilling processes—traditional landfill and hygienic landfill. Traditional landfill means centralized stockpiling of garbage in ponds, pits, or depressions under natural conditions, without cover or scientific treatment. Hygienic landfill is the approach of adopting engineering and technical measures to prevent pollution or environmental and land damages. Featuring reliable technology, simple process, easy management, and low investment and operating cost, the second landfilling approach has become one of the main garbage disposal approaches in various countries. But its biggest hazard lies in liquid dialysis and methane leakage, which are currently solved through waterproof measures and methane recovery and reuse.

(IV) Garbage compost

Garbage composting technology means, by the way of microorganisms, turning unstable organic substances into stable ones and reducing the content, mitigating the odor, and improving the physical properties of volatile substances in garbage. High-temperature composting can kill the pathogenic bacteria, eggs, and grass seeds in garbage, and compost products can serve as soil conditioners and source of nutrition for plants. Garbage composting is one of the effective ways of garbage recycling in China.

Domestic waste in China has high water content, low heat value, complex components, and high content of organic matters and is collected in mixture, so biological composting treatment is a suitable disposing approach. China implemented rural water-logged compost in the 1950s–1960s, promoted small-scale experimental compost in the 1970s–1980s, and is now in the stage of mechanized, large-scale, and automatic composting treatment. After all these years, the garbage composting technology has been improved. There are several composting processes to treat urban domestic garbage, either with silo or trough or other devices, which apply intermittent dynamic high-temperature aerobic fermentation, static high-temperature aerobic fermentation and dynamic high-temperature aerobic fermentation, respectively.

## 10.4 Roadmap of Low-Carbon Technology Development

Based on the technical maturity and cost-effectiveness of different low-carbon technologies, and the development of relevant industries, the roadmap of China's low-carbon technology development and application in the future is shown in Table 10.1. Before 2020, the focus will be on existing industries, infrastructure, and relatively mature technologies, such as high-speed railway and clean and efficient coal utilization. Efforts will be made to promote the large-scale application of energy-saving technologies, equipment, and production processes, with energy conservation being the keyword. We will develop renewable energies in an orderly manner and place equal emphasis on distributed and centralized energy within the bearing capacity of the grid and infrastructure. We will improve the energy supply system, improve energy management, and set up the platform to consume electricity generated from renewable energies. After 2020, we will boost the industrialization and large-scale application of key low-carbon technologies on the basis of making breakthroughs, such as the Gen-IV nuclear power technology and spent fuel treatment technology, smart grid, thin-film PV battery, etc. At that time, China will be a global leader of low-carbon technology.

**Table 10.1** Roadmap of future development and application of low-carbon technologies

| 2010 | 2020 | | 2030 | |
|---|---|---|---|---|
| Energy exploitation | Clean, efficient, safe coal production technology; Oil-based secondary resource processing technology, oil shale utilization technology; CBM, shale gas exploitation technology | Polysilicon, monosilicon cell; Onshore wind power; Gen-IV nuclear power; Medium/low-temperature geothermal power generation, GSHP | Full-automatic coal mining and symbiotic resource mining technology; Prospecting and exploitation technology for unconventional oil and gas resources, such as shale gas and combustible ice | Thin film PV battery; Offshore wind power; Spent fuel treatment technology; Hydrogen utilization technology |
| Energy conversion | SC, USC power generation, gangue power generation; Co-generation Natural gas CCHP | Modern coal chemistry: coal-based multi-generation, coal-to-olefin, coal-to-methanol | IGCC | Coal-to-liquid, coal-to-gas |
| Energy consumption | Industrial residual heat, pressure recovery technology; Advanced energy-saving production process, technology and equipment; Energy-saving household appliances | High-speed railway, pure EV, dual-fuel vehicle; Passive buildings; Construction insulation, thermal insulation, ventilation technology | Coordinated disposal of urban sewage, garbage and wastes by industrial enterprises; High-efficiency, intelligent household appliances | Buildings of renewable energies; EV, fuel cell vehicle |
| Energy system | UHV grid; Pumped storage, air energy storage | Distributed heating, power generation | Smart grid; Chemical energy storage, electromagnetic energy storage | Flexible power transmission; Demand response |

# Chapter 11
# Policy Guidance of Low-Carbon Development

The modernization of China is necessarily a low-carbon one. Low-carbon development is a long-term vision critical to people's well-being and national future. It is required for the "two centenary goals," and marks a great opportunity for accelerating the transformation of economic development mode, restructuring the economy, and promoting new industrial revolution. As is mentioned above, however, the transition toward low-carbon mode does not occur, evolve, and complete automatically. It needs strong policies to guide the transformation of socioeconomic development, raise people's awareness, identify low-carbon development paths, and translate these into actions.

## 11.1 Low-Carbon Development is a Prevalent Trend in International Socioeconomic Development

Major developed countries and regions have incorporated the low-carbon concept into their state socioeconomic development strategies and guide the society to transit to low-carbon mode by releasing and implementing a package of strategies and policies.

### *11.1.1 Low-Carbon Development as State Strategy*

The United Kingdom pioneers in low-carbon economy. With the release of the white paper *Our Energy Future—Creating a Low Carbon Economy* in 2003, the UK was the first to escalate low-carbon development to a state strategy. In 2008, the UK founded the Department of Energy and Climate Change, which became the world's first government department focusing on climate change.


X. Du et al., *Overview of Low-Carbon Development*,
https://doi.org/10.1007/978-981-13-9250-4_11

The White House released *The President's Climate Action Plan* in 2013, which put forward comprehensive arrangements on the basis of three major goals: energy security, economic growth, and reaction to climate change; and released *All-of-the-Above Energy Strategy* in 2014, which acted as a guideline of energy system low-carbon transition.

The European Union issued its climate change and renewable energy package in 2008, aiming to drive the transition of EU economy toward a high energy efficiency and low carbon mode and to lead the world into "post-industrial-revolution" era. Then, in 2011, it released *2050 Low Carbon Roadmap*, in which it proposed that the EU could win competitiveness in technology and the whole economy through high emission reduction targets and appropriate policies.

Japan advocates low-carbon development. In 2008, it issued "Fukuda Blueprint" to encourage the efforts for low-carbon society, and released multiple state-level action plans, such as *Japan Basic Energy Plan*, *State New Energy Strategy*, *Low-carbon Society Action Plan,* and *New Economic Growth Strategy*, in order to guarantee the reaction to climate change and develop energy. After the Fukushima Daiichi nuclear disaster, the Japanese government adopted more measures, including accelerating the development of renewable energy and increasing natural gas supply, to continue its low-carbon energy state strategy.

### 11.1.2 Clear Emission Targets to Cope with Climate Change and Develop Energy

Clear mid- and long-term targets and specific action plans promoted low-carbon development in developed countries and regions, and secured their green energy development, emission reduction, and energy conservation.

The European Union sets the goal of capping greenhouse gas emission in low-carbon development and energy–climate integration. Step by step, it proposed several energy transition goals—"reduce greenhouse emission by 20% by 2020 compared to 1990 levels and raise the proportion of renewable energy to 20%," "reduce greenhouse emission by 80–95% by 2050 compared to 1990 levels," and "reduce greenhouse emission by 40% by 2030 compared to 1990 levels, raise the proportion of renewable energy to 27%, and increase energy efficiency by 30%." It also identified the paths to hit these targets through a series of energy–climate integration strategies—*Europe 2020 Strategy* (2010), *Energy Roadmap 2050* (2011), *2030 Climate and Energy Framework* (2014), etc.

Japan proposed in "Fukuda Blueprint" and "New Countermeasures Against Climate Warming" released in 2008 to "reduce greenhouse gas emission by 60–80% by 2050 compared to the current level," and identified mid- and long-term targets in *Basic Act on Global Warming Countermeasures*—"reduce greenhouse gas emission by 25% by 2020 and by 80% by 2050 compared to 1990 levels." Japan has released a number of state-level plans, including *Japan Basic Energy Plan*, *State*

*New Energy Strategy, Low-carbon Society Action Plan,* and *New Economic Growth Strategy*, all of which have supported its measures to cope with climate change and develop energy.

### 11.1.3 Sound Laws and Regulations System to Secure Low-Carbon Development

Developed countries and regions pay high attention to related legal system development. They adjust and amend existing laws and regulations to meet the demand for low-carbon development, and meanwhile, expand and make special laws to relegate emerging issues and situations in low-carbon development.

For example, the United States Supreme Court identified greenhouse gases like carbon dioxide as air pollutants in 2007, and ordered EPA to make emission standards as soon as possible. The country also applies the policy tools for controlling other pollutants to greenhouse gases.

The United Kingdom adopted the world's first domestic emission reduction law—*The Climate Change Act*—to combine greenhouse gas emission reduction, adaptation of climate change, and low-carbon transition and to stipulate the system, procedures, and institutions.

The European Union has adopted a series directives—2001/77/EC (on renewable energy), 2003/30/EC (on biodiesel), 2003/96/EC (on energy taxation), 2003/54/EC (on power market liberalization), and 2003/87/EC (on greenhouse emission trade)—since 2000, which formed the greenhouse gas emission law system. This legal system encourages EU members to improve energy efficiency and develop renewable energy, in order to reduce greenhouse gas emission.

### 11.1.4 Low-Carbon Technology R&D and More Innovation Input to Establish Low-Carbon Technological System

In low-carbon era, the technology system also needs to turn to the track of green development. All countries are pursuing industrial upgrading through conceptual and technological innovation.

The United States, which invests more resources in low-carbon economy R&D than any other countries in the world, regards the development and application of energy-saving and low-carbon technology as the starting point of "green revolution." Through *American Recovery and Reinvestment Act of 2009*, it invested USD 23 billion in renewable energy production, in order to deepen and expand smart grid research plan, make technical roadmaps, improve technical innovation in wind turbines, solar cell, biofuel, and other renewable energies, and raise the capacity and competitiveness of manufacturing advanced energy equipment. After 2010, the EPA updated the emission standards for power plants, oil and gas exploitation, automobile and biofuel, and established a full set of technical standard guidelines for all sectors. *The President's Climate Action Plan* released by the White House in 2013 regards low-carbon economy as the new engine for future development, upgrades traditional strong industries (such as energy and automobile), and implements new green development policies through incentives including subsidies, preferential taxes, and government procurement. As a result, more jobs have been created and the competitive of the United States in low-carbon technology has been improved.

The European Union considers low-carbon economy as a new economic engine and a cradle of jobs, and thus incorporated low-carbon economy into its future strategic plan. *The 2050 Low Carbon Roadmap,* an outcome of the collaboration between the business community and researchers, aims to develop low-carbon technologies in six potential sectors: wind power, solar power, bioenergy, CCS, and other two. *The Europe 2020 Strategy* released in March 2010 proposed to enlarge the input in energy conservation, emission reduction, and clean energy, making low-carbon sectors pillar industries of future economic development.

Germany put forward "Industry 4.0" led by smart manufacturing, and spares no efforts to implement this plan to maintain the international competitiveness of its industry. Currently, in low-carbon energy technology research, Germany has formed an intensive funding system and has taken the lead in the world in terms of energy conservation and new energy technology. Energy conservation and environmental protection industries have become Germany's pillars.

## 11.2 China's Low-Carbon Actions

For China, low-carbon development is a must for identifying its own development demand, transforming economic development pattern, and pursuing sustainable development. In recent years, China has released, implemented, and strengthened a series of plans and policies to reduce greenhouse gas emission and realize green, low-carbon, and sustainable development.

### *11.2.1 Raise Awareness and Gradually Elaborate Low-Carbon Development Philosophy*

With stronger economy and changes in domestic situation, China further elaborated and clarified the philosophy of pursuing ecological progress and sustainable development, and put forward green development and low-carbon development step by step.

(1) In 2005, Hu Jintao said on the National Population, Resources and Environment Work Conference that one of the priorities in environmental work now

was "to improve the law and policy system promoting ecological progress, make national ecological protection plans, and educate the public on ecological progress."

(2) At the end of 2005, *the Decision of the State Council on Implementing the Scientific Development View and Strengthening the Environmental Protection* mentioned that the environmental protection work shall, under the guidance of Scientific Outlook on Development, "rely on scientific and technological progress to develop recycling economy, advocate ecological progress, strengthen environmental law, improve regulation system, and establish long-term mechanism."

(3) In 2007, the Report on the 17th National Congress of the CPC further clarified the new requirements of ecological progress, and set "become a ecologically developed country by 2020" as one of the important requirements of building a well-off society in an all-round way.

(4) In 2010, the *12th Five-Year Plan* regarded climate change reaction and green and low-carbon development as an important policy direction. "Green Development" was incorporated into the *Plan* as an independent chapter.

(5) In 2011, the Report on the 18th National Congress of the CPC elaborated ecological progress in a more systematic, complete, and philosophical manner, and proposed to integrate ecological progress into all aspects and whole process of economic, political, cultural, and social development. Green, recycling, and low-carbon development would be promoted.

(6) In 2013, the 3rd Plenary Session of the 18th Central Committee of the CPC proposed to "deepen institutional reforms related to ecological progress with a view to building a beautiful China."

(7) In March 2015, the Political Bureau of CPC Central Committee reviewed and approved *Opinions of the CPC Central Committee and the State Council on Accelerating the Ecological Civilization Construction*, which proposed that "we need to incorporate the ecological progress into all aspects of the whole process of economic, political, cultural and social progress, stay committed to the principle of giving high priority to conserving resources, protecting the environment and promoting its natural restoration, and strive for green, circular, and low-carbon development. Driven by reform and innovation-driven development, we pursue ecological progress by making comprehensive advancement while focusing on priorities".

With richer content and connotation, ecological progress has become a major state strategy, and low-carbon development is a critical measure.

### *11.2.2 Establish Low-Carbon Development Administration Institution*

The CPC Central Committee and State Council attach great importance to climate change and low-carbon development. In 2007, National Leading Group to Address

Climate Change and Energy Conservation and Pollutant Discharge Reduction was founded under the State Council, as a national deliberation and coordination organ on addressing climate change, conserving energy, and reducing emission. The head of the Leading Group is the Premier, Vice Premier, and State Councilors act as deputy heads, and group members include the heads and persons-in-charge in all the Ministries and Commissions. The Leading Group Office was affiliated to the NDRC to undertake the regular work of the Group.

NDRC, as the principal coordinator in addressing climate change, established the Department of Climate Changes to undertake the specific work of the Leading Group in terms of addressing climate change, to analyze the impacts of climate change on socioeconomic development, and to organize the drafting of major strategies, plans, and policies on climate change and low-carbon development.

At the provincial level, all the 31 provinces, autonomous regions, and municipalities (not including Hongkong, Macao, and Taiwan) have established leading organs to address climate change, with the chief of provincial government as the head. Accordingly, departments under provincial Development and Reform Commission were established to coordinate the work on climate change.

This top-level design and administrative institution development shows the importance attached by the government to climate change and reflects the profound influence of low-carbon development on China's economy and people's well-being.

### *11.2.3 Form a Policy System with Clear and Complete Hierarchy*

Low-carbon policies and administration of China are not entirely separated. Instead, based on current economic, energy, and environmental administration system, a policy system with clear and complete hierarchy has been formed.

Take a series of policies addressing climate change released during the 12th Five-year Plan period as examples. From macroscopic plans to specific policy tools, a framework with clear and complete hierarchy was formed and improved. As shown in Fig. 11.1, this system has four levels: a. Under the state policy of ecological progress, *National Climate Change Program (2014–2020)*, a. mid- and long-term guideline, proposed the goals for 2020 in five aspects: greenhouse gas emission control, low-carbon pilot projects, climate change adaptation, capacity building, and international exchanges; b. Under the *National Climate Change Program*, in terms of climate change mitigation, the *Work Plan for Greenhouse Gas Emission Control during the 12th Five-Year Plan Period* stipulates specific targets and plans, and aligns with related energy conservation and emission reduction plans; in terms of climate change adaptation, *National Climate Change Adaptation Strategy* was released; c. In terms of climate change mitigation, energy, industry, construction, and transportation authorities have made sectoral special plans of addressing climate change, or have incorporated addressing climate change into their development plans and energy conservation and emission reduction plans, elaborating sectoral targets and priorities more specifically; d. Under the sectoral plans, specific policy measures, such as subsidies and mandatory standards, were issued to meet the goals.

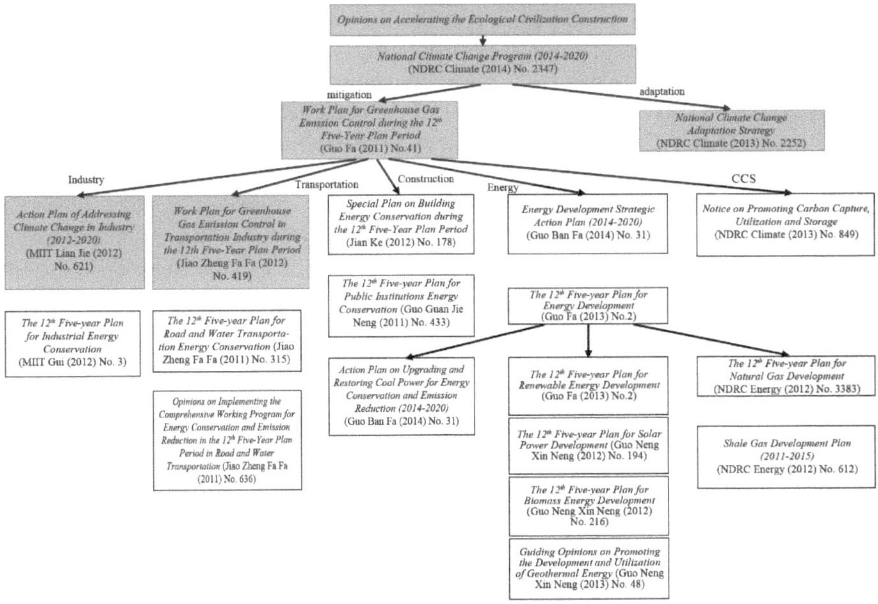


**Fig. 11.1** 中国控制温室气体排放的政策体系. China's greenhouse gas emission control policy system. *Note* Gray rectangulars refer to plans specially for climate change or greenhouse gas emission control. White rectangulars refer to plans closely related to climate change

## *11.2.4 Promote Low-Carbon Transition Under Sectoral Emission Reduction Targets*

*National Climate Change Program (2014–2020)* (NDRC Climate (2014) No. 2347) issued in September 2014 is a mid- and long-term guideline, which proposed major goals in five aspects: greenhouse gas emission control, low-carbon pilot projects, climate change adaptation, capacity building, and international exchanges.

In 2009, China proposed a target on addressing climate change for the first time—reducing carbon emission per unit of GDP by 40–45% by 2020 compared to 2005 levels. The general targets during the 12th Five-year Plan period are reducing carbon dioxide emission per unit of GDP by 17%, and raising the share of nonfossil fuels in primary energy consumption to 11.4%. These two indicators were incorporated in the Five-Year Plan for the first time, which shows that China has started to shift from single-target control of energy intensity to a more strict double-target control of energy intensity and carbon intensity.

In addition, aggregated energy consumption target was introduced for the first time during the 12th Five-Year Plan period: by 2015, national total energy consumption and power consumption shall be controlled at 4 billion tons of standard coal and 6.15 trillion kW · h, respectively; and by 2020, total primary energy consumption shall be controlled at 4.8 billion tons of standard coal.

In 2014, China made peak year target public for the first time. Carbon dioxide emission is planned to peak around 2030 and efforts would be made peak early. The share of nonfossil fuels in primary energy consumption is planned to rise to around 20%. It is also the first time to set a target of the peak year of absolute emission.

In June 2015, the Chinese government submitted *Enhanced Actions on Climate Change: China's Intended Nationally Determined Contributions* to the Secretariat of UNFCCC. This INDC proposed to achieve the peaking of carbon dioxide emissions around 2030 and making best efforts to peak early; to lower carbon dioxide emissions per unit of GDP by 60–65% from the 2005 level; to increase the share of nonfossil fuels in primary energy consumption to around 20%; and to increase the forest stock volume by around 4.5 billion $m^3$ on the 2005 level. This document identifies the goals to strengthen the actions of addressing climate change after 2020, the paths and policy measures to achieve these goals, and the targets of controlling greenhouse gas emission by 2030.

Under the state master goals, sectoral special plans propose more specific targets. As shown in Table 11.1, these targets, on the basis of the policy system shown in Fig. 11.1, have become the policy guidance on low-carbon development in all sectors.

### 11.2.5 Comprehensively Take Use of Policies Tools Such as Orders, Regulations, Economic Incentives, Market Mechanisms, and Information Release

i. Laws and regulations

Laws, as norms adjusting power relations and social relations, not only provide incentives and guarantees for low-carbon economic development but demonstrated China's resolution to address climate change and legal process of implementing international treaties. Currently, China has about 30 laws, 90 administrative regulations and a large number of environmental protection standards offering a primary legal support to low-carbon development. Domestic laws closely related to low-carbon development include *Energy Conservation Law, Law on Promoting Clean Production, Renewable Energy Law, Circular Economy Promotion Law, Forestry Law, Grassland Law, Environmental Protection Law*, etc.

In terms of top-level laws and regulations design, NDRC, NPC Environmental and Resources Committee, NPC Legislative Affairs Commission, State Council Legal Office, and related departments jointly founded a climate law drafting work leading group. So far, a legal framework has come into being. Meanwhile, the government is coordinating research institutions to study low-carbon development law-making. National legislation can learn lessons from local low-carbon development pilot projects, so that a legal framework will be established.

**Table 11.1** Targets of controlling greenhouse gas emission in major sectors in China

| | Content | 2015 | 2020 |
|---|---|---|---|
| General | $CO_2$ per unit of GDP | Lower by 17% in five years | Lower by 40–45% from 2005 level |
| Energy | Share of nonfossil fuels in energy consumption/% | 11.4 | 15 |
| | Total primary energy consumption/tons of standard coal | 40 | Around 48 |
| | Share of nonfossil fuels installed capacity/% | 30 | – |
| | Share of gas consumption in primary energy consumption/% | 7.5 | >10 |
| | Share of coal consumption in primary energy consumption/% | 65 | 62 |
| | Total coal consumption/100 million tons | – | 42 |
| | Conventional hydropower installed capacity/100 million kW | 2.6 | 3.5 |
| | Conventional hydropower generating capacity/1 trillion kW · h | 0.91 | 1.2 |
| | Nuclear power installed capacity/10,000 kW | 4000 | 5800 |
| | Grid-connected wind power installed capacity/100 million kW | 1.0 | 2 |
| | Solar power installed capacity/100 million kW | 0.21 | 1 |

(continued)

**Table 11.1** (continued)

| | Content | 2015 | 2020 |
|---|---|---|---|
| Industry | $CO_2$ emission per unit of industrial value added lowered from 2005 level/% | – | Around 50 |
| | Reduction rate in five years of energy consumption of industries beyond a designated scale/% | 21 | – |
| | Share of value added of strategic emerging industries in GDP/% | 8 | Around 15 |
| | Share of value added of service industry in GDP/% | 47 | 52 |
| Construction | Share of urban green buildings in new buildings/% | 20 | 50 |
| | Energy-conserving design standard implementation rate/% | >95 | – |
| | Reduction rate of energy consumption per unit area of public buildings/% | 10 | – |
| Transportation | Share of buses use in transportation large- and medium-sized cities/% | – | 30 |
| | $CO_2$ emission per passenger person-kilometers in road transportation | Down by 7% from 2005 level | Down by 5% from 2010 level |
| | $CO_2$ emission per rotation volume of goods transport in road transportation | Down by 13% from 2005 level | Down by 13% from 2010 level |
| | $CO_2$ emission per unit of rail transport volume reduced from 2010 level/% | – | Down by 15% from 2010 level |

(continued)

**Table 11.1** (continued)

| | Content | 2015 | 2020 |
|---|---|---|---|
| | $CO_2$ emission per unit of water transport volume/% | Down by 15% from 2005 level | Down by 13% from 2010 level |
| | $CO_2$ emission per unit of civil aviation transport volume/% | Down by 3% from 2010 level | Down by 11% from 2010 level |
| Agriculture and forestry | Forest coverage/% | 21.66 | – |
| | Forest area increased from 2005 level/10,000 $hm^2$ | 3000 | 4000 |
| | Forest growing stock | Up by 600 million $m^3$ from 2010 level | Up by 1.3 billion $m^3$ from 2005 level |

*Data sources* National Climate Change Program (2014–2020), The 12th Five-year Plan for Energy Development (Guo Fa (2013) No.2), Energy Development Strategic Action Plan (2014–2020) (Guo Ban Fa (2014) No. 31), The 12th Five-year Plan for Industrial Energy Conservation, Action Plan of Addressing Climate Change in Industry (2012–2020) (MIIT Lian Jie (2012) No. 621), Special Plan on Building Energy Conservation during the 12th Five-year Plan Period (Jian Ke (2012) No. 72), The 12th Five-year Plan for Public Institutions Energy Conservation (Guo Guan Jie Neng (2011) No. 433), Work Plan for Greenhouse Gas Emission Control in Transportation Industry during the 12th Five-Year Plan Period (Jiao Zheng Fa Fa (2012) No. 419), The 12th Five-year Plan for Road and Water Transportation Energy Conservation (Jiao Zheng Fa Fa (2011) No. 315), Opinions on Implementing the Comprehensive Working Program for Energy Conservation and Emission Reduction in the 12th Five-Year Plan Period in Road and Water Transportation (Jiao Zheng Fa Fa (2011) No. 636), Key Points on Addressing Climate Change of Forestry during the 12th Five-year Plan Period (Ban Zao Zi (2011) No. 241), The 12th Five-year Plan for China Civil Aviation Development

ii. Administrative orders

Administrative orders are commonly used policy tools in low-carbon development. It is a mandatory order issued by the government to government agencies, enterprises, the public, and individuals on energy conservation and emission reduction. Classified by types of policy tools, they include mandatory standards, mandatory tasks, and industry access system; by contents, they cover outdated capacity phaseout, transportation, construction and green government procurement, etc. For example, the NDRC issued *Ten Thousand Enterprises Action Plan for Energy Conservation and Low-carbon Development* for nine key energy consumption industries: steel and iron, nonferrous metals, coal, power, petroleum and petrochemical, chemical engineering, building materials, papermaking, and textile. According to the Action Plan, key enterprises were required to sign letters of responsibility of energy conservation targets and establish energy management system, and energy auditing was organized for these enterprises. In the action, NDRC played a coordinative and managing role; National Bureau of Statistics follows and collects the energy consumption data of enterprises, Administration of Quality Supervision, Inspection and Quarantine inspects the enterprises' energy measurement meters; State-owned Assets Supervision and Administration Commission evaluates the energy-saving performance of state-owned enterprises directly administered by it; and provincial energy conservation authorities supervise energy-saving progress of regional enterprises.

iii. Economic incentives

Economic incentives include price tools, tax and fee policies and fiscal fund support, etc. For example, China uses price tools to promote the low-carbon transition in industries with high energy consumption and high carbon emission. In 2006, China applied differential tariff to eight high-energy-consuming industries: electrolytic aluminum, ferroalloy, calcium carbide, caustic soda, cement, steel and iron, yellow phosphorus and zinc smelting. In 2007, NDRC released four policy documents, demanding the implementation of differential tariff. In 2010, NDRC required to cancel the preferential power price provided by some local governments for high-energy-consuming enterprises after the international financial crisis. The differential tariff for the above eight industries was continued, and the differences between rates were increased. Punitive tariff was imposed on the products that exceed national and local energy consumption standards.

By adjusting low-carbon development-related taxation policy and thus guiding ways of production and consumption, the government encourages low-carbon production methods and lifestyle. Currently, the government's taxation measures for low-carbon development include resource tax, vehicle purchase tax, consumption tax, corporate income tax, value-added tax and pollutant discharge fee, etc.

So far, China has multiple categories of low-carbon-related fiscal supporting funds. Major fund incentive mechanisms include fiscal fund for urban energy conservation and emission reduction pilot projects, earmarked fund for renewable energy development, contract energy management incentive fund, earmarked fund for circular economy development, energy conservation technological retrofits incentive fund, earmarked fund for low-carbon development, earmarked fund for emerging industry development, building energy efficiency subsidy and earmarked fund for transportation energy conservation, and emission reduction. For example, to promote ten key energy-saving projects, the NDRC issued the *Interim Administrative Measures for the Financial Incentive Funds for Energy Conservation Technology Retrofits*, replacing subsidies with rewards. For technology retrofits projects saved more than 10,000 tons of standard coals, in eastern China, reward was given at RMB 200 per ton of standard coal, and in western China, 250 per ton of standard coal.

iv. Market mechanism

Carbon trading is a market mechanism conducive to low-carbon development. In the 12th Five-year Plan, China proposed "step by step establishing carbon trading market" as one of the measures to control greenhouse gas emission for the first time. Meanwhile, it put forward "making a master plan of carbon trading market development." This means that China's carbon trading policy follows the concept of "from pilot projects to larger scale." In October 2011, Notice on Carrying out Carbon Emissions Trading Pilots approved the pilot projects in two provinces and five municipalities, including Beijing and Shanghai. All the pilot regions have studied and established basic system of carbon trading, and kicked off the project in 2013–2014. The last contract fulfillment was by 2016. Based on the practices and lessons of pilot projects, China would accelerate to develop a nationwide carbon trading market, and planned to run it in 2016.

v. Pilot projects

On July 19, 2010, NDRC issued a notice to identify the five provinces (Guangdong, Liaoning, Hubei, Shaanxi, and Yunnan) and eight municipalities and cities (Tianjin, Chongqing, Shenzhen, Xiamen, Hangzhou, Nanchang, Guiyang, and Baoding) as the first batch of pilot regions. On November 29, 2012, NDRC issued another notice to identify 29 provinces, municipalities, and cities (including Beijing, Shanghai, Shijiazhuang, etc.) as the second batch. The central government proposed several requirements to these pilot regions: first, clarify the direction and principles of work; second, make low-carbon development plan; third, establish low-carbon industry system featuring low-carbon, green, environmental-friendly, and circular economy, including promoting green and energy-saving buildings, establishing low-carbon transportation network, etc.; fourth, establish greenhouse gas emission data collection and management system; fifth, establish accountability system of controlling greenhouse gas emission targets; sixth, encourage low-carbon and green lifestyle and consumption model. The two batches of pilot projects covered six provinces and 36 municipalities and cities. In 2010, pilot regions accounted for 57% of national GDP, 42% of national population, 58% of energy consumption, and 56% of national carbon dioxide emission based on fossil fuels consumption. By geographical distribution, these pilot regions cover regions in different development phases and with different features: three more developed regions in China—North China, East China, and South China, rapid growing region—Southwest China and underdeveloped regions—Middle West China and Northeast old industrial area. Low-carbon development pilot projects were a bottom-up attempt to explore low-carbon development model. They will provide best practices for regions with all fundamentals and features in China.

## 11.3 Difficulties and Challenges in China's Low-Carbon Transition

### *11.3.1 Lack of Interdepartmental Coordination Among Low-Carbon, Energy, and Environment Policies*

Addressing climate change is closely related to economic development, energy development, and environmental protection. Currently, however, these fields are fragmented, yet to be coordinated and integrated. At the state level, despite that National Leading Group to Address Climate Change and Energy Conservation and Pollutant Discharge Reduction was founded as early as in 2007 and the NDRC was identified to undertake specific work, economic development, energy development, environmental protection, and climate change addressing are still separated, managed by different departments. Lack of holistic philosophy led to interest conflicts between these departments, and as a result, a large amount of coordination was needed, and synergy was difficult to foster. So far, China has neither related laws and regulations nor special fund on climate change, which resulted in lack of legal foundation, means of restraint, and strength of implementation.

At regional level, in many parts of China, energy conservation is in the charge of regional Economic and Information Commission, pollutant discharge reduction is in the charge of local environmental protection authorities, and low-carbon development is in the charge of local development and reform commission. Related evaluation indicators are also distributed to different departments. Out of the consideration of departmental interests, these departments may have difficulties in coordination with each other, and thus, the pace of work is dragged. For example, the local government needs to reach the obligatory targets on greenhouse gas emission control and energy consumption reduction. Among these, targets on greenhouse gas emission control can be reached by trading, while energy conservation targets, in the current stage, are mainly reached by government-led administrative or fiscal means. As a result, some conflicts may emerge in specific work.

### 11.3.2 Vacancy in Special Legislations on Low-Carbon Development

The rule of law is necessary for the modernization of state governance system and capability. China's existing legislation related to low-carbon development cannot meet current demand. Although some of the existing laws play a critical role in promoting low-carbon development, no laws, regulations, and norms directly stipulate carbon emission control. Existing related laws are not able to adequately adjust the social relationship in terms of low-carbon development, and this leads to restraints in design and effect of related institutions and policies. It means that existing government rules directly stipulating greenhouse gas emission in China all lack host law base. Local related laws and regulations are few in number and have not formed a complete system, failing to ensure and support the pilot projects and experiments. Under the circumstances of "doing nothing unless the law provides basis," vacancy in local regulations leads, to a large part, to difficulties in forming long-term mechanisms and foundation for fulfilling low-carbon development targets.

### 11.3.3 Lack of Low-Carbon Development Systems

As is pointed by the 4th Plenary Session of the 18th CPC Central Committee, strict legal system is needed to protect ecological environment. Thus, a legal system pursuing ecological progress should be formed at a quicker pace, in order to restrain development activities and promote green, circular, and low-carbon development. Based on government rules of NDRC and other departments, China has kicked off system for voluntarily reducing and trading greenhouse gas emission, low-carbon product certification system and greenhouse gas emission statistics accounting and reporting system nationwide, and started the pilot project of carbon trading system in some provinces and municipalities. These systems, however, are implemented separately through issuing policies, instead of forming a united and coordinative framework and covering all effective measures to control greenhouse gas emission. And most of them are still in the period of trial implementation. China need to combine top-level design with pilot projects, as well as form an institutional framework for low-carbon development, strengthen the design of major systems including total greenhouse gas emission control, carbon trading, product labeling and certification, etc., improve greenhouse gas emission statistics accounting system, and follow-up the practices and lessons of low-carbon development pilot projects. In this way, China is to pursue a low-carbon development path with both bottom-up and top-down features.

### 11.3.4 Incomplete Fiscal Policies

In recent years, the central government established earmarked funds, such as government earmarked funds for energy conservation and emission reduction, to support low-carbon urban development through budgetary work. These earmarked funds are mainly used for pilot projects, demonstration areas, institution and standards research, and design and basic scientific research. China's low-carbon fiscal policy, however, has not rationalized the interest relationships among actors and the government–market relationship, nor formed synergies among regions or departments. For example, most of the pilot low-carbon cities did not adequately plan the input and output values of major low-carbon projects and lack complete and clear mechanisms to secure low-carbon investment and financing. Moreover, the government is not the only actor in green and low-carbon investment; rather, the central government, local governments, enterprises, public institutions, foreign investors, and individuals are part of it. But now, the government fails to drive large sum of nongovernmental investments. The leading and leveraging roles of fiscal funds are yet to take full use of, and the financing channels in the financial market are yet to be accessible.

## 11.4 Suggestions on Strengthening the Guidance of Low-Carbon Policy

By 2020, China is going to be built into a moderately prosperous society completely. After 2030, industrialization and urbanization will be realized and people's level of living will also be improved significantly. Domestic and international situations ask China to take the responsibilities of dealing with climate change, which are consistent with China's development level, capabilities, and international status, so as to realize low-carbon development. By then, people's income level, public awareness, and regulation system will have changed. China needs to comprehensively consider its future changes in society, economy, energy, and environment and improve its policy system of low-carbon development on the basis of evaluating major policies.

### *11.4.1 Realize Balanced Development Through Low-Carbon Development, and Coordinate Overall Development*

First, the realization of low-carbon development needs the long-term revolution of economic development model and energy system. The transformation of development model and energy system is not a fleeting one. It takes dozens of years to materialize. Hence, related work must be deployed and arranged beforehand. China needs to make determination, decide early on, perceive from a strategic, holistic and long-term view, and balance the development of energy and economy through low-carbon development and ecological civilization establishment. China needs to study, demonstrate, and decide China's mid- and long-term target for low-carbon development, specify policy expectation and propose the roadmap and timeline for economic transformation and economic restructuring as soon as possible. Meanwhile, top-level design should be reinforced, knowledge and actions should be aligned and development philosophy, arrangement of strategy, setting of objective and path, development of technology and funding and policies, etc. should be secured.

Second, China should establish a clear philosophy of scientific development and incorporate greenhouse gas emission index into economic development plan and evaluation system. Government at all levels would align their development philosophy, put low-carbon development and greenhouse gas emission control on the top of the agenda, and regard low-carbon development as the important driving force of motivating economic development model. Government responsibilities and policy orientation should be strengthened, major tasks of controlling greenhouse gas emission should be completed, and local government's work of controlling greenhouse gas emission should be effectively pushed so as to guarantee the realization of emission target and improve low-carbon development.

Third, China's coordination and leadership in tackling climate change and energy development should be planned as a whole. National low-carbon development strategy drafting shall be aligned and major topics regarding national low-carbon development and international climate change cooperation shall be balanced and coordinated, and unified leadership on low-carbon development shall be strengthened. What's more, China's low-carbon development involves multiple areas and levels. Hence, it is very difficult to solely rely on central governing authority to manage comprehensively, organize, and coordinate with strategic management as the core. Under the mechanism of multilevel management of low-carbon development, management, and coordination potentials of local governing bodies at provincial, municipal and prefecture levels should be given full play to. Local governing bodies at all levels, especially provincial governing authority, should take more responsibilities, and strengthen their comprehensive management and coordination capability by deepening system reform.

Last, establish the working mentality with controlling total carbon emission as the guidance. Effectively manage total greenhouse emission amount by setting the target of total greenhouse emission amount for each region and key industry in different phases: China can pilot trial emission control in key regions and industries with mature conditions on the basis of guaranteeing the realization of carbon emission intensity target; nationwide control on total greenhouse gas emission amount can be realized, and simultaneous control on carbon emission intensity and total amount can be promoted in parallel after policy system is relatively complete. After low-carbon development mentality achieves overall popularization and low-carbon policy and mechanism are comprehensively established, total carbon emission amount can be used as the objective, energy utilization efficiency shall be comprehensively improved, and carbon emission index shall be used to lead energy development.

### 11.4.2 Improve Policy System by Guaranteeing Low-Carbon Development

Low-carbon development should be strengthened from the following aspects on the basis of coordinating the current energy, environment, and climate change policies.

First, complete the low-carbon development legal system. Specific law on promoting low-carbon development should be formulated and related legal regulations should be drafted in a complete, systematic, and orderly manner. China's legal regulations related to low-carbon development should be reviewed from the level of ecological civilization development, legislation, and consolidation of regulations shall proceed strategically, and law shall play the role of leading, regulating, promoting, and securing in low-carbon development. The legal system should be one that cooperates with each other and has a full coverage. It shall not only provide legal foundation for policy-making and practical activities, but also provide effective system security for future development, and break the system bottleneck of low-carbon development. Low-carbon development legislation shall effectively coordinate related regulations related to low-carbon development such as *Law on Prevention and Control of Atmospheric Pollution, Circular Economy Promotion Law, Clean Production Promotion*

*Law, Energy Conservation Law,* and *Renewable Energy Law*, which forms a relatively complete legal regulation system of low-carbon development.

Second, low-carbon development system shall be established. Low-carbon development should be promoted by establishing a complete low-carbon regulation system. Low-carbon development is the important vehicle of promoting the orderly execution of low-carbon development, and the necessary guarantee of the effective execution of low-carbon development. China's development reality should be the foundation, international proven system shall be referred to, and the relationship between government and market in improving low-carbon development shall be tackled properly. Market should play a decisive role in allocating low-carbon resources. Further, improve and strengthen related system establishment with building key system as the core and enhance the authenticity and effectiveness of low-carbon development. In the future, China should insist on combining top-level design and pilot demonstration, establish the regulation system that is consistent with low-carbon development, improve the design of significant low-carbon development mechanism of greenhouse gas emission control, carbon credit trading, product logo, and accreditation, and improve greenhouse gas emission statistical calculation system.

Secondly, actively explore leveraging free-market method and achieve emission control target with a relatively low cost. China should establish carbon credit trading system and gradually build domestic carbon credit trading markets. It shall further give full play to market mechanism, actively explore leveraging free-market method, and achieve emission control target with a low cost. Meanwhile, actively formulate and build the related accreditation and standard system from a strategic level, establish a set of legal regulation systems that is related to developing a united nationwide carbon market, improve the transparency of trading markets, ensure the delivery of information, and provide effective guarantee for the trading on carbon market.

Last, improve economic incentive mechanism, and lead the formation of low-carbon producing and living style and consumption model. a. Letting price play the adjustment role, speeding up energy pricing mechanism reform, and reflecting carbon's external cost into the price of product. b. Establishing green taxation system, promoting resource tax and environment tax reform and studying, and formulating carbon tax plan. c. Optimizing government investment, improving government's investment areas, strengthening government investment, improving government's investment patterns, and intensifying the leading and exemplary role played by government procurement. d. Optimizing the implementation environment of low-carbon investment policy, realizing the diversity of the investment sources of low-carbon development, including building the government investment and financing platform, leading financial institutions to engage in financial businesses, improving market entrance and elimination mechanism, cultivating investment and financing intermediaries, improving financing guarantee system, and building low-carbon investment and financing information platform.

### 11.4.3 Develop Low-Carbon Industry by Leveraging Carbon Innovation

Technological system featured with low carbon has become the future trend of global development, and all countries have increased their investment and support into R&D in low-carbon industry, so as to reach the high point of low-carbon competition. China should follow and lead the trend, facilitate low-carbon innovation, strengthen technological storage, facilitate industrial transformation, establish low-carbon technological system, and improve national competitiveness amidst new situations.

First, establish a standard system with carbon index as the direction, formulate national standard of greenhouse gas emission in key industries, establish the standard, logo, and accreditation system of low-carbon products, and encourage enterprises to establish corporate greenhouse gas emission standard that is more strict than national and local standard so as to support the development of low-carbon technologies.

Second, take a series of policies and measures to stir up the investment into low-carbon technologies, facilitate R&D, demonstration, promotion, and spreading of low-carbon technologies, further enhance innovation capabilities and technical levels, and increase the comparative advantage of low-carbon economy and global competitiveness.

Third, take different policy tools and financing instruments for different types of technologies in different development phases. For technologies in R&D phase, the major role of government is providing funding for R&D and demonstration and improving related infrastructure construction. For technologies in demonstration phase, the major role of government is setting sound incentive policies for specific technologies, such as on-grid price discounts, tax exemption and reduction, loan guarantee, etc. For technologies in promotion phase, related technology standard should be set up and a greenhouse trading system should be established. For technologies in deployment phase, China should drive the promotion and application of technologies by removing market barriers.

Fourth, strongly promote the transfer and cooperation on international technologies, draft special preferential policies, attract advanced overseas technologies and funding to China, including keeping track of the updated progress of advanced low-carbon technologies, promoting technology transfer and joint R&D, improving the understanding and application of transferred technologies, and jointly demonstrate, share achievements, and achieve a win-win result.

### 11.4.4 Build Low-Carbon Cities Featuring Climate Wisdom

The idea of building low-carbon cities has been widely promoted and recognized in China. Since the launch of pilot endeavors in two batches of low-carbon provinces, districts, and cities in 2010, they began to explore the low-carbon development models, systems, and mechanisms, as well as policies and measures according to their local conditions. The *US–China Joint Announcement on Climate Change* issued jointly by China and the United States by the end of 2014 has specific proposals on building low-carbon cities: "In response to growing urbanization and increasingly significant greenhouse gas emissions from cities and recognizing the potential for local leaders to undertake significant climate action, the United States and China will establish a new initiative on Climate-Smart/Low-Carbon Cities under the Climate Change Working Group (CCWG)."

Cities are the main places in which people work and live and are main carbon emitter. Building low-carbon cities is an important way to realize low-carbon development.

First, the building of low-carbon cities should highlight the key points of policy guidance according to different stages of development. There are over 600 cities in China with different development stages. Among them, Beijing and eastern coastal cities such as Shanghai and Shenzhen boast per capita GDP of over 8000 USD, making them middle-to-upper developed areas. The per capita carbon emission in those cities is higher than the national average because those cities are in the later stage of urbanization and industrialization. In the future, the emission growth will mainly come from construction and transportation. We should strictly control the general volume of carbon emission by emphasizing the control of excessive growth of emission in transportation, construction, and living sectors. Some fast-growing cities in central and eastern China are in the process of industrialization and urbanization and they have large room for growth in industrial, construction, and transportation sectors. We should carry out dual constraints of carbon emission intensity and total carbon emission, adhere to the low-carbon development path, and strive to achieve leap-forward development. Some cities in western China are relatively backward in economic development, and their carbon emissions remain quite low. The key point of low-carbon development is to rationally arrange the industrial and energy systems in the process of urbanization and industrialization in the future. Besides, they should develop related industries and strengthen policy guidance for new industrialization construction and low-carbon urbanization based on their endowment of resources.

Second, encourage cities to explore innovative and effective models and create regional highlights. Cities should be encouraged to innovate based on local foundations and capabilities, for example, exploring market-oriented mechanisms beneficial to emission reduction, such as carbon emission trading, voluntary emission reduction agreement, low-carbon certification, regional carbon sink ecological compensation, etc; exploring greenhouse gas emission quota management, emission index decomposition assessment, carbon assessment of major engineering construction projects, and "Internet + low-carbon cities" and other low-carbon development models with local characteristics; and encouraging cities to carry out pilot demonstrations of low-carbon industrial parks, low-carbon transportation, low-carbon buildings, communities, businesses, products, etc., in a bottom-up manner.

Third, build smart cities featuring climate wisdom that are efficient, livable, and low-carbon. Work out top-level design and build low-carbon cities with next-generation information technologies such as the Internet of things, the Internet, cloud computing, big data, etc. Create smart and livable cities by carrying out digitalized city construction management and using information technology to improve the overall level of urban water, gas, and heat supply systems. Raise the operational efficiency of cities by building a public platform, providing intelligent government services, public services, community operations, and residents' living facilities. Promote innovation and the development of smart urban industries and economic transformation.

### *11.4.5 Strengthen International Cooperation by Tackling Climate Change*

Responding to climate change and developing a low-carbon economy will become an important area for China to participate in the formulation of new international rules, build an external exchange platform, and seize the high ground of industrial technology. Specific measures to strengthen international cooperation in the development of low-carbon may include the following:

First, promote climate diplomacy in a comprehensive and in-depth manner. We should integrate the international cooperation on climate change and our diplomatic strategies for a new era, further enhance the cooperation with the United States, the European Union, "Basic Four", etc., in climate change, encourage and promote localities, enterprises, think tanks, and other organizations to have broader participation and strive to build the climate diplomacy into a new highlight of our diplomatic work under the new situation. This will make new contributions to speeding up the building of new relations between big powers, deepening mutually beneficial cooperation and connectivity with neighboring countries, consolidating traditional friendship, and strengthening cooperation between China and developing countries.

Second, actively participate in and guide the construction of the international climate change system. We should adhere to the United Nations Framework Convention on Climate Change as the main channel for international cooperation on climate change, adhere to the principle of "common but differentiated responsibilities," the principle of fairness and the principle of respective capabilities, and actively participate in the international climate negotiations on new agreements and follow-up institutional arrangements, as well as make a constructive contribution and play a guiding role. We should promote the improvement and implementation of international climate regime in a fair and effective way and through cooperation and win-win.

Third, promote the cooperation and green development with relevant countries by combining the Belt and Road strategy. Countries along the Belt and Road have different resource endowments, strong economic complementarities, great potential for cooperation and space, and the technical and construction needs for green and low-carbon development. We should build the Belt and Road into a low-carbon corridor and a model for global low-carbon cooperation by combining with the Belt and Road initiative, taking the low-carbon industry and infrastructure as the starting point, carrying forward China's status as a large developing country, providing low-carbon development experiences, technologies, and services for less-developed countries and regions through policy communication, facility connectivity, smooth trade, and economic integration.

Fourth, actively carry out "South–South cooperation" and promote "going global" of low-carbon enterprises and products in China. We should continue to increase the financial input into the "South–South cooperation" in low-carbon field, launch the "South–South Cooperation" fund operation for climate change as soon as possible, gradually expand the scale of funds, and improve the efficiency of fund utility. We should actively participate in the design and improvement of the systems, rules, and management of the Green Climate Fund, explore the use of multilateral mechanisms to expand channels and ways to deal with climate change "South–South cooperation." We should encourage international institutions to support the "South–South cooperation" in low-carbon fields, encourage and guide enterprises, research institutions, universities and social groups to increase investment in international cooperation, establish diversified investment channels, and form a multichannel international cooperation input system involving central investment, local support and social capital integration. We should increase the proportion of low-carbon assistance in the national foreign aid funding, give full play to climate foreign aid, and raise China's radiation influence on neighboring and developing countries. We should encourage and support enterprises, universities, and scientific research institutions to obtain investment from foreign funds in joint research and development through various channels.

Fifth, improve the research capacity of think tanks, strengthen their researches on basic theories, emphasize the accumulation of data and information, make targeted policy research, and establish long-term strategic partnership with domestic and international first-class think tanks and regular exchange and discussion mechanisms on global low-carbon development issues. This will continuously enhance the research level and thus provide necessary research and consultation suggestions for formulating and adjusting the internationalization strategy of China's scientific and technological development. We should strengthen national research and carry out systematic strategic analysis on different countries' advantageous areas, their China-related policies and science, and technology management system. We should strengthen the strategic research on major issues related to low-carbon development in the world and make full use of overseas research channels to track and grasp the latest trends in international low-carbon development in a timely manner.

### *11.4.6 Improve Governance Model Based on the Philosophy of Social Coordination*

Low-carbon development is immediately related to the fundamental interests of the people. It cannot do without any actors, including the government, enterprises, social organizations, and the public. Therefore, to achieve low-carbon development, not only government policy guidance but also a coordinative social governance mechanism and model is needed, in order to make use of the capability of all the actors.

First, the government restrains or guides the activities of enterprises and the public. The government needs to use administrative tools, legal measures, economic incentives, moral restrains, market mechanism, and information release to lead the low-carbon transition of the economy and promote the building of social awareness. Meanwhile, it needs to set low-carbon development goals for the whole society and establish and improve institutional channels and platforms for communication and engagement, in order to foster mature nongovernmental actors. These measures can help advance low-carbon development.

Second, social organizations need to play a more important role by enjoying more spaces of activity and building low-carbon capacity. So far, all over the world, nearly 10,000 social organizations aim for environmental protection, green and sustainable development, and these organizations have become a mainstream force promoting green and low-carbon development. From the "Earth Hour", a global campaign calling for attention to carbon footprints in daily life, to An Inconvenient Truth, the famous speech by former US Vice President Al Gore, more and more social forces have been dedicated to publicity and activities, shaping a new social culture. China needs to enhance it social governance, establish a holistic network, improve the government's service procurement mechanism, incorporate social organizations into public service system, and encourage these organizations to localize their publicity and make it continuous.

Third, social culture and consumer psychology need to be guided, in order to raise the low-carbon awareness and public engagement in the whole society. People is the master of the society as well as the actor and critical link in low-carbon practice. It is imperative to raise people's awareness and build low-carbon production and lifestyle through demonstration and publicity. By guiding people's consumption behaviors, we can encourage low-carbon production of companies. By building incentive institutions, we can engage more citizens and enterprises in low-carbon development, and make low-carbon lifestyle a voluntary choice of people. For example, all forms of campaigns on low-carbon life can be held on Earth Day, Environment Day, Low Carbon Day, Ozone Day, Car-free Day, Energy Conservation Week, Public Transportation Week, etc. Media can be mobilized to cover low-carbon life to promote the understanding and personal engagement of the public. Basic-level communities can be encouraged to hold low-carbon activities.

# Appendix

See Figs. A.1, A.2, A.3, A.4, A.5, A.6 and A.7.

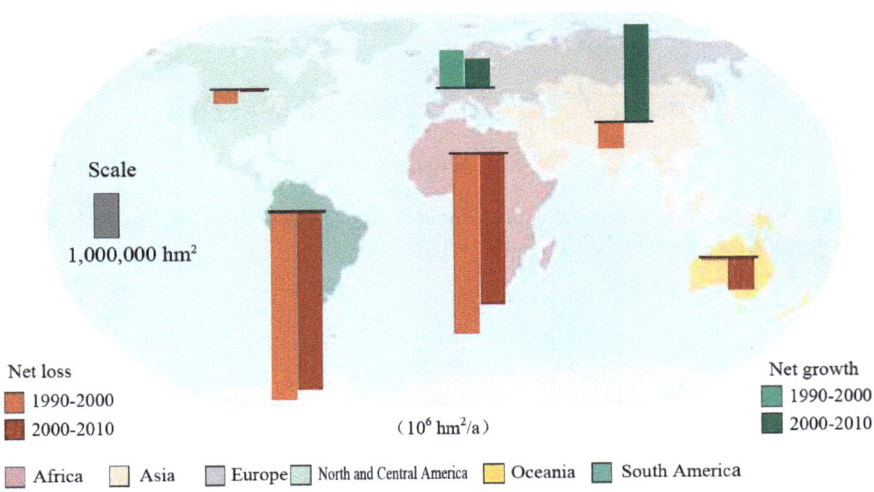


**Fig. A.1** Annual changes in forest area by region form 1990 to 2010


X. Du et al., *Overview of Low-Carbon Development*,
https://doi.org/10.1007/978-981-13-9250-4

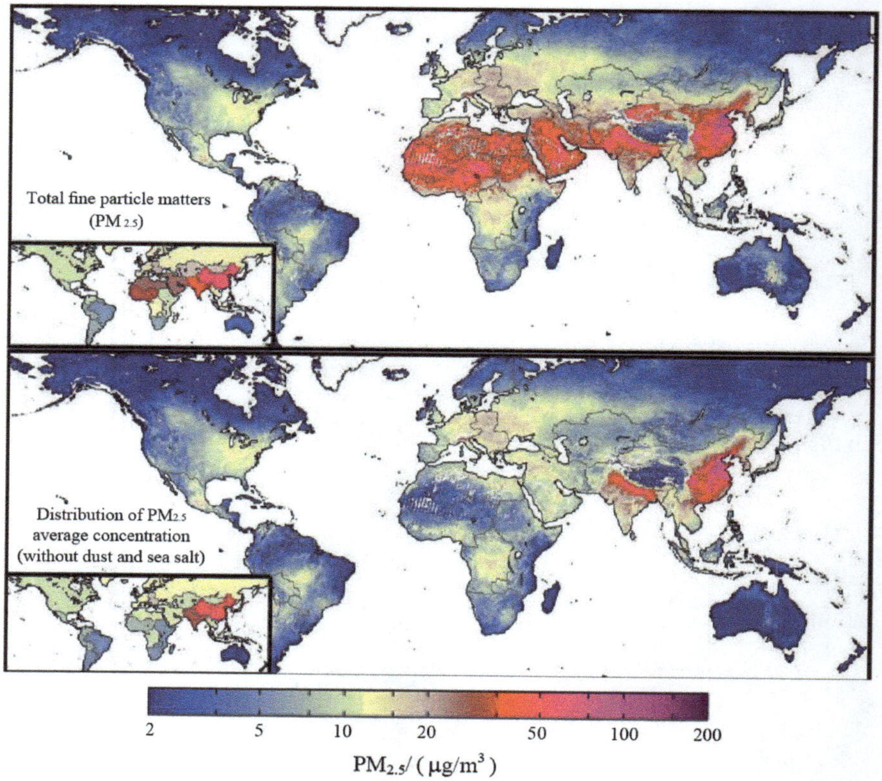


**Fig. A.2** $PM_{2.5}$ average concentration distribution across the globe from 2001 to 2010 (*Source Environmental Health Outlook*, 2015, v. 123)

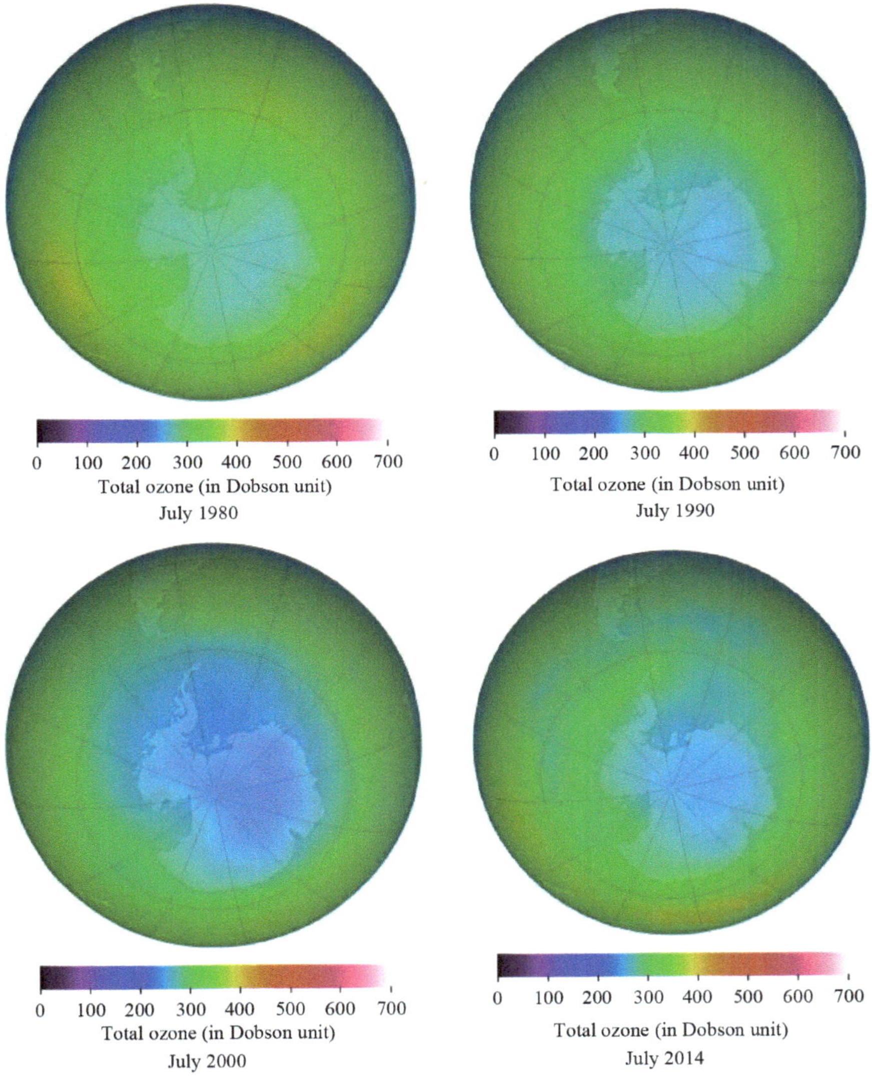


**Fig. A.3** Distribution of monthly average concentration of ozone layer over Antarctica (Blue-violet indicates areas with low concentration of ozone, and red-yellow indicates areas with high concentration of ozone) (*Source* NASA: Goddard Space Flight Center)

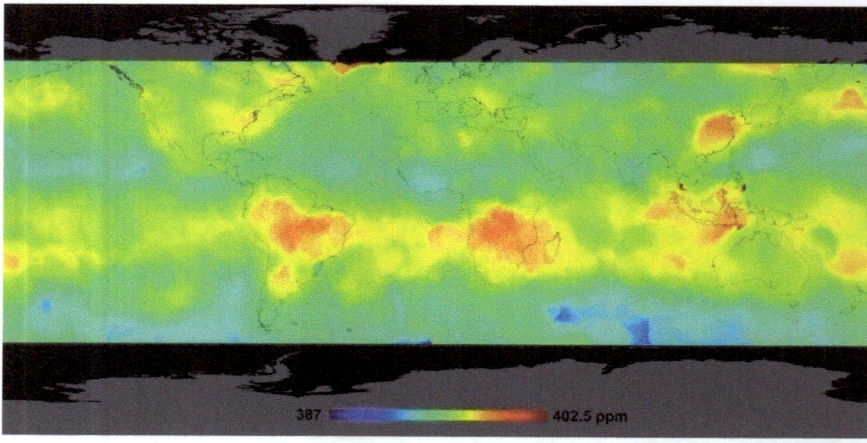


**Fig. A.4** NASA satellite data: Global carbon dioxide concentration map for the fall of 2014 (from October 1–November 11)

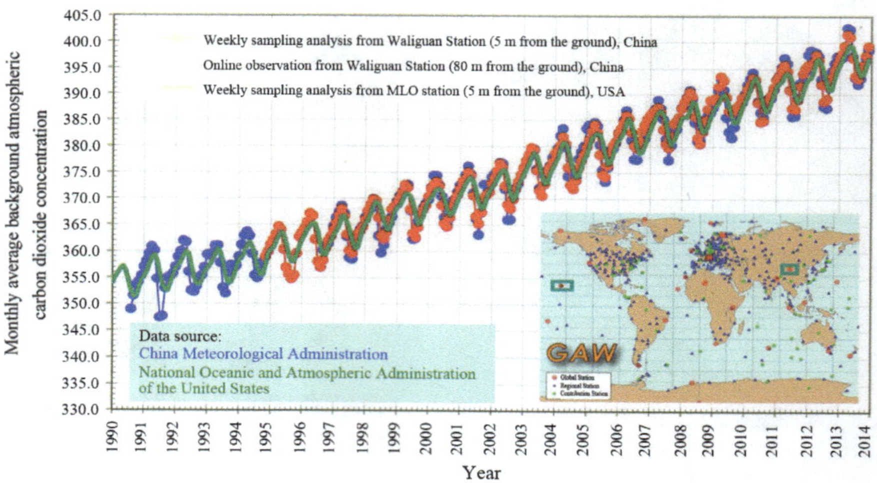


**Fig. A.5** Monthly average background variations of atmospheric carbon dioxide concentration from 1990 to 2014

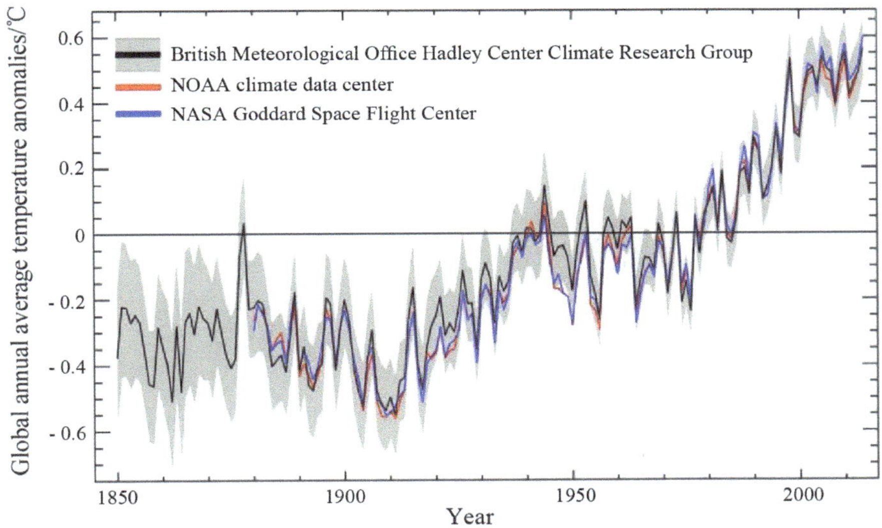


**Fig. A.6** Global annual average temperature anomalies from 1850 to 2014 (against the 1961–1990 average) (*Source* The World Meteorological Organization)

**Fig. A.7** Contrast showing smog in Beijing

# References


1. Xi Jinping. Report to the 18th National Congress of the Communist Party of China [R]. 2012.
2. Wang Anjian. Statistics of the Report Meeting of the Ministry of Land and Resources [R]. 2015.
3. Fu Bojie. Report on the 2013 Annual Conference of the China Association for Science and Technology [R]. 2013.
4. Research team of the Chinese Academy of Engineering. Research Report on the Consultative Project of the Scientific and Technological Issues When Dealing with Climate Change [R]. 2014.
5. Du Xiangwan, et al. The International Comparative Study of the Room for Development of China's Energy [J]. Engineering Sciences, 2013, 15(6): 4–10.
6. US-China Joint Statement on Climate Change. 2014.
7. Zhao Jianbo, Tian Yuanbing, Wang Qiang, et al. The Philosophical Speculation on Global Changes and Civilization Evolution [J]. The Science Education Article Collects, 2013(30): 199–200.
8. Ding Ming. On the Ecological Civilization Dimensionality of Human Development [D]. Dalian: Liaoning Normal University, 2010.
9. Xu Chun. The Position of Ecological Civilization in Human Civilization [J]. Journal of Renmin University of China, 2010(2): 37–45.
10. Liu Jing. The Study on Building Ecological Civilization of the Socialism with Chinese Characteristics [D]. Beijing: Central Party School of the Communist Party of China, 2011.
11. Qian Yi, Tang Xiaoyan. Environmental Protection and Sustainable Development [M]. Second Edition. Beijing: Higher Education Press, 2010.
12. Ministry of Water Resources. China Water Resources Bulletin(1997–2013) [R].
13. World Steel Association. Steel Statistical Yearbook 2014 [M]. 2015.
14. Cui Feng, Tang Guojian. Environmental Sociology [M]. Beijing: Beijing Normal University Press, 2010: 76–82.
15. Gong Ke. The World's Eight Big Pollution Incidents and Green GDP [J]. Journal of Shenyang University, 2005(4): 3–6.
16. Yang Guifang. Analysis on the Connotations of Ecological Civilization [J]. Ecological Economy, 2010(12): 185–188.
17. Yang Hejiao. The Contemporary Thinking of Building Ecological Civilization [J]. Theory Monthly, 2014(3): 157–161.
18. Liu Yanchun. A few Thoughts on Ecological Civilization [J]. Chinese Cadres Tribune, 2003(3): 56–57.
19. Bao Shuangye. Study on Ecological Civilization amid the Social Transformation in Today's China [D]. Shanghai: East China Normal University. 2012.
20. Yu Mouchang. Ecological Civilization: the New Form of Human Civilization [J]. Changbai Journal, 2007(2): 138–140.




X. Du et al., *Overview of Low-Carbon Development*,
https://doi.org/10.1007/978-981-13-9250-4

21. World Economic Development Report [R]. Shanghai: Shanghai University of Finance and Economics Press, 2009.
22. Zhang Guo. Ecological Civilization Building and Low-carbon Economic Development—Study on China's Case [J]. Frontiers, 2011(3): 168–170.
23. Ba Yongqing. Low Carbon, the Key to Ecological Civilization—A Basis of Building Ecological Civilization in Contemporary Society [J]. Journal of Shengli College, China University of Petroleum, 2014(9): 73–76.
24. Xu Chun. Theoretical Explanations of the Concept of Ecological Civilization [J]. Journal of Peking University (Philosophy and Social Sciences Page), 2010(1): 61–63.
25. Frederick F. Cartwright, Michael Biddiss. Disease & History [M]. Translated by Chen Zhongdan, Zhou Xiaozheng. Jinan: Shandong Pictorial Publishing House, 2004.
26. Cheng Faliang, Sun Chengfang. Environmental Protection and Sustainable Development [M]. Second Edition. Beijing: Tsinghua University Press, 2009.
27. Liu Shaokang. An Introduction to Environment and Environmental Protection [M]. Beijing: Tsinghua University Press, 2002.
28. Editorial Board of Atmospheric Sciences Dictionary. Atmospheric Sciences Dictionary [M]. Beijing: China Meteorological Press, 1994: 408–677.
29. Yang Xin, Chen Yizhen, Liu Houfeng, et al. The Features and Causes Analysis of the Continuous Heavy Smog in January 2013, Beijing [J]. China Environmental Science, 2014, 34(2): 282–288.
30. Li Zhanwu. Integrated Measures Are Required to Tackle the PM 2.5 Pollution [J]. Sino-Global Energy, 2013, 18(9): 7–13.
31. BP. BP Statistical Review of World Energy [M]. 2013.
32. BP. BP Statistical Review of World Energy [M]. 2014.
33. National Bureau of Statistics of China. China Energy Statistical Yearbook [M]. Beijing: China Statistics Press, 2014.
34. Food and Agriculture Organization of the United Nations. Global Forest Resources Assessment 2010 Main Report [R]. 2011.
35. Du Xiangwan. The Essence and Way out of International Climate Negotiations [J]. Advances in Climate Change Research, 2014, 10(5): 384–387.
36. Global International. Global International Study on Legislations of Climate—A Review of Legislations of Climate in 33 Countries. 2012.
37. Xu Dongqing. Practices and Experience of Low-carbon Economic Growth in Developed Countries [J]. Forum of World Economics & Politics, 2009(6): 112–116.f.
38. Gou Hongyang. Low Carbon Plot [M]. Taiyuan: Shanxi Economic Press, 2010.
39. Long Yushi. Plot or Science [J]. New Economy, 2010(9): 90.
40. Wang Qian, et al. Theoretical Analysis of "Carbon Plot" and Policy Directions [J]. Economist, 2011(10): 45–50.
41. Zhang Kunmin, Pan Jiahua, Cui Dapeng. Low-carbon Economy Theory [M]. Beijing: China Environmental Science Press, 2008.
42. China Council for International Cooperation on Environment and Development. Research Report on Global Experience and China's Practices of Low-carbon Economy [R]. 2008.
43. He Jiankun. The Key to Low-carbon Economy is Low-carbon Technological Innovation [J]. Green Leaf, 2009(1): 46–50.
44. Pan Jiahua, Zhuang Guiyang, et al. Internal Report on Distinguishment of Concepts and Evaluation Criteria System of Low-carbon Economy [R]. 2009.
45. Yuan Nanyou. The Conceptual Connotations of Low-carbon Economy [J]. Urban Environment and Urban Ecology, 2010(1): 43–46.
46. Zhuang Guiyang. The Connotations of Low-carbon Economy and the Building of Integrated Evaluation Criteria System [J]. Economic Perspectives, 2011(1): 132–136.
47. Li Junfeng, Ma Lingjuan. The Still Nascent Low-carbon Economy: Low-carbon Economy is the New Rule Outlining the World Development Landscape [J]. World Environment, 2008(2): 17–20.

48. Hao Wensheng. The Progressive Path of China's Low-carbon Economic Development [J]. Urban Environment and Urban Ecology, 2010(6): 17–23.
49. Ji Kunsen. Coordinated Development of Circular Economy and Low-carbon Economy [R]. The 12th Annual Meeting of China Association for Science and Technology, 2010.
50. Wang Anjian, Wang Gaoshang, Chen Qishen, et al. Energy and National Economic Development [M]. Beijing: China Land Press, 2008.
51. Zhu Xiaoying. Analysis of Canada's Environmental Policies from Its Attitudes towards the Kyoto Protocol [D]. Hebei Normal University, 2007: 34–35.
52. State Grid Energy Research Institute. Global Energy and Power Statistical Book [M]. State Grid Corporation of China, 2012.
53. Fang Jia. Research on Issues Concerning Industrial Structure [M]. Beijing: China Renmin University Press, 1997.
54. Ding Chun. Transformation of the Economic Growth Pattern of the Federal Republic of Germany and Analysis of its Causes [J]. Deutschland-studien, 2000(3).
55. Zhang Wei. Experience and Inspirations of Energy Saving and Emissions Reduction Practices of Germany [J]. World Economy, 2008(3): 64–68.
56. Huang Wenjiie. The Energy Issues and Government's Energy Policies of France [J]. World Economy, 1988(8).
57. Li Weijun. Comparative Study of Environmental Governance of the United States, Japan and the United Kingdom and What can be Learned for China—System, Policy and Action [D]. Shanghai: Fudan University, 2008.
58. Liang Rui. Study on the Clean Air Act of the United States [D]. Qingdao: Ocean University of China, 2010.
59. Xue Wenbo, Fu Fei, Wang Jinnan, et al. Simulations of the Capacity of Atmosphere Environment Under the Condition of a Healthy Level of PM 2.5 in all Chinese Cities [J]. China Environmental Science, 2014, 34(10): 2490–2496.
60. Du Xiangwan, et al. Study of China's Medium-to-long Term Energy Development Strategy (2030-2050) [M]. Beijing: Science Press, 2011.
61. Peng Jinxin. Human beings Go from Dealing with Climate Change to Low-carbon Economy [J]. Environmental Science & Technology, 2009, 32(12): 1–8.
62. Cui Yi, Hao Shouyi, Zhang Lixin. How to Transform a High-carbon Economy to a Low-carbon One [J]. Ecological Economy, 2010(4): 30–38.
63. Department of Integrated Affairs, National Bureau of Statistics. Grand Reform, Immense Opening-up and Great Development—One of the Series of Reports on the Achievements of China's Economic and Social Development in the Past 30 Years since Reform and Opening Up [R]. 2008.
64. Pang Changwei. Developing Low-carbon Economy and Guarding Ecological Security [J]. Journal of China Jinggangshan Cadre College, 2009, 2(1): 125–128.
65. Yang Ting. The Thoughts on Chinese Residents' Low-carbon Life in the Context of Low-carbon Economy [J]. Journal of Heze College, 2010, 32(5): 31–34.
66. Zhou Ran. Review and Reflections on China's Democracy and Legislation since Reform and Opening Up [J]. Shandong Archives, 2008(5): 56–57.
67. Du Xiangwan, Zhou Dadi. The Scientific, Green and Low-carbon Energy Strategy of China [J]. Engineering Science, 2011, 13(6): 3–10.
68. Cheng Huaiwen, et al. The Theory and Empirical Evidence of the Society's Capacity to Adapt to Water Resources Shortage [J]. Acta Ecologica Sinica, 2011, 31(5): 1430–1439.
69. Song Tingting. Analysis of Issues Regarding Sustainable Resource and Environmental Development of China [J]. Resources and Environment, 2014(18): 113.
70. Chen Canping. Analysis of Causes of China's Water Resources Problem and Study on Solutions [J]. Journal of Southwest Minzu University, 2014, 6: 141–144.
71. United Nations Development Program, et al. World Resources 2000–2001: People and Ecosystems: The Fraying Web of Life [M]. Beijing: China Environmental Science Press, 2002.
72. Shen Manhong. Study on Methods of Environmental Economy [M]. Beijing: China Environmental Science Press, 2001.

73. Li Keguo, et al. Environmental Economy [M]. Beijing: China Environmental Science Press, 2003.
74. National Bureau of Statistics. China Statistical Yearbook, China Industrial Economy Statistical Yearbook, China Energy Statistical Yearbook 2001–2014 [J]. Beijing: China Statistics Press.
75. Zhao Yaru. Win-win Results of Economy and Environment—Experience of Germany's "Social Ecological Market Economy" [J]. China WTO TRibune, 2008(62): 75–76.
76. Chen Zhiguo. Experience of Japan's Transformation of Environmental Management Strategy [J]. Environmental Protection, 2013, 41(15): 73–75.
77. Du Xiangwan. Energy Revolution—For a Sustainable Future [J]. Journal of Beijing Institute of Technology, 2014, 16(5): 1–8.
78. The Energy Research Institute of the National Development and Reform Commission (NDRC). Disquisition Report on the Medium-to-long Term Potential for Energy Saving and Emission Reduction for Industries and How to Unleash It [R]. 2014.
79. NDRC. National Plan of Adapting to Climate Change (2014–2020) [R]. 2014.
80. NDRC, Ministry of Finance, Ministry of Housing and Urban-Rural Development, et al. National Strategy to Adapt to Climate Change [R]. 2013.
81. NDRC. Strengthen Actions Against Climate Change—China's Nationally Determined Contributions [R]. 2015.
82. National Energy Administration. Review of the World Nuclear Power Development and Technological Growth in 2013 [N]. http://www.nea.gov.cn/2014-01/03/c_133015069.htm, 2014-06-20.
83. Zhang Bo, Pang Shuya. Green House Gas Emissions in the United Kingdom: Break-down of Causes and Study on Ways of Emission Reduction [J]. Journal of European Studies, 2012(1): 100–114.
84. Pan Jiahua, Chen Ying, Zhuang Guiyang, et al. Incentives for Low-carbon Development and Lessons Learned [J]. China Economic & Trade Herald, 2006(18): 51–52.
85. Li Yuyan. Potential of Shale Gas Resource and Progress in its Exploration and Extraction [R]. 2014.
86. Cai Fang. Speech on the 2011 Annual Conference of the Chinese Economists 50 Forum [R]. 2012.
87. Ministry of Commerce: A Survey of Thousands of Enterprises by the Ministry of Commerce: the Average Export Profit Rate was less than 3% [N]. http://money.163.com/13/0618/10/91L753BC00252G50.html.
88. World Bank, World Trade Organization Database, et al.
89. WTO. International Trade Statistics 2012.
90. Gao Shixian, Ren Dongming. Study on Promoting Energy Production and Consumption Revolution [M]. Beijing: China Economic Publishing House, 2014.
91. Dai Yande, Xiong Huawen, Jiao Jian. Research on the Achievement of Energy Saving Potential of China's Industries and Technological Roadmap in 2020 [M]. Beijing: China Science and Technology Press, 2013.
92. Strategy Research Team of the Energy Field, Chinese Academy of Sciences. China's Energy and Technological Development Roadmap by 2050 [M]. Beijing: Science Press, 2009.
93. Han Wenke, Zhang Jian'guo, Gu Lijing. Green Building: China is Taking Action [M]. Beijing: China Economic Publishing House, 2013.
94. Dai Yande, Hu Xiulian. Study on China's Technological Potential and Costs of Carbon Dioxide Emission Reduction [M]. Beijing: China Environmental Science Press, 2013.
95. Academy of Macroeconomic Research, NDRC. Research on the Low-carbon Development in Chinese Cities. 2012.
96. Zhou Peng, Zhou Dequn, Yuan Hu. Low-carbon Development Policies: International Experience and Chinese Strategy [M]. Beijing: Economic Science Press, 2012.
97. Editorial Board of China Low-carbon Yearbook. China Low-carbon Yearbook 2010-2014 [M]. Beijing: Metallurgical Industry Press.

98. Qi Ye. China Low-carbon Development Report 2011–2012 [M]. Beijing: Social Sciences Academic Press, 2011.
99. Qi Ye. China Low-carbon Development Report 2013 [M]. Beijing: Social Sciences Academic Press, 2013.
100. Du Xiangwan. The Depth of Climate Change: Addressing Climate Change and Transformation in Development [J]. Chinese Journal of Population Resources and Environment, 2013(9).
101. He Jiankun. The Situation and Solutions of Energy Development in China and Climate Change [N]. 2014–05–10.
102. NDRC. China's Policy and Actions to Tackle Climate Change 2007–2014 [EB].
103. CPC Central Committee, State Council. On Accelerating the Ecological Civilization Construction [EB/OL]. 2015.
104. State Council. Energy Development Plan of the 12th Five-Year Plan (No. 2 (2013) Issued by the State Council) [EB/OL]. 2013.
105. General Office of the State Council. Program of Action for the Energy Development Strategy (2014–2020) (No. 31 (2014) Issued by the General Office of the State Council) [EB/OL]. 2014.
106. Ministry of Industry and Information Technology. Action Plan of Industries Addressing Climate Change (2012–2020) (No. 621 (2012) of the Ministry of Industry and Information Technology) [EB/OL]. 2012.
107. Ministry of Housing and Urban-Rural Development. Special Planning for the Energy Conservation of Buildings of the 12th Five-Year Plan (No. 72 (2012) of Jian Ke) [EB/OL]. 2012.
108. National Government Offices Administration. Plan for Public Institutions Energy Conservation During the 12th Five-Year Plan Period (No. 433 (2011) of the Energy Saving Sector of the National Government Offices Administration) [EB/OL]. 2011.
109. Ministry of Transport. Work Plan for Greenhouse Gases Emission Control of Traffic and Transport During the 12th Five-Year Plan Period (No. 419 (2012) Issued by the Ministry of Transport) [EB/OL]. 2012.
110. Ministry of Transport. Plan for Energy Conservation and Emission Reduction in Road and Water Transport During the 12th Five-Year Plan Period (No. 315 (2011) Issued by the Ministry of Transport) [EB/OL]. 2011.
111. State Forestry and Grassland Administration. Key Points in Forestry's Action to Address Climate Change During the 12th Five-Year Plan Period (No. 241 (2011) of the State Forestry and Grassland Administration) [EB/OL]. 2011.
112. State Council. Made in China 2025 (No. 28 (2015) Issued by the State Council) [EB/OL]. 2015.
113. Chen Ningling. Analysis on the Consumption of Animal Food and Predictions of Demands for Pig and Chicken Feed in China [D]. Southwest University, 2011.
114. Chen Jun, Dan Bin. Loss of Fresh Agricultural Products During Shipping and Its Solutions [J]. Modernization of Management, 2008(4): 19–21.
115. DTI. Energy White Paper: Creating a Low Carbon Economy. 2003.
116. BP. Statistical Review of World Energy. 2012.
117. The World Bank. World Development Indicators. http://data.worldbank.org.cn/indicator/all.
118. Stojanovic D. Pejovic S., Milosevic Z. Greenhouse Gases and Means of Prevention. Acta Medic Medianae, 2013, 52(3):49–54.
119. Staehelin J, Harris N R P, Appenzeller C, Eberhard J. Ozone Trends: A Review. Reviews of Geophysics, 2010, 39: 231–290.
120. Donkelaar A, Martin R, Brauer M, Boys B. Use of Satellite Observations for Long-Term Exposure Assessment of Global Concentrations of Fine Particulate Matter. Environmental Health Perspective, 2015, 23: 135–143.
121. Arrhenius S. On the Influence of Carbonic Acid in the Air upon the Temperature of the Ground. Philosophical Magazine and Journal of Science, 1896, 5, 41: 237–276.
122. Berger A and Tricot C. The greenhouse effect. Surveys in Geophysics, 1992, 13: 523–549.
123. Charney J, Coauthors G. Carbon dioxide and climate: A scientific assessment. National Academy of Science, 1979.
124. Hansen J, et al. Climate impact of increasing atmospheric carbon dioxide. Science, 1981, 213: 957–966.

125. IEA. 2012. Energy policies of IEA Countries: The United Kingdom[R]. 2012.
126. IPCC. Climate change 2014: Synthesis Report[M]. Cambridge: Cambridge University Press, 2014.
127. Manabe S, Wetherald R. Thermal Equilibrium of the Atmosphere with a Given Distribution of Relative Humidity. Journal of the Atmospheric Sciences, 1967, 24(3): 242–259.
128. Steinman B, Mann M, Miller S. Atlantic and Pacific multidecadal oscillations and Northern Hemisphere temperatures. Science, 2015, 347: 988–991. http://doi.org/10.1126/science.1257856.
129. Department of Energy & Climate Change. 2003. Planning our electric future: a white paper for secure, affordable and low-carbon energy. Accessed at: www.gov.uk/government/publications/planning-our-electric-future-a-white-paper-for-secure-affordable-and-low-carbon-energy.
130. Ministry of the Environment, Japan. Low-Carbon Society Scenarios towards 2050. http://2050.nies.go.jp/.
131. Ann P. Kinzig, Daniel M. Kammen. National Trajectories of Carbon Emissions: analysis of proposals to foster the transition to low-carbon economies. Global Environmental Change, 1998, 8.
132. Pearce, D. Green Economics. Environmental Values, 1992, 1: 3–13.
133. Ellis, Baker, et al. Policies for Low Carbon Growth. Overseas Development Institute, accessed at: http://www.odi.org.uk/sites/odi.org.uk/files/odi-assets/publications-opinion-files/5571.pdf. 2009.
134. Funder, Fjalland, et al. Low Carbon Development and Poverty Alleviation. DIIS Report, accessed at: http://www.diis.dk/graphics/Publications/Reports2009/DIIS_Report_2009-20_Low_Carbon_Development_and_Poverty_Alleviation.pdf. 2009.
135. UNDESA. A Guidebook to the Green Economy. Accessed at: http://www.uncsd2012.org/content/documents/528Green%20Economy%20Guidebook_100912_FINAL.pdf. 2012.
136. Wlokas, Rennkamp, et al. Low Carbon Development and Poverty. Energy Research Centre, University of Cape Town, accessed at: http://www.erc.uct.ac.za/Research/publications/12-Wlokas-etal-Development_and_poverty.pdf. 2012.
137. Sen, A. Development as Freedom. New York, Random House. 1999.
138. Christa, Briner, et al. Low-Emission Development Strategies (LEDS): Technical, Institutional and Policy Lessons. OECD Report. Accessed at: http://www.oecd.org/env/cc/46553489.pdf. 2010.
139. Foxon T, Fujino J. Policy implications from the Low-Carbon Society. 2008.
140. UNDESA. A Guidebook to the Green Economy. Accessed at: http://sustainabledevelopment.un.org/content/documents/GE%20Guidebook.pdf. 2012.
141. Bruyn S M, Opschoor J B. Development in the throughput-income relationship: Theoretical and empirical observations [J]. Ecological economics, 1997, 20(3): 255–268.
142. OECD. Environmental indicators-Development, measurement and use [R]. Paris OECD, 2003:13.
143. UK Trade & Investment. Our energy future-creating a low carbon economy. 2003.
144. Executive Office of the President. The President's Climate Action Plan. June 2013.
145. GDP and its breakdown at current prices in US Dollars. NU. http://unstats.un.org/unsd/default.htm.
146. IEA. $CO_2$ emissions statistics 2013. http://www.iea.org/statistics/topics/CO2emissions/.
147. Thomas Lindhqvist, Reid Lifset. Getting the Goal Right: EPR and DfE. Journal of Industrial Ecology, 1998, 2(1): 6–8.
148. IEA. $CO_2$ Emissions From Fuel Combustion Highlights. 2014.
149. EU commission. Energy Roadmap 2050. 2013.
150. IPCC. 2006 IPCC Guidelines for National Greenhouse Gas Inventories, Prepared by the National Greenhouse Gas Inventories Programme, Eggleston H.S., Buendia L., Miwa K., Ngara T. and Tanabe K. (eds). Published: IGES, Japan. 2006. Chapter 1, 2: 123–124.
151. BP. BP Statistical Review of World Energy June 2014. http://www.bp.com/statisticalreview,2014-06-20.

152. IEA. Technology Roadmap-Hydropower. Paris: International Energy Agency. 2012.
153. IHA. 2013 IHA Hydropower Report. London: International Hydropower Association. 2013.
154. Melissa Lott. Bridging the gap between supply and demand. The Journal of International Energy Agency, (6): 34–35; IHA, 2013.
155. EIA. Annual Energy Outlook 1998 - With Projections to 2020. http://www.eia.gov/forecasts/aeo/archive.cfm. 1998.
156. Julie M. Carey. Forbes: Surprise Side Effect of Shale Gas Boom: A Plunge In U.S. Greenhouse Gas Emissions. http://www.forbes.com/sites/energysource/2012/12/07/surprise-side-effect-of-shale-gas-boom-a-plunge-in-u-s-greenhouse-gas-emissions/.2012.
157. EIA. AEO2014 Early Release Overview. http://www.eia.gov/forecasts/aeo/er/early_carbonemiss.cfm. 2013.
158. EIA. Annual Energy Outlook. 2014.
159. EPA. Clean Power Plan Proposed Rule. http://www2.epa.gov/carbon-pollution-standards/clean-power-plan-proposed-rule. 2014.
160. Council of the European Union. Council Directive of 16 December 1980 relating to the carbon dioxide emissions and the fuel consumption of motor vehicles. 80/1268/EEC. 1980.
161. European Parliament and the Council of the European Union. Directive 1999/94/EC of the European Parliament and of the Council of 13 December 1999 relating to the availability of consumer information on fuel economy and $CO_2$ emissions in respect of the marketing of new passenger cars. 1999.
162. European Parliament and the Council of the European Union. Directive 2009/29/EC of the European Parliament and of the Council of 23 April 2009 amending Directive 2003/87/EC so as to improve and extend the greenhouse gas emission allowance trading scheme of the Community. 2009.
163. European Parliament and the Council of the European Union. Regulation(EU) No. 2011. 510/2011 of the European Parliament and of the Council of 11 May 1999 setting emission performance standards for new light commercial vehicles as part of the Union's integrated approach to reduce $CO_2$ emissions from light-duty vehicles.
164. Phylipsen G, Bode J, Blok K, et al. A Triptych sectoral approach to burden differentiation: GHG emissions in the European Bubble. Energy Policy, 1998(26): 929–943.
165. European Parliament and the Council of the European Union. Decision No 406/2009/EC of the European Parliament and of the Council of 23 April 2009 on the effort of Member States to reduce their greenhouse gas emissions to? 2009.
166. European Parliament and the Council of the European Union. Directive 2009/28/EC of the European Parliament and of the Council of 23 April 2009 on the promotion of the use of energy from renewable sources and amending and subsequently repealing Directive 2001/77/EC and 2003/30/EC. 2009.
167. European Council. Conclusion on 2030 Climate and Energy Policy Framework. SN79/14. http://www.consilium.europa.eu/uedocs/cms_data/docs/pressdata/en/ec/145356.pdf.2014.
168. HM Treasuty. Carbon price floor consultation: the Government response. 2011.
169. Executive Office of the US President. The President's Climate Action Plan [EB/OL]. 2013.
170. European Commission. 2030 Framework for Climate and Energy Policies [EB/OL]. 2014.
171. Germany. The new High-Tech Strategy[EB/OL]. 2014.

Kolibri 360 GmbH
Ferdinand-Jühlke-Straße 7
99095 Erfurt, Deutschland
produktsicherheit@kolibri360.de